Annals of the International Society of Dynamic Games Volume 6

Series Editor
Tamer Başar

Annals of the International Society of Dynamic Games

Advances in Dynamic Games and Applications

Eitan Altman
Odile Pourtallier
Editors

Birkhäuser
Boston • Basel • Berlin

Eitan Altman
Odile Pourtallier
INRIA
06902 Sophia Antipolis
France

Library of Congress Cataloging-in-Publication Data
Advances in dynamic games and applications / editors, Eitan Altman, Odile Pourtallier.
p. cm.—(Annals of the International Society of Dynamic Games ; 6)
Includes bibliographical references.
ISBN 0-8176-4202-1 (acid-free paper)
1. Game theory—Congresses. I. Altman, Eitan. II. Pourtallier, Odile.
III. Annals of the International Society of Dynamic Games; v. 6.
QA269.A364 2001
519.3—dc21 00-069774

AMS Subject Classifications: 91A05, 91A06, 91A10, 91A15, 91A23, 49N70, 91A25, 91A28, 91A65, 91A80, 49N90, 49N30

Printed on acid-free paper.

Birkhäuser

ISBN 0-8176-4202-1
ISBN 3-7643-4202-1 SPIN 10771035

Production managed by Louise Farkas; manufacturing supervised by Erica Bresler.
Typeset by The Bartlett Press, Inc., Marietta, GA.
Printed and bound by Maple-Vail Book Manufacturing Group, York, PA.
Printed in the United States of America.

9 8 7 6 5 4 3 2 1

Contents

Preface

Game theory is a rich and active area of research of which this new volume of the Annals of the International Society of Dynamic Games is yet fresh evidence. Since the second half of the 20th century, the area of dynamic games has managed to attract outstanding mathematicians, who found exciting open questions requiring tools from a wide variety of mathematical disciplines; economists, social and political scientists, who used game theory to model and study competition and cooperative behavior; and engineers, who used games in computer sciences, telecommunications, and other areas.

The contents of this volume are primarily based on selected presentation made at the 8th International Symposium of Dynamic Games and Applications, held in Chateau Vaalsbroek, Maastricht, the Netherlands, July 5–8, 1998; this conference took place under the auspices of the International Society of Dynamic Games (ISDG), established in 1990. The conference has been cosponsored by the Control Systems Society of the IEEE, IFAC (International Federation of Automatic Control), INRIA (Institute National de Recherche en Informatique et Automatique), and the University of Maastricht. One of the activities of the ISDG is the publication of the Annals. Every paper that appears in this volume has passed through a stringent reviewing process, as is the case with publications for archival journals.

The volume has four parts. The first two parts are theoretic ones: the first (6 papers) on dynamic games (involving deterministic dynamics) and the second (5 papers) on stochastic games. These parts contain mostly novel research results, but also a survey paper on cooperative differential games. The third part of the volume (4 papers) is dedicated to algorithms and numerical solution approaches for dynamic games, and the final part (3 papers) concerns several applications of dynamic games.

We wish to thank especially all the associate editors and the referees for their valuable contributions that made this volume possible.

Sophia Antipolis, France

Eitan Altman
Odile Pourtallier

Contributors

Eitan Altman, INRIA, 2004 Route des lucioles, BP93, 06902 Sophia Antipolis, France

Pierre Bernhard, I3S CNRS-UNSA, 200 Route des lucioles, BP121, 06902 Sophia Antipolis, France

Michael H. Breitner, Technische Universität Clausthal, Fachbereich Mathematik und Informatik, Erzstraße 1, D-38678 Clausthal-Zellerfeld, Germany

Dean A. Carlson, Department of Mathematics, University of Toledo, Toledo, OH 43606-3390, USA

Stephane Crepey, INRIA BP 93, 2004 Route des Lucioles, 06902 Sophia-Antipolis Cedex, France

Maurizio Falcone, Department of Mathematics, University of Rome, "La Sapienza," P. Aldo Moro 2, 00185 Rome, Italy

Eugene A. Feinberg, Department of Applied Mathematics, SUNY at Stony Brook, Stony Brook, NY 11794-3600, USA

Jerzy Filar, School of Mathematics, University of South Australia, The Levels 5095, Australia

Vladimir Gaitsgory, School of Mathematics, University of South Australia, The Levels 5095, Australia

Alan J. Goldman, Department of Mathematical Sciences, The John Hopkins University, Baltimore, MD 21218-2682, USA

Alain Haurie, LOGILAB/HEC, University of Geneva, 102 Boulevard Carl-Vogt, CH-1211, Geneva-4, Switzerland

Naira Hovakimyan, Institute of Mechanics, National Academy of Sciences of Armenia, Yerevan, Armenia

Heinz-Uwe Küenle, Institut für Mathematik, Brandenburgische Technische Universistät Cottbus, Postfach 10 13 44, 03013 Cottbus, Germany

Sergey S. Kumkov, Institute of Mathematics and Mechanics, Ural Branch of RAS, S. Kovalevskaya Str. 16, Ekaterinburg 620219, Russia

Peiro Lanucara, Department of Mathematics, University of Rome "La Sapienza," P. Aldo Moro 2, 00185 Rome, Italy

Lina Mallozzi, Dipartimento di Matemaatica e Applicazioni, Università degli Studi di Napoli "Federico II," Via Claudio 21, 80125 Napoli, Italy

Monica Marinucci, CASPUR, P. Aldo Moro 2, 00185 Rome, Italy

Francesco Moresino, LOGILAB/HEC, University of Geneva, 102 Boulevard Carl-Vogt, CH-1211, Geneva-4, Switzerland

Jacqueline Morgan, Dipartimento di Matematica e Applicazioni, Università degli Studi di Napoli "Federico II," Via Claudio 21, 80125 Napoli, Italy

Minh-Tuan Nguyen, DSTO, P.O. Box 15000, Salisbury, Australia

Valery S. Patsko, Institute of Mathematics and Mechanics, Ural Branch of RAS, S. Kovalevskaya Str. 16, Ekaterinburg 620219, Russia

Tadeusz Radzik, Technical University, Institute of Mathematics, Wybreze Wyspianskiego 27, 50-370 Wroclaw, Poland

Alain Rapaport, INRIA, 2 Place Viola, 34060 Montpelier Cedex, France

Uwe Rettig, Technische Universität München, Zentrum Mathematik, D-80290 München, Germany

Piercesare Secchi, Dipartimento di Matematica, Politecnico di Milano, Piazza Leonardo da Vinci 32, I-20133 Milano, Italy

William D. Sudderth, School of Statistics, University of Minnesota, Minneapolis, MN 55455-0488, USA

V.L. Turova, Center of Advanced European Studies and Research, Friedensplatz 16, 53111 Bonn, Germany

Gisèle Umbhauer, Bureau d'Economie Thèorique et Appliquèe, 62 Avenue de la Forêt Noire, 67085 Strasbourg Cedex, France

Nicolas Vieille, HEC, Departement Finance et Economie, 78351 Jouy en Josas, France

Oskar von Stryk, Technische Universität München, Zentrum Mathematik, D-80290 München, Germany

Agnieszka Wisniewska-Matyszkeil, Institute of Applied Mathematics and Mechanics, Warsaw University, 02-097 Warsaw, Poland

PART I
Dynamic Games: Theory

On Problems with Information in Some Games: Modelling the Strategies in Some Dynamic Games

Tadeusz Radzik
Technical University
Institute of Mathematics
Wrocław, Poland

Alan J. Goldman
The Johns Hopkins University
Department of Mathematical Sciences
Baltimore, Maryland, USA

Abstract

This paper proposes a new framework for modelling certain duel-like games played in continuous time. The key difficulty is that the continuum with the natural order is not a well-ordered set, so problems arise while constructing normal forms for such games. The paper deals with zero-sum two-person games in which each player expends an initial stock of "resources" during a time interval. Such an expenditure can be "silent" (not observable by the opponent) or "noisy" (at each moment the opponent has full information about the history of the game), giving rise to three possible types of games, depending on whether the number of "noisy players" is 0, 1, or 2. Here, with the help of Zorn's lemma, the normal forms of these games are precisely formulated, a modelling effort that includes identification and resolution of a technical difficulty in describing the strategy spaces for the noisy case.

1 Introduction

This paper deals with zero-sum games in which each of two players P_i ($i = 1, 2$) with an initial endowment of $E_i > 0$ units of some homogeneous resource must distribute some or all of that endowment over a common one-dimensional continuum. For definiteness and naturalness of the subsequent description, we take that continuum to be a *time interval*, specifically $I = [0, 1]$, so that each player's distribution progresses over time. In specifying and analyzing such a game, one must obviously consider whether P_i's opponent constantly knows of P_i's expenditures so far (and thus, for example, can determine P_i's remaining supply): if this is the case, P_i is called a *noisy player*, while at the opposite extreme (where the opponent is totally ignorant of P_i's past distribution) P_i is called *silent*. (A somewhat different set of ideas [and variant models] results from replacing "noisy" and "silent" by "deaf" and "hearing" [or perhaps "sharp-eared"], respectively.)

Thus, a situation of the indicated type actually gives rise to three types of games: Γ_{ss}, in which both players are assumed silent; Γ_{nn}, in which both are noisy; and Γ_{ns}, in which P_1 is noisy and P_2 is silent. (Renumbering the players turns Γ_{sn} into Γ_{ns}, so the former does not require a separate analysis.) In our analysis the basic game is Γ_{ss}, and its normal form is designated $\Gamma = \langle M_1, M_2, K \rangle$, where

(i) the *strategy space* M_i of P_i is a specified subset of the set of all measures μ_i on (the Borel sets) of [0, 1] satisfying $\mu_i([0, 1]) <= E_i$,
(ii) $K: M_1 \times M_2 \to R$ is the *payoff function* (from P_2 to P_1).

Thus, for example, $\mu_1([0, t))$ gives the quantity of P_1's resource expended up to time t. Obviously, in this model, the two players act independently; once chosen by P_i, μ_i cannot be modified at an intermediate time to exploit knowledge about the opponent's previous actions. So this game indeed corresponds to Γ_{ss}.

The terms "noisy" and "silent" (rather than, say, "observable" and "secret") come from the early literature on applications to "duels": here the resources are "ammunition," and μ_i describes the time pattern of firing by P_i as the two opponents approach each other during the time interval [0, 1]. Many different particular cases of such games have been analyzed in the literature. For special choices of M_1, M_2, and K, the game to be discussed becomes some of the games of timing on the unit square of Karlin [9, Ch. 5], or various versions of discrete duels [there are many papers here, e.g., [2], [3], [7], [11], [16], [18]] or of nondiscrete duels [the most important here are [12], [14], [22]]. In turn, the works [9, Chs. 8.3 and 8.4], and [15], contain an analysis of mixtures of discrete and nondiscrete duels. (A full review and discussion are given in [17]). Also, some deterministic duels (where the players are equipped with fixed strategies) were analyzed (e.g., [1]) in the literature, where they are called "stochastic duels" and have close relationships with the duels discussed above. Among other things, our results will permit correcting a subtle error in the analysis of a nondiscrete noisy duel given in [13].

Other very essential counterparts of the games Γ, Γ_{ns}, Γ_{sn}, and Γ_{nn} were studied in the theory of differential games. Instead of measures μ_i from M_i in Γ, the players' basic strategies in such games are control functions. Also here, the problems with description of the players' strategy spaces, particularly in "noisy" differential games, were widely discussed, and many various constructions for these spaces were proposed. The main motivation for those constructions was to find such strategy spaces to ensure the existence of values of differential games. For several important papers related to this topic, readers are referred to [4], [5], [6], [8], [19], [20], and [21].

Our main purpose is somewhat different. We are interested in carefully modelling the players' strategy spaces (in Γ_{nn}) possessing some specified properties, providing explicit discussion while aiming at simplicity consistent with accuracy. In particular, we are interested in the question of how to define the strategy spaces

M_1^n and M_2^n in the game Γ_{nn} (independently of the payoff function K) to preserve the following two properties:

(i) upon applying any two strategies $\xi_1 \in M_1^n$ and $\xi_2 \in M_2^n$ by the players, the game Γ_{nn} "realizes" with a unique outcome strategy-pair (μ_1, μ_2) from the basic game Γ, that is, with $(\mu_1, \mu_2) \in M_1 \times M_2$;
(ii) the strategy spaces M_1^n and M_2^n should be as "large as possible."

The property of uniqueness in (i) can be very essential in some models. It is, for example, in the case of Γ when the spaces M_1 and M_2 describe possible ways of distributing "resources." Then a very natural question for the game Γ_{nn} arises: what will be the actual distribution of the players' resources from applying any two player strategies ξ_1 and ξ_2 from M_1^n and M_2^n, respectively? The analogous question can also arise in a differential game: with what actual trajectory will the game exhibit when the players apply strategies from its "noisy version?" Therefore, assuring the existence of a value of the game Γ_{nn} does not have be the only motivation in constructing the spaces M_1^n and M_2^n. The achievement of (ii)'s "large as possible" aim is more problematic; see our concluding Remark 4.3.

The second purpose of the present paper is to give a rigorous definition of a player's strategies in noisy duels of type Γ_{nn}. The lack of just such a rigorism was the reason for only "partially valid" conlusions in [13]. Also, many other papers on games of timing deal with strategies from Γ_{nn} without the necessary exact definitions.

Different constructions of the counterparts of the strategy spaces M_1^m and M_2^n for differential games are analyzed in the literature (the papers earlier listed). They are of the form of "delayed-reaction strategies," where the delay does not depend on the behavior of the opponent during the "noisy" differential game. The construction presented in the present paper is more general. The delay functions for players whose choices depend essentially on their opponents' strategies at all moments are basic in our construction of M_1^n and M_2^n. We give a careful theoretical treatment (with the help of Zorn's lemma) of how the normal forms of Γ_{nn} can be obtained from that of $\Gamma = \Gamma_{ss}$, which we regard as basic. In addition, we investigate general relationships among these games. We will show, in particular, that formulating a mathematical model for the (certainly realistic) situation Γ_{nn} is a rather delicate matter, where casual acceptance of a natural-looking description can lead to misleading conclusions. It is shown that the normal form of Γ_{ns} is completely determined by its basic game Γ, but that the same is true only for some cases of Γ_{nn}. In general, the problem of mutual relationships between Γ_{ss} and Γ_{nn} is very complex, and rather far from being completely solved. The constructions presented in this paper seem to be easily transferred to models of n-person games.

The organization of this paper is as follows: Section 2 contains a short analysis of the strategy spaces in games Γ_{ns} and Γ_{sn}. In Section 3 we define and discuss the strategy spaces M_1^n and M_2^n for the game Γ_{nn}. Section 4 contains properties of those spaces, which is the main result of the paper.

In concluding this section, we introduce some necessary notation. It will be convenient to have symbols both for the distribution function and for the restrictions (initial truncations) of any particular measure μ_i: we let

$$I := [0, 1],$$
$$F(\mu_i, t) = \mu_i([0, t]), \quad t \in I,$$
$$\mu_{it} \text{ — the restriction of } \mu_i \text{ defined by } \mu_{it} = \mu_i \text{ on } [0, t) \text{ and } \mu_{it} \equiv 0 \text{ on } [t, 1].$$

(By definition, $[0, 0) = \emptyset$.)

Also, it will be convenient at times to refer to the players (in either order) as P_i and P_j, with $i = 1, 2$ and $j = 3 - i$; the indices i and j will be reserved for this usage.

2 Strategy Spaces $\tilde{M}_1$, $\tilde{M}_2$, and the Games Γ_{ns} and Γ_{sn}

In this section we give a rigorous construction of the strategies $\tilde{M}_1$ and $\tilde{M}_2$ arising in the games Γ_{sn} and Γ_{ns}, respectively. These kinds of strategy spaces are well known in differential games theory as the class of *nonanticipating* (or *causal*) *strategies*. They were first introduced in [19] and next studied in a series of papers (see [4], [5], and [21]). Although this construction is quite simple, we present it explicitly (in a game of timing context) because it is closely related to the subsequent and more difficult analysis of the game Γ_{nn}.

If P_i is the silent player and P_j the noisy one, then P_j cannot exploit information about P_i's previous activities. Therefore, it is natural to take P_j's strategy space to be the same set, M_j, as in the basic game.

On the other hand, since P_j is noisy, P_i at time t is aware of μ_{jt} if P_j has chosen μ_j. It is therefore natural to define a strategy for P_i to be a measure whose distribution function $\xi_i(t)$ depends, at each $t \in I$, on μ_{jt}. More rigorously, ξ_i will be called a strategy for P_i if it is a function of the form

$$\xi_i = \xi_i(\mu_j, t), \qquad (\mu_j, t) \in M_j \times I, \tag{1}$$

where

(i) for each $\mu_j \in M_j$, there exists $\mu_i \in M_i$ (necessarily unique) such that $\xi_i(\mu_j, t) = F(\mu_i, t)$ for all $t \in I$,

(ii) for each $t \in I$,

$$\xi_i(\mu_j, u) = \xi_i(\bar{\mu}_j, u), \qquad u \in [0, t],$$

whenever $\mu_{jt} = \bar{\mu}_{jt}$.

Therefore, for any ξ_i and $\mu_j \in M_j$, the function $\xi_i(\mu_j, \cdot)$ is the distribution function of some measure $\mu_i \in M_i$. In this connection, in the sequel, the function $\xi_i(\mu_j, \cdot)$ will be identified with that measure μ_i. Also, when $\xi_i(\mu_j, \cdot)$ is independent of μ_j, ξ_i will be identified with that μ_i in M_i.

The *strategy space* $\tilde{M}_i$ for P_i is taken to consist of all the functions ξ_i of the form (1) and satisfying conditions (i) and (ii); clearly, this set contains M_i, and properly in general but with trivial-case exceptions when M_i is a singleton or when all measures in M_j coincide on [0, 1).

Now we can give the *normal form of game* Γ_{ns}; it is the triple

$$\Gamma_{ns} = \langle M_1, \tilde{M}_2, \tilde{K} \rangle$$

where the payoff function $\tilde{K}: M_1 \times \tilde{M}_2 \mapsto R$ is defined by the formula

$$\tilde{K}(\mu_1, \xi_2) = K(\mu_1, \xi_2(\mu_1, \cdot)), \tag{2}$$

with M_1 and K defined in the previous section. We have also defined a set $\tilde{M}_1$ of strategies for P_1, analogous to $\tilde{M}_2$, and so we can obtain the normal form of Γ_{sn} using the obvious analog of (2).

To clarify the relationship between games Γ_{ss} and Γ_{ns}, assume that Γ_{ss} has value v. If μ_1^* is an optimal strategy for P_1 in Γ_{ss}, then it follows from (2) that $\tilde{K}(\mu_1^*, \xi_2) \geq v$ holds for all $\xi_2 \in \tilde{M}_2$, whereas if $\mu_2^* \in M_2 \subset \tilde{M}_2$ is an optimal strategy for P_2 in Γ_{ss}, then (2) yields $\tilde{K}(\mu_1, \mu_2^*) \leq v$ for all $\mu_1 \in M_1$. The same logic applies if "optimal" is replaced by "ϵ-optimal" and v is replaced by $v - \epsilon$ or $v + \epsilon$ as appropriate. This argument, which formalizes an observation made for the special case of nondiscrete duels in Sect. 2 of [10], immediately yields the following three corollaries.

Corollary 2.1. *If game Γ_{ss} has value v, then v is also the value of the games Γ_{ns} and Γ_{sn}. Furthermore, if strategy μ_i^* is optimal (ϵ-optimal) for P_i in Γ_{ss}, then it is also optimal (ϵ-optimal) for P_i in Γ_{ns} and Γ_{sn}.*

Corollary 2.2. *The sets of optimal strategies for P_1 in the games Γ_{ss} and Γ_{ns} are identical.*

Corollary 2.3. *If games Γ_{sn} and Γ_{ns} have values, then*

$$\text{val}(\Gamma_{sn}) \geq \text{val}(\Gamma_{ns}).$$

For basic games $\Gamma = \Gamma_{ss}$, some of the games $\Gamma_{ss}, \Gamma_{sn}, \Gamma_{ns}$, or Γ_{nn} may not have a value. In such cases, probability extensions $\bar{\Gamma}_{ss}, \bar{\Gamma}_{sn}, \bar{\Gamma}_{ns}$, or $\bar{\Gamma}_{nn}$, appropriately, can be considered in order to ensure, usually successfully, the existence of the value. (For a game given in a normal form $G = \langle A, B, K \rangle$, the game $\bar{G} = \langle \bar{A}, \bar{B}, \bar{K} \rangle$ is a probability extension if $\bar{A}$ and $\bar{B}$ are sets of probability measures on some σ-fields of A and B, respectively, and $\bar{K}$ is the expected-payoff function defined on $\bar{A} \times \bar{B}$ in the usual way as the double integral of K.) It is worth mentioning here that, in general, only Corollary 2.3 remains true when games Γ_{ss}, Γ_{sn}, and Γ_{ns} are replaced by their respective probability extensions $\bar{\Gamma}_{ss}$, $\bar{\Gamma}_{sn}$, and $\bar{\Gamma}_{ns}$. Moreover, game Γ_{ns} and/or Γ_{sn} may have values (not necessary equal to each other) even when Γ_{ss}, with its smaller strategy space for one player, does not. Furthermore, Corollary 2.2 is not true for P_2: in general, P_2 has a larger set of optimal strategies in Γ_{ns} than in Γ_{ss}.

3 Strategy Spaces M_1^n and M_2^n for Games Γ_{nn}

We have already mentioned a comment in Sect. 2 of [10]. Its point is that a player in Γ_{ss} who is about to use an optimal strategy can announce that strategy in advance without risking a payoff less than the game value; since such an announcement sacrifices the advantage of silence, the passage goes on to infer that a player's optimal strategies in Γ_{ss} remain optimal in Γ_{ns} (or Γ_{sn}) and Γ_{nn}, which must therefore have the same value as Γ_{ss} if the latter has a value. Full justification for such a claim of course requires rigorous specification of strategy spaces M_1^n and M_2^n for players P_1 and P_2, respectively, in game Γ_{nn}. In particular, any two strategies $(\xi_1, \xi_2) \in M_1^n \times M_2^n$ should be "realized" in the form of an outcome pair $(\mu_1, \mu_2) \in M_1 \times M_2$ from the basic game Γ, that is, they should satisfy the equalities

$$\xi_1(\mu_2, t) = F(\mu_1, t), \qquad \xi_2(\mu_1, t) = F(\mu_2, t), \qquad \text{for } t \in I. \tag{3}$$

Unfortunately, after taking $M_1^n = \tilde{M}_1$ and $M_2^n = \tilde{M}_2$, system (3) may have either no solution (in μ_1 and μ_2) or more than one solution. The second possibility can occur, for example, in the case of $M_1 = M_2$ consisting only of continuous measures, when ξ_1 and ξ_2 are "imitative" strategies to each other. Then each pair $(\mu_1, \mu_2) \in M_1 \times M_2$ with $\mu_1 = \mu_2$ constitutes a solution of (3).

The problems (described above) with spaces M_1^n and M_2^n have counterparts in differential game theory ([4], [5], [6], [8], [20], and [21]). One of the approaches proposed in those publications, was to ensure the existence of solutions in the counterpart of (3) (without the uniqueness) for all nonanticipating strategies by adding some regularity assumptions on differential games. Then after adopting various modifications of the standard definition of a game value, the existence of "saddle points" could be shown. The second approach in differential games was based on modifications of nonanticipating strategies to various types of delayed-reaction strategies, in order to get the existence of a value of a game. However, the common feature of all these constructions was that the adopted delays did not depend on the behavior of the opponents during the game.

Our approach in constructing spaces M_1^n and M_2^n for game Γ^{nn} is different. We want to construct those two sets in such a way to have for any two strategies $(\xi_1, \xi_2) \in M_1^n \times M_2^n$ a unique solution of system (3) in $(\mu_1, \mu_2) \in M_1 \times M_2$, and would like to have M_1^n and M_2^n as large as possible. To this end, we adopt more general delay strategies (in the form of a modification of the strategies from $\tilde{M}_1$ and $\tilde{M}_2$), by allowing delay functions to depend on the behavior of the opponents at all previous moments during the game. This is done in the rest of the section.

The general conclusion from our discussion about (3) is that sets $\tilde{M}_1$ and $\tilde{M}_2$ are too large to serve together as the strategy spaces of Γ_{nn} (this is the subtle point overlooked in [13]). These sets must be replaced by suitable subsets M_1^n and M_2^n, for which a natural payoff function K_n: $M_1^n \times M_2^n \mapsto R$ can be defined based on the payoff function K of Γ_{ss}. Our particular proposal for the set M_i^n, presented below, will weaken in a slight but decisive way P_i's ability to utilize instantaneous full knowledge of his opponent's past behavior; in particular, it will exclude the troublesome imitative strategy $\tilde{\xi}_i$ described above.

Consider, then, player P_i and his opponent P_j. We regard P_i as making, at any time $r \in I$, the following pair of decisions. First, he chooses (or he may be forced by limited abilities for rapid response to accept) a positive value for a *forward delay* $\hat{\epsilon}_i = \hat{\epsilon}_i(\mu_{jr}) > 0$ that depends on μ_j at most via the manner of distribution μ_{jr} of the opponent's resource during $[0, r)$. This value $\hat{\epsilon}_i(\mu_{jr})$ will be interpreted as follows: player P_i can fully react to his opponent's behavior in $[0, r)$, described by μ_{jr}, at moment $t = r + \hat{\epsilon}_i(\mu_{jr})$ but not earlier. [For example, that would happen if, by moment r, P_i's sensors have received information that determines μ_{jr}, but the processing of that information by P_i as a basis for responding requires, for its completion, the length $\hat{\epsilon}_i(\mu_{jr})$ of time.] So, at any time $t \in I$, P_i has at his "effective disposal" only the information about μ_{ju} [the opponent's behavior during $[0, u)$], where moment u, which depends on t and (via μ_{jt}) on μ_j, is determined by

$$u = \sup\{r \in I : r + \hat{\epsilon}_i(\mu_{jr}) \leq t\}$$

[here, $\sup(\emptyset) = 0$]; P_i, knowing μ_{ju}, reaches and acts on a decision about his strategy at this moment t. In other words, at any moment $t \in I$, player P_i "effectively knows" the history of his opponent's behavior only up to just before time $t - \epsilon_i(\mu_j, t)$, where

$$\epsilon_i(\mu_j, t) = t - \sup\{r \in I : r + \hat{\epsilon}_i(\mu_{jr}) \leq t\},$$

and he can react to P_j's behavior only to within this *backward delay* $\epsilon_i(\mu_j, t)$. In the sequel, it will be more convenient to use the backward delay function $\epsilon_i(\mu_j, t)$, $(\mu_j, t) \in M_j \times I$ (called shortly a *delay*), instead of the forward delay function $\hat{\epsilon}_i(\mu_{jr})$. To give rigorous expression to this intuitive idea, we first define the class D_i of those delay functions $\epsilon_i : M_j \times I \mapsto R$ that have the following four properties.

(i) For all $(\mu_j, t) \in M_j \times I$,

$$0 \leq \epsilon_i(\mu_j, t) \leq t, \tag{4}$$

(ii) Let $(\mu_j, t) \in M_j \times I$ and $\epsilon_i(\mu_j, t) = \epsilon$. Then for any $\bar{\mu}_j \in M_j$ satisfying $\mu_{jt-\epsilon} = \bar{\mu}_{jt-\epsilon}$,

$$\epsilon_i(\mu_j, t) = \epsilon_i(\bar{\mu}_j, t). \tag{5}$$

(iii) For each $\mu_j \in M_j$, the function

$$h(t) := t - \epsilon_i(\mu_j, t), \qquad t \in I, \tag{6}$$

which corresponds to the previous "u," is nondecreasing in t.

(iv) For any $r \in [0, 1)$ and $\bar{\mu}_j \in M_j$, there exists $t \in (r, 1]$ such that for all $\mu_j \in M_j$ with $\mu_{jr} = \bar{\mu}_{jr}$,

$$t - \epsilon_i(\mu_j, t) \leq r. \tag{7}$$

Note that for such a t, the delay appearing in (7) cannot be zero.

The class D_i is always nonempty. It is enough to note that, for instance, the delay function $\epsilon_i(\mu_j, t) = t$, for all μ_j and t, satisfies (4)–(7) and thereby belongs to D_i. Note that with this delay function, P_i is in effect made deaf (or, P_j is made silent), since it is never able to react to the opponent's behavior. In view of the "reaction delay" interpretation given earlier, properties (i)–(iii) are natural enough. Property (iv), needed for the later theoretical development, can also be easily interpreted. Namely, it says [jointly with condition (iii)] that for for any moment $r < 1$, there is at least a short subsequent interval during which P_i is unable to observe and react to all of the opponent's behavior after that moment r. In other words, P_i cannot react instantly to P_j's behavior, and near the start of the play (take $r = 0$) must resort to preplanned behavior rather than "reacting" at all.

In the sequel, to simplify formulas, we shall also use the notation

$$\epsilon(i) := \epsilon_i.$$

Next, for any ϵ_i in D_i, we define, similarly to the definition of $\tilde{M}_i$, the subset $M_i^n(\epsilon_i)$ of $\tilde{M}_i$—nonempty since it contains M_i—to consist of all strategies for P_i of the form

$$\xi_{\epsilon(i)} = \xi_{\epsilon(i)}(\mu_j, t), \qquad (\mu_j, t) \in M_j \times I, \tag{8}$$

where the functions $\xi_{\epsilon(i)}$ satisfy the following two conditions, of which the first repeats (i) below (1), while the second is analogous to (ii) there.

(i) For each $\mu_j \in M_j$, there exists $\mu_i \in M_i$ such that $\xi_{\epsilon(i)}(\mu_j, t) = F(\mu_i, t)$ for all $t \in I$,

(ii) For each $t \in I$ and each μ_j and $\bar{\mu}_j$ in M_j,

$$\xi_{\epsilon(i)}(\mu_j, u) = \xi_{\epsilon(i)}(\bar{\mu}_j, u), \qquad u \in [0, t],$$

whenever, for some r, $0 \le r \le t$, $\mu_{jr} = \bar{\mu}_{jr}$ and

$$t - \epsilon_i(\mu_j, t) \le r, \qquad t - \epsilon_i(\bar{\mu}_j, t) \le r.$$

(To justify this property, it can be shown with the aid of (5) and (6) that at any time $u \in [0, t]$, P_i's "effective information" about the opponent's behavior is the same versus $\bar{\mu}_j$ as versus μ_j.)

With these definitions completed, we now define the strategy space for player P_i in game Γ_{nn} to be

$$M_i^n = \bigcup \{M_i^n(\epsilon_i) : \epsilon_i \in D_i\}. \tag{9}$$

It follows that $M_i \subset M_i^n \subset \tilde{M}_i$ (strict inclusions in general).

Remark 3.1. For the strategy space of a player P_i facing a noisy opponent in Γ_{nn}, we have motivated the restriction from $\tilde{M}_i$ to M_i^n by the need to avoid the paradox of "mutual instantaneous imitation." Another motivation for this restriction, valid for some applications but perhaps not for all, is that modelling P_i as capable of "instantaneous reaction" would exaggerate his real powers of *response* to an unacceptable degree. But then this would hold whether P_i is himself noisy or silent, i.e., whether P_i is involved in a noisy duel or as the "silent" player in a mixed duel. Such considerations suggest analyzing games in which the players' strategy spaces are M_i^n and M_j.

Remark 3.2. The strategy space defined by (9) may still be "too large," but for an entirely different reason. Its size reflects the idea that every delay function in D_i is a plausible choice for player P_i. But wouldn't the player want to choose from D_i a delay function allowing reactions "as fast as possible," so as to base his moment-by-moment decisions and actions on maximal information about the opponent's past behavior? Formalizing this idea may show that many of the strategies in (9) are dominated, and therefore eliminable, in detailed analyses of specific models.

An implicit assumption in the last comment, and indeed throughout the present paper, is that the activities of a noisy player, once detected and reactable-to by the opponent, play that role without error—ironically, they would be called "noise free" in the language of the "signal-versus-noise" distinction. In a model enlarged by relaxing such assumptions, P_i might prefer longer delays if they permit sharper estimation of the opponent's past behavior and hence of his current resource stock.

Exploring the ideas in the two previous paragraphs appears worthwhile. But we choose here to tackle the complications of the present formulation, arising from the research's original motivation, before (and, we think, as a necessary basis for) introducing still-further complexities.

Remark 3.3. Since the structure of strategies in M_i^n can be rather complex, it is useful to describe a relatively simple construction that yields a wide variety of such strategies. Namely, consider any $\epsilon_i \in D_i$ such that, for each $\mu_j \in M_j$, $\epsilon_i(\mu_j, t)$ is right-hand-continuous in t. Then, for any $\xi_i \in \tilde{M}_i$, it is easy to verify that the strategy $\hat{\xi}_{\epsilon(i)}$ defined by

$$\hat{\xi}_{\epsilon(i)}(\mu_j, t) = \xi_i(\mu_j, t - \epsilon_i(\mu_j, t)), \qquad (\mu_j, t) \in M_j \times I, \tag{10}$$

with the property that (as a function of t) it lies in M_i for all $\mu_j \in M_j$, also lies in $M_i^n(\epsilon_i)$. (Membership in M_i seems plausible, since it would be strange if P_i, reacting to some μ_j with an allowable delay function, became able to perform some distribution he could not carry out in the basic game.) Denote the set of all these $\hat{\xi}_{\epsilon(i)}$ by $\hat{M}_i^n(\epsilon_i)$. One could ask here whether the set $\hat{M}_i^n = \bigcup\{\hat{M}_i^n(\epsilon_i)\}$, where the union runs through all right-hand-continuous (in t) functions $\epsilon_i \in D_i$, coincides with M_i^n. Unfortunately, in general, the answer is no, which is not difficult to show. Namely, first note that after setting $r = 0$ in (7), (4) and (6) imply the following: for any ϵ_i, there exists $t_0 > 0$ such that $\epsilon_i(\mu_j, t) = t$ for all $t \in [0, t_0)$ and all $\mu_j \in M_j$. Therefore, any strategy $\hat{\xi}_{\epsilon(i)}$ of the form (10) prescribes player P_i not to distribute any of his resources in an open interval of the form $(0, t_0)$ with $t_0 > 0$. Since, in general, not all members in M_i, and thereby also in M_i^n, possess this property [see the inclusion below (9)], the asserted strict inclusion $\hat{M}_i^n \subset M_i^n$ follows.

We conclude this section by noting the subsets $M_i^n(\epsilon_i)$ of M_i^n that arise from certain natural and important choices of $\epsilon_i \in D_i$. We consider the following three cases.

Case 1: For a constant $\epsilon > 0$, let $\epsilon_i(\mu_j, t) = \min(\epsilon, t)$ for all $t \in I$ and $\mu_j \in M_j$. Then, obviously, $\epsilon_i \in D_i$. The strategies in $M_i^n(\epsilon_i)$ are those available to P_i if he can observe and react to the opponent's behavior only with the constant delay ϵ throughout the game.

Case 2: Here the member ϵ_i of D_i is a function $\epsilon_i = \epsilon_i(t)$ of only the one variable t. Thus, the delay in P_i's current information about the opponent's past actions depends only on the current moment t, and not on those past actions themselves. Here (5) is automatically satisfied, but the delay function must still be constrained by (4), (6), and (7).

Case 3: This is an interesting subcase of case 2. For some natural number m, we take $\epsilon_i(\mu_j, 1) = 0$, and for $k = 0, 1, \ldots, m-1$,

$$\epsilon_i(\mu_j, t) = t - k/m \qquad \text{if } k/m \le t < (k+1)/m$$

for all $\mu_j \in M_j$. It is easy to see that (4)–(7) hold, so that $\epsilon_i \in D_i$. The idea here is that P_i can observe P_j's (past) behavior only at the discrete set of moments $0, 1/m, 2/m, \ldots, 1$.

4 The Main Theorem

In this section we shall show that, under an additional but not very restrictive assumption on $M_i, i = 1, 2$, sets M_1^n and M_2^n constructed in Section 3 will indeed serve well to describe the strategy spaces for the players in the game Γ_{nn}. The basic result is Theorem 4.1 below, which states that the choice of such strategies by both players will indeed lead to a definite pair of distributions of resources, this pair lying in the domain $M_1 \times M_2$ of K. This result will be proven under the following assumption, which will also (Remark 4.2) be shown to be necessary for the result to hold.

Assumption: For any sequence $\{\mu_i^m\}$ in M_i with $\mu_{i,t(m)}^m = \mu_{i,t(m)}^{m+1}$ for some $0 \le t(1) < t(2) < \cdots < 1$ ($i = 1, 2, m = 1, 2, \ldots$), there exists a measure $\bar{\mu}_i \in M_i$ such that $\bar{\mu}_{i,t(m)} = \mu_{i,t(m)}^m$ for all $m \ge 1$.

Remark 4.1. It may be helpful to illustrate both what this assumption permits and what it forbids. Let t^* denote the limit of its ascending sequence $\{t(m)\}$, let l_i be the limit of the (bounded, nondecreasing) sequence $\{\mu_i^m([0, t(m)))\}$, and let $e_i \ge 0$ denote the excess of P_i's initial resources E_i over l_i. In the situation described, P_i can follow the distribution μ_i^1 in $[0, t(1))$, then follow μ_i^2 in $[t(1), t(2))$, etc., so that for all m the distribution up to just before $t(m)$ is exactly that of μ_i^m (element of M_i). Thus, in $[0, t^*)$, P_i has distributed a total of l_i. The claim of the assumption is that he can distribute some (maybe all or none) of the remainder e_i during $[t^*, 1]$

in at least one way, which, according to whatever criteria define M_i, is allowable both "in itself" and in how it fits with the previous distribution over $[0, t^*)$.

Suppose, for example, that beyond the total-resources limitation the only requirement for membership in M_i is the "smoothness" condition of absolute continuity, so that failure of differentiability for the cumulative distribution at t^* is "excusable." Then P_i can meet the assumption's claim simply by choosing a ceasefire during $[t^*, 1]$. On the other hand, suppose membership in M_i requires not only absolute continuity, but also expenditure of all of P_i's resources. To see that the assumption fails in this case, take the sequence $\{t(m)\}$ to be such that $t^* = 1$, and—as is possible—choose each μ_i^m in M_i to expend no resources during $[0, t(m))$. Then the desired $\bar{\mu}_i$ would need a burst of activity at $t = 1$, which is incompatibile with absolute continuity.

Theorem 4.1. *For any* $\xi_{\epsilon(1)} \in M_1^n$ *and* $\xi_{\epsilon(2)} \in M_2^n$ *given by* (8)*, there are unique measures* $\mu_1^* \in M_1$ *and* $\mu_2^* \in M_2$ *such that*

$$\xi_{\epsilon(1)}(\mu_2^*, u) = F(\mu_1^*, u), \qquad \xi_{\epsilon(2)}(\mu_1^*, u) = F(\mu_2^*, u), \qquad \textit{for } u \in I. \tag{11}$$

Proof. To demonstrate uniqueness, assume that the pairs (μ_1^*, μ_2^*), $(\mu_1^0, \mu_2^0) \in M_1 \times M_2$ both satisfy (11). Let

$$r = \inf\{u : \mu_{1u}^* \neq \mu_{1u}^0 \quad \text{or} \quad \mu_{2u}^* \neq \mu_{2u}^0\}. \tag{12}$$

(We take $\inf(\emptyset) = 1$.)

Then for $i = 1, 2$, we have $\mu_{iu}^* = \mu_{iu}^0$ for all $u < r$, which implies

$$\mu_{ir}^* = \mu_{ir}^0 \qquad \text{for } i = 1, 2. \tag{13}$$

Now, using condition (ii) below (8) with $t = r$, for $i = 1, 2$ we get

$$\xi_{\epsilon(i)}(\mu_j^*, u) = \xi_{\epsilon(i)}(\mu_j^0, u) \qquad \text{for } u \in [0, r]. \tag{14}$$

Hence, if $r = 1$, (11) implies that $\mu_i^* = \mu_i^0$ for $i = 1, 2$, the desired assertion of uniqueness.

Suppose, therefore, for the rest of this argument, that $r < 1$. Then, by (13) and (5–7), there exists t, $r < t \leq 1$, such that for $i = 1, 2$,

$$t - \epsilon_i(\mu_j^*, t) \leq r, \qquad t - \epsilon_i(\mu_j^0, t) \leq r. \tag{15}$$

Therefore, by (13) and (15), the assumptions of condition (ii) below (8) are satisfied with $\mu_i = \mu_i^*$ and $\bar{\mu}_i = \mu_i^0$, so that we have

$$\xi_{\epsilon(i)}(\mu_j^*, u) = \xi_{\epsilon(i)}(\mu_j^0, u) \qquad \text{for } u \in [0, t],$$

for $i = 1, 2$. But this equality and (11) imply that $\mu_{it}^* = \mu_{it}^0$ for $i = 1, 2$, contradicting (12) because of $r < t$. Thus, the proof of uniqueness is complete (without use of the assumption preceding Theorem 4.1). We now free the symbol "r" for other uses.

For the demonstration of existence, we shall construct a certain partially ordered system $(N, \leq)$, prove that any of its maximal elements yields a pair (μ_1^*, μ_2^*)

satisfying (11), and then use the assumption to show that $(N, \leq)$ satisfies the hypotheses of Zorn's lemma (a basic axiom that in an equivalent form generalizes to "larger infinities" the familiar principle of mathematical induction), which will then assure the existence of a maximal element.

For each $t \in I$, let N^t consist of all $(\mu_1, \mu_2) \in M_1 \times M_2$ such that

$$\xi_{\epsilon(1)}(\mu_2, u) = F(\mu_1, u), \qquad \xi_{\epsilon(2)}(\mu_1, u) = F(\mu_2, u), \qquad \text{for } u \in [0, t). \quad (16)$$

[we omit condition (16) when $t = 0$].

Finally, let $N = \bigcup\{\{t\} \times N^t : t \in I\}$. Note that N is nonempty because $N^0 = M_1 \times M_2 \neq \emptyset$.

It can be easily seen that, for each $t \in I$, the set N^t has the following property: for $\mu_i, \bar{\mu}_i \in M_i$, $i = 1, 2$,

$$\left[(\mu_1, \mu_2) \in N^t,\ \mu_{it} = \bar{\mu}_{it} \text{ for } i = 1, 2\right] \Rightarrow (\bar{\mu}_1, \bar{\mu}_2) \in N^t. \quad (17)$$

The partial order is defined as follows: $(t, \mu_1, \mu_2) \leq (r, \nu_1, \nu_2)$ holds if $[\, t < r$ and $\mu_{it} = \nu_{it}$ for $i = 1, 2\,]$ or $[\, t = r$ and $\mu_i = \nu_i$ for $i = 1, 2\,]$. Verifying that this binary relation has the defining properties of a partial order is straightforward: reflexive, transitive, and antisymmetric.

Suppose (τ, μ_1, μ_2) is a maximal element of N under this partial order. Then (16) holds for all $t \leq \tau$. By (i) below (8), for $i = 1, 2$ there exists $\bar{\mu}_i \in M_i$ for which

$$\xi_{\epsilon(1)}(\mu_2, u) = F(\bar{\mu}_1, u), \qquad \xi_{\epsilon(2)}(\mu_1, u) = F(\bar{\mu}_2, u), \qquad \text{for all } u \in I. \quad (18)$$

Therefore, $F(\mu_i, \cdot) = F(\bar{\mu}_i, \cdot)$ on $[0, \tau)$ and, consequently,

$$\mu_{i\tau} = \bar{\mu}_{i\tau}, \qquad i = 1, 2. \quad (19)$$

We next show that $\tau = 1$. If $\tau < 1$, then (6), (7) with its $r = \tau$, and (19) imply that for some ρ, $\tau < \rho \leq 1$, and $i = 1, 2$,

$$\rho - \epsilon_i(\mu_j, \rho) \leq \tau, \qquad \rho - \epsilon_i(\bar{\mu}_j, \rho) \leq \tau. \quad (20)$$

We easily see that (19) and (20) are the assumptions of condition (ii) below (8) with $r = \tau$ and $t = \rho$, so that

$$\xi_{\epsilon(i)}(\mu_j, u) = \xi_{\epsilon(i)}(\bar{\mu}_j, u), \qquad u \in [0, \rho], \quad (21)$$

for $i = 1, 2$.

But (18) and (21) lead to the equalities

$$\xi_{\epsilon(1)}(\bar{\mu}_2, u) = F(\bar{\mu}_1, u), \qquad \xi_{\epsilon(2)}(\bar{\mu}_1, u) = F(\bar{\mu}_2, u), \qquad \text{for all } u \in [0, \rho),$$

whence $(\rho, \bar{\mu}_1, \bar{\mu}_2) \in N$.

Therefore, in view of (19) and the inequality $\tau < \rho$, we get $(\tau, \mu_1, \mu_2) < (\rho, \bar{\mu}_1, \bar{\mu}_2)$, contradicting the maximality of (τ, μ_1, μ_2). Therefore, $\tau = 1$.

Suppose, therefore, that $(1, \bar{\mu}_1, \bar{\mu}_2)$ is a maximal element in $(N, \leq)$. Then, (19) and (20) hold with $\tau = \rho = 1$, so reasoning as before, (21) holds with $\rho = 1$. Hence, by (18), the pair $(\bar{\mu}_1, \bar{\mu}_2)$ provides the desired solution of (11).

In view of the above, to complete the proof it suffices to show that $(N, \leq)$ has at least one maximal element. By Zorn's lemma, we need only to show that any linearly ordered subset $\{(t^\alpha, \mu_1^\alpha, \mu_2^\alpha) : \alpha \in A\}$ has an upper bound in N. Note that $(t^\alpha, \mu_1^\alpha, \mu_2^\alpha) < (t^\beta, \mu_1^\beta, \mu_2^\beta)$ implies $t^\alpha < t^\beta$. Let $\bar{t} = \sup\{t^\alpha : \alpha \in A\}$. If $\bar{t} = t^\gamma$ for some $\gamma \in A$, then $(t^\gamma, \mu_1^\gamma, \mu_2^\gamma)$ is easily seen to provide the desired upper bound. So assume this is not the case.

Take any sequence $\{\alpha(m)\}_{m=1}^\infty$ of elements from A such that $\{t^{\alpha(m)}\}$ is an increasing sequence convergent to $\bar{t}$. Hence, it follows, by the assumption and the definition of the order in N, that there exists a pair $(\bar{\mu}_1, \bar{\mu}_2) \in M_1 \times M_2$ such that for $i = 1, 2$,

$$\bar{\mu}_{i,t(m)} = \mu_{i,t(m)}^{\alpha(m)}, \qquad m = 1, 2, \ldots \tag{22}$$

[here $t(m) = t^{\alpha(m)}$].

Since for all m, $(\mu_1^{\alpha(m)}, \mu_2^{\alpha(m)}) \in N^{t(m)}$, therefore, by (17) and (22), also $(\bar{\mu}_1, \bar{\mu}_2) \in N^{t(m)}$ for all m. Now taking into account the easily seen fact that $\bigcap_{m=1}^\infty N^{t(m)} = N^t$ for $t = \bar{t}$, we finally deduce that $(\bar{t}, \bar{\mu}_1, \bar{\mu}_2)$ is an upper bound in N of the linearly ordered subset considered above. Thus, the theorem has been proved. □

Remark 4.2. We have seen that the assumption stated before Theorem 4.1 was sufficient for the truth of that theorem, and it was certainly necessary for the particular proof we gave. For logical completeness, we now show that it is actually necessary in order to ensure the truth of the theorem in general. Consider the assumption-violating situation at the end of Remark 4.1: let $M_1 = M_2$ consist of all absolutely continuous distributions that exhaust the player's resources. Take any sequence $t(0) = 0 < t(1) < t(2) < \ldots$ convergent to 1. Generalizing case 3 in Remark 3.3, we take the delay-function class $D_1 = D_2$ to consist of the single function given by $\epsilon(t) = t$ in $[0, t(1))$, $\epsilon(t) = t - t(k)$ for $t \in [t(k), t(k+1))$ for each $k \geq 1$, and $\epsilon(1) = 0$. We can now define member ξ_ϵ of M_i^n (for both $i = 1$ and $i = 2$) as follows. In reaction to any μ_j of M_j:

(1) do nothing during $[0, t(1))$;
(2) during interval $[t(k), t(k+1))$, where $k \geq 1$, expend at a uniform rate the same total resource-quantity $\mu_j([t(k-1), t(k))$ that the opponent expended over the preceding interval. The action at $t = 1$ is determined by the resource-exhaustion requirement. If both players adopt ξ_ϵ in game Γ_{nn}, then they will expend no resources during $[0, 1)$, and so in order to to exhaust the resources, the μ_1^* and μ_2^* of (11) would require a burst of activity at $t = 1$, contradicting the absolute continuity. Thus, the theorem's conclusion does not hold for this case.

In view of the result of Theorem 4.1, we can define a map $C = (C_1, C_2)$: $M_1^n \times M_2^n \mapsto M_1 \times M_2$ by

$$C_1(\xi_1, \xi_2) = \mu_1, \qquad C_2(\xi_1, \xi_2) = \mu_2,$$

where (μ_1, μ_2) is the solution of (11) with $\xi_{\epsilon(i)} = \xi_i$. Now the payoff function K^n: $M_1^n \times M_2^n \mapsto R$ for Γ_{nn} can be defined by the natural

$$K_n(\xi_1, \xi_2) = K(C(\xi_1, \xi_2)), \tag{23}$$

so that we have the normal form $\Gamma = \langle M_1^n, M_2^n, K^n \rangle$.

To summarize our progress so far, we can observe that the payoff functions $\tilde{K}$ in Γ_{ns} and K^n in Γ_{nn} are indeed uniquely determined by the payoff function K in the game Γ_{ss} and are related to each other by

$$K^n(\xi_1, \xi_2) = \tilde{K}(\xi_1, \xi_2) \quad \text{if } (\xi_1, \xi_2) \in M_1 \times M_2^n, \tag{24}$$

$$K^n(\xi_1, \xi_2) = K(\xi_1, \xi_2) \quad \text{if } (\xi_1, \xi_2) \in M_1 \times M_2. \tag{25}$$

In addition, we also have

$$K^n(\xi_1, \xi_2) = K^n(\mu_1, \xi_2) \quad \text{if } C_1(\xi_1, \xi_2) = \mu_1. \tag{26}$$

Namely, eqs. (24) and (25) are a simple consequence of the two statements

$$C_k(\xi_1, \xi_2) = \xi_k, \quad \text{whenever } \xi_k \in M_k, \quad k = 1 \text{ or } 2,$$

$$C_2(\xi_1, \xi_2) = \xi_2(\xi_1, \cdot), \quad \text{whenever } (\xi_1, \xi_2) \in M_1 \times M_2^n,$$

which follow from the uniqueness result in Theorem 4.1 and then, together with (11) and (24), yield (26).

Remark 4.3. Theorem 4.1, together with the observation that the strategy spaces in duel Γ_{ss} satisfy the assumption, justifies taking M_1^n and M_2^n as strategy spaces for the players in Γ_{nn}. In this connection, one could ask whether these spaces are "maximal," in the sense that there are no sets greater than M_1^n and M_2^n for which that theorem remains true. The considerations of the first part of Section 4 show that it is not possible to replace the sets M_1^n, M_2^n by $\tilde{M}_1$ and $\tilde{M}_2$ since we would then lose the uniqueness in Theorem 4.1, and thereby the payoff function K^n in game Γ_{nn} would not be well defined [see (23)]. The set M_i^n is described with the help of the set D_i of delay functions ϵ_i satisfying (4)–(7). If we constructed the set D_i in another way to have $\epsilon_i \equiv 0 \in D_i$, then the definition of M_i^n [conditions (i) and (ii) below eq. (7)] would imply the equality $M_i^n = \tilde{M}_i$, which is not satisfactory. On the other hand, $\epsilon_i \equiv 0$ satisfies (4)–(6), though not (7). In view of that, the question set above can be replaced by the next one: is it possible in some interesting or natural way to replace condition (7) by a weaker one (this would make the set D_i greater, enlarging thereby the space M_i^n) in such a way to preserve both the existence and uniqueness of solutions in Theorem 4.1? Unfortunately, we do not know the answer.

Acknowledgments

The authors thank the referees for several useful comments and suggestions. The research of T. Radzik was supported by KBN Grant no. 211589101 under contract 664/2/91. A. J. Goldman's research was supported in part by NSF Grant ECS-9111548.

REFERENCES

[1] Barfoot, C. B. Continuous-time markov duels: Theory and application, *Naval Res. Logistics*, **36**(3): 243–253, 1989.

[2] Blackwell, D. and Girshick, M. A. *Theory of Games and Statistical Decisions*, John Wiley, New York, 1954.

[3] Cegielski, A. Tactical problems involving uncertain actions, *J. Opt. Th. Appl.* **49**(1):81–105, 1986.

[4] Elliott, R. J. and Kalton, N. J. The existence of value in differential games, *Mem. Amer. Math. Soc*, 1972.

[5] Elliott, R. J. and Kalton, N. J. The existence of value in differential games of pursuit and evasion, *J. Diff. Eqs.*, **12**(3):504–523, 1972.

[6] Fleming, W. H. The convergence problem for differential games, *J. Math. Anal. Appl.*, **3**, pp. 102–116, 1961.

[7] Fox, M. and Kimeldorf, G. S. Noisy duels, *SIAM J. Appl. Math.*, **17**(2):353–361, 1969.

[8] Friedman, A. On the definition of differential games and the existence of value and of saddle points, *J. Differential Eqs.* **7**, pp. 69–91, 1970.

[9] Karlin, S. *Mathematical Methods and Theory in Games, Programming, and Economics*, Vol. 2, Reading, MA, 1959.

[10] Kimeldorf, G. S. and Lang, J. P. Asymptotic properties of non-discrete duels, *J. Appl. Prob.* **14**, pp. 153–161, 1977.

[11] Kurisu, T. On a duel with time lag and arbitrary accuracy functions, *Intl. J. Game Theory*, **19**, pp. 43–56, 1991.

[12] Lang, J. P. and Kimeldorf, G. S. Silent duels with non-discrete firing, *SIAM J. Appl. Math.* **31**, pp. 99–110, 1976.

[13] Positielskaya, L. N. Non-discrete noisy duels, *Eng. Cybernetics*, **2**, pp. 40–44, 1984. Engl. trans. **3**, pp. 25–29, 1984.

[14] Radzik, T. Games of timing related to distribution of resources, *J. Opt. Th. Appl.*, **58**(3):443–471, 1998.

[15] Radzik, T. Games of timing with resources of mixed type, *J. Opt. Th. Appl.*, **58**(3): 473–500, 1998.

[16] Radzik, T. General noisy duels, *Mathematica Japonica*, **36**(5):827–857, 1991.

[17] Radzik, T. Results and problems in games of timing, In T.S. Ferguson, L.S. Shapley, and J. MacQueen, eds., *Statistics, Probability and Game Theory. Papers in Honor of David Blackwell*, Lecture Notes-Monograph Series, v. 30, Institute of Mathematical Sciences, Hayward, CA, pp. 269–292, 1996.

[18] Restrepo, R. Tactical problems involving several actions, In Contributions to the Theory of Games III, *Ann. Math. Stud.*, **39**, Princeton Univ. Press, Princeton; pp. 313–335, 1957.

[19] Roxin, E. The axiomatic approach in differential games, *J. Opt. Th. Appl.*, **3**, 153–163, 1969.

[20] Ryll-Nardzewski, C. A theory of pursuit and evasion, In *Advances in Game Theory, Ann. Math. Studies*, **52**, Princeton Univ. Press, Princeton, pp. 113–126, 1964.

[21] Varaiya, P. and Lin, J. Existence of saddle points in differential games, *SIAM J. Cont.*, **7**(1):141–157, 1969.

[22] Yanovskaya, Y. B. Duel-type games with continuous firing, *Eng. Cybernetics*, **1**, pp. 15–18, 1969.

A Historical Perspective on Cooperative Differential Games

Alain Haurie
University of Geneva
LOGILAB/HEC
Geneva, Switzerland

Abstract

This paper provides a brief historical perspective on the use of cooperative solution concepts in the theory of differential and dynamic games. The paper surveys the publications that appeared in control journals, like *Automatica*, the *Journal of Optimization and Applications* (JOTA), the *SIAM Journal on Control and Optimization*, the *IEEE Transactions on Automatic Control*, and the proceedings of the successive Differential Games Symposia, during a period that spans from the origin of differential game theory, in the early 1960s, until 1990. The survey concentrates on solution concepts that imply Pareto optimality with respect to the reward criteria of the different players. The survey does not include the topic of team theory.

1 Introduction

From the very beginning of the development of the theory of games, it has been recognized that the players could be placed in a situation where the mode of play does not correspond to a confrontation of interests but rather to a cooperation where they may enter into a bargaining process for dividing the benefits of a coordination of their actions. The theory would then predict the outcome that could result from a fair sharing of the dividends of the cooperation. The development of the theory of differential games in the late 1950s and early 1960s permitted the explicit consideration of time in the representation of the rules of the game. This provided a new set of tools for the study of the cooperation between agents. This brief paper surveys the main contributions in this domain. To have an idea of the time span during which these developments occurred, we may recall that

- Rufus Issacs' book, [15], totally devoted to two-player, zero-sum differential games, became available in 1964;

- in 1969 Jim Case published the first paper in English[1] introducing m-player differential games [4], while Starr and Ho [47] and [46], were the first to introduce cooperative solutions in differential games;
- Typical cooperative game solution concepts, like the *Nash bargaining solution* or *the Core*, have been studied in a dynamic and differential game context in the mid 1970s, [33], [20], [21], [18], [44], [49];
- Finally, in the 1980s, the first papers dealing with a solution concept that has sometimes been called *cooperative equilibrium* appeared: a subgame perfect equilibrium that includes a preplay agreement among the players on the coordinated use of a threat–punishment mechanism.

This paper has been written as an attempt to capture, in a broad perspective, the main development of a theory of cooperative dynamic games using the paradigm of differential games or *generalized optimal control problems*. I apologize in advance for the possible omissions of significant contributions, the list of references having been built only to the best of my (necessarily limited) knowledge.[2] The listing of references is by order of first appearance in the text. I have decided that *team theory* is not a branch of the theory of cooperative games.[3] I have also decided arbitrarily that "history" ended in 1990 ...

The paper is organized as follows: in Section 2 the cooperative game concepts introduced in the early days of the theory of games are briefly reviewed; in Section 3 the cooperative solution concepts of the "classical" theory of games, which were going to be adapted to the theory of cooperative differential games, are also briefly reviewed; in Section 4 the contribution of mathematical programming and control theory to the theory of cooperative solutions is reviewed; in Section 5 the cooperative solutions in dynamic games are considered; in conclusion, the potential future development of the theory of cooperative differential games is assessed.

2 The Early Days

The theory of cooperation between economic agents takes its origin from economic analysis. I believe that it is appropriate to refer to [9] and [37] for the first "operational" and "formal" definition of a cooperative outcome. Both used the formalism of ordinal utility functions.

---

[1]It seems that Petrosjan and Murzov had already published a paper, in Russian [38], dealing with noncooperative differential games.

[2]In particular, I have had very limited access to the publications written in Russian, so I don't pretend that a correct representation of the early contributions to dynamic cooperative game made in the former USSR is given in this paper.

[3]The main reason for this decision was only to limit the scope of my work. Indeed there are many similarities between the results of team theory and those of cooperative game theory.

2.1 The Edgeworth Contract Curve

F. Y. Edgeworth, in 1881 [9], considers the case of two agents deriving utility from the allocation of two goods that they exchange in the amount x and y from initial holdings $(a, 0)$ and $(0, b)$, respectively. The utility functions of the two agents are $F(x, y)$ and $\Phi(x, y)$, respectively. He arrives at the conclusion that the settlements they will consent to are determined as the locus of points (x, y) verifying

$$F_x(x, y)\Phi_y(x, y) - F_y(x, y)\Phi_x(x, y) = 0. \tag{1}$$

This locus is called the *contract curve*. It is also the locus of tangency points between pairs of indifference curves. Edgeworth gave three justifications for condition (1). The third one shows that a settlement means Pareto efficiency: "... motion is possible so long as, one party not losing, the other gains ..."

2.2 Pareto Efficiency

An allocation is efficient in the sense of Pareto [37] if it is impossible to improve the utility (satisfaction) of one agent without decreasing the utility (satisfaction) of at least one other agent. Consider again two agents and represent in a two-dimensional space the possible utilities the agents can obtain. The *Pareto frontier* is the set of achievable utility pairs that are not dominated. More formally, let $\mathcal{F} \subset \mathbf{R}^2$ be the set of achievable utility pairs (ω_1, ω_2). A pair (ω_1^*, ω_2^*) is undominated in $\mathcal{F}$ if

$$\begin{aligned} (\omega_1, \omega_2) &\in \mathcal{F} \\ (\omega_1, \omega_2) &\geq (\omega_1^*, \omega_2^*) \\ &\Longrightarrow \\ (\omega_1, \omega_2) &= (\omega_1^*, \omega_2^*). \end{aligned} \tag{2}$$

3 Classical Game Theory

3.1 von Neuman and Morgenstern

In their seminal book, founding the classical theory of games in 1944, von Neumann and Morgenstern [51] go through a rationale that I would summarize as follows:

1. Start with a game in extensive form involving a set M of players $j \in M$. Define the strategies as mappings from information to moves and associate a matrix game with the strategic form, also called the *normal form* of the game.
2. Introduce a cardinal utility function so that the players can always resort to randomized (mixed) strategies $u = (u_j)_{j \in M}$ and strive to maximize their respective utilities $J_j(u)$.
3. Prove existence of saddle points in the zero-sum matrix games played with mixed strategies.

4. Assume that the players can transfer their utilities, so the utility of a group S of players, called a *coalition*, is described by

$$J_S(u) = \sum_{j \in S} J_j(u).$$

5. Define the *characteristic function* of the game as the mapping $v : S \mapsto \mathbf{R}$ that associates with every coalition $S \subset M$ the guaranteed utility[4]

$$v(S) = \max_{u_S} \min_{u_{M-S}} J_S(u).$$

6. Introduce the concept of an *imputation* $\omega = (\omega_j)_{j \in M}$ as a proposed payment to each player and the idea of "blocking":
a coalition blocks an imputation ω if

$$\sum_{j \in S} \omega_j < v(S).$$

Now a variety of cooperative solution concepts can be introduced with the help of the concept of blocking. von Neumann and Morgenstern proposed the concept of *stable set*. I will not elaborate on this concept, which found no adepts among the developers of the theory of differential games. The solution concept known as *the Core* has been more successful in this regard.

3.2 The Core

An imputation is in *the Core* if no coalition, including the M-set, can block it. This implies, in particular, that the imputation is an undominated, Pareto-efficient outcome.

3.3 The Core in the Case of Nontransferable Utilities

The concept of a coalition blocking an imputation, and the associated concept of the *Core*, has been extended to the case where the players cannot transfer utility to each other (see [2]). This yielded two concepts, the $\alpha - $ *Core* and the $\beta - $ *Core*. We shall only refer here to the $\alpha - $ *Core*.

The characteristic function is now defined as a mapping from 2^M to $2^{\mathbf{R}^m}$ that associates with each possible coalition $S \subset M$ the closure $v(S)$ of the set of imputations blocked by this coalition. So one defines $v(S)$ by

$$\omega \in v(S) \text{ if there exists } u_S \in U_S \quad \text{s.t. } \forall j \in S,\ \forall u_{M-S} \in U_{M-S},$$
$$J_j(u_S, u_{M-S}) \geq \omega_j.$$

---

[4]We use the obvious notation $u_S = 3D(u_j)_{j \in S}$.

3.4 A Plethora of Cooperative Solution Concepts

In addition to the *stable set* and the *Core*, there are many other cooperative solution concepts that exhibit different sorts of stability with respect to the possible coalitions formed by the players. In order to keep this survey very short, I will not enter into these developments. Furthermore, there are very few (actually, I don't know of a single one) uses of these solution concepts to the domain of differential games.

3.5 The Nash Bargaining Solution

The Nash bargaining solution has been introduced by J. Nash ([35] and [36]), as a way to identify a fair sharing of the benefits of cooperation. The solution concept was based on the following ingredients:

- an axiomatic description of fairness;
- the possibility to change cardinal utilities via affine transformations without changing the preferences;
- a status quo describing the outcome $(\hat{\omega}_1, \hat{\omega}_2) \in \mathcal{F}$ if the bargaining process fails to reach an agreement.[5]

Nash's remarkable result was that the players should aim at the settlement $({\omega_1}^*, {\omega_2}^*) \in \mathcal{F}$ that satisfies

$$({\omega_1}^*, {\omega_2}^*) = \arg \max_{(\omega_1, \omega_2) \in \mathcal{F}} (\omega_1 - \hat{\omega}_1)(\omega_2 - \hat{\omega}_2). \tag{3}$$

4 Computing Pareto Outcomes and Finding Elements of the Core; Mathematical Programming and Control

4.1 Vector Optimization in Mathematical Programming and Control

Pareto optimality according to criterion (2) leads very naturally to the theory of mathematical programming with a vector-valued objective function. The optimization of systems with performances evaluated along several criteria has a relatively recent history. L. Zadeh [53] initiated the research toward this direction in the control and optimization community. He has been closely followed by Klinger [29]. The basic idea is the scalarization approach, where the vector criterion $(J_j(u))_{j \in M}$ is replaced with the scalarized criterion $\sum_{j \in M} \alpha_j J_j(u)$ with a positive weighting scheme, $\alpha_j > 0,\ j \in M$.

Da Cunha and Polak ([6] and [7]) proposed the first precise set of optimality conditions for a system with a vector performance index.

[5] To simplify the exposition, I consider, here again, the two-player case.

4.2 Pareto Optimality in Differential Games

R. W. Reid and S. J. Citron [40] study the geometry of the noninferior set via the functional relationship between noninferior solution and the weighting vectors. Y. C. Ho [25] shows the interplay between control theory and cooperative differential games and he coins the term *generalized control theory*. In [3], [50], [31], [45], [41], [18], [32], and [30], the development of a theory of vector-valued optimization for dynamical systems is pursued and the problem of selecting a particular Pareto outcome in the set of undominated solutions is posed in terms of *compromise solution* ([41] and [18]). This concept is very close to Nash's bargaining solution, the only difference being the engineering flavor of compromise solution.

Another favorite exercise at this time has been to find examples where solutions would be simultaneously a Nash equilibrium and a Pareto solution ([4]).

4.3 Characteristic Functions and the Core

The characterization of the $\alpha - Core$ calls for the computation of the characteristic function $v(S)$. In [21], [18], [44], and [49], it is recognized that the computation of the set $v(S)$ reduces to the characterization of a set of vector maxmin-solution. Necessary conditions extending the results of nondifferentiable optimization were proved in these papers. Some approaches were based on the guaranteed reachability for dynamical systems ([15], [21], and [18]), while in [44] and [49] a more classical differential game framework was used.

5 The Development of a Theory of Cooperative Differential Games; Dynamic Games

The developments I referred to in the previous sections have often been presented in a control-theoretic context. However, the problems were, most of the time, akin to the paradigm of optimization in vector (Banach) space. The time dimension was not genuinely exploited to convey a different view or interpretation to the idea of a cooperative solution. In this last part of the survey I shall try to address the concepts that have really benefited from the new light shed by the differential (or more generally dynamic) game formalism. In a general formalism, we consider a dynamical system described by the state equation

$$\begin{aligned} \dot{x}(t) = \quad & f(x(t), u_1(t), \ldots, u_m(t), t), \\ x(0) = \quad & x^0 \end{aligned} \tag{4}$$

and a payoff (reward) functional

$$J_j(x^0, u(\cdot)) = \int_0^T L_j(x(t), u_1(t), \ldots, u_m(t), t)\, dt. \tag{5}$$

5.1 Variations Around Pursuit–Evasion Games

The class of *pursuit–evasion games* provides the archetype of differential games. They were called *games of kind* by Isaacs [15]. Medanic and Andjelic [34] proposed an interesting framework mixing the theory of cooperative games and the paradigm of zero-sum differential games. In their model a player controls a dynamic system, while the opponent controls the "target site," that is, selects one cost functional among several possible ones. The problem can be formulated as follows:

$$\min_u \max_i J_i(x_0, u(\cdot)).$$

The authors introduce the relaxed problem

$$J_m(u(\cdot), c, x_0) = \sum_i c_i J_i(x_0, u(\cdot)),$$

where the weighting $c_i \geq 0$ satisfies $\sum_i c_i = 1$ and show that the solution of the previous problem is also obtained from the solution of

$$\min_u \max_c J_m(u(\cdot), c, x_0).$$

Ho, [26], in a comment on a claim of [34], provided an interesting cooperative game interpretation. In [27], Hogan studies a differential game with two pursuers and a single evader. The two pursuers use the Nash bargaining point to determine their cooperative pursuit strategy.

5.2 Bargaining Solution

In 1973 Liu [33], proposed a direct extension of the Nash bargaining solution to the context of nonzero-sum differential games with feedback strategies. It was then shown in [20] that the solution concept was lacking stability and would not benefit, in most cases from the property of *time consistency*. More precisely, in applying directly the Nash bargaining solution concept to a differential game or, more generally, a dynamic game, the solution would be valid only for the initial state. This comes from the fact that the status-quo point $(\hat{\omega_1}, \hat{\omega_2})$ will change along the Pareto optimal trajectory $(t, x(t))$ generated by the Nash bargaining solution obtained at the initial state $t = 0, x(0) = x^0$. The same type of time inconsistencies were to be found in *the Core* concept; see [19]. Petrosjan has made many publications, in Russian, on this problem of time inconsistency and on the possible regularization techniques.

5.3 Coalition Dynamics

The "state" of a cooperative game could include, in its description of the current situation, a representation of the coalition structure. This is the idea sketched, e.g., in [39]. However, the endogenous dynamics of coalitions is not straightforward, and very little has been done in this direction of research.

5.4 Equilibria with Trigger Strategies

A celebrated result in the classical theory of games is known as the *Folk theorem*; see, e.g., [13]. This result shows that some Pareto optimal settlements can be transformed into subgame perfect equilibria in a repeated game context, through the use of *trigger strategies*. Other similar results have been obtained for stochastic games with imperfect information, where an equilibrium dominating the Nash–Cournot one could be sustained by trigger strategies. These results had an immediate impact on the theory of cooperative differential and dynamic games. First, the use of trigger strategies is permitted in games where each player can recall the whole past trajectory of the system under control. In fact, the class of memory strategies had already been introduced in a differential game framework when Friedman [12] and Krassovski and Subbotin [28] proposed a rigorous theory of zero-sum differential games. In cooperative differential games one generally assumes that the players have agreed on a cooperative mood of play in a preplay negotiation. The use of trigger strategies permits them to implement this cooperative play as an equilibrium. Papers [23], [24], and [48] have looked at the definition of equilibria in the class of memory strategies and have shown the possible uses of the concept in typical economic applications like fishery management [16], [17] or capitalist-worker competition [22].

6 Conclusion

There is still considerable room for developing a dynamic theory of cooperation. Differential games represent very well the changing of the state of the system that results from the players' actions. On the one hand, most of the theories of cooperative behavior in m-person games completely bypass the time dimension or reduce it to the paradigm of repeated games. On the other hand, the attempts for implementing cooperative solutions in differential games have consisted mainly in translating the axiomatic definition of the solution into an ad-hoc optimization problem and solve it. Maybe an interesting future lies in the definition of an axiomatic that would include precise assumptions on the use of time by the players involved in the cooperative process.

Acknowledgments

This research was supported by FNRS.

REFERENCES

[1] Aumann, R. Acceptable points in general cooperative n-person games, *Annals of Mathematical Studies*, **40**, 287–324, 1954.

[2] Aumann, R. A survey of cooperative games without side payments, in M. Shubik, ed., *Essays in Mathematical Economics*, Princeton University Press, 1969.

[3] Basile, G. and Vincent, T. L. Absolutely cooperative solution for a linear, multiplayer differential game, *JOTA*, **6** (1):41–46, 1970.

[4] Case, J. Toward a theory of many-player differential games, *SIAM J. Cont. and Opt.*, **7** (2):179–197, 1969.

[5] Case, J. A class of games having Pareto optimal nash equilibria, *JOTA*, **13** (3):379–386, 1974.

[6] DaCunha, N. O. and Polak, E. *Constrained minimization under vector-valued criterion in finite-dimensional spaces*, UC Berkeley, Electronics Research Laboratory, Memorandum no. ERL-M188, 1966.

[7] DaCunha, N. O. and Polak, E. Constrained minimization under vector-valued criterion in linear topological spaces, in A. V. Balakrishnan and L. W. Neustadt, eds., *Mathematical Theory of Control*, Academic Press, New York, 1967.

[8] Isaacs, R. *Differential Games*, John Wiley, New York, 1964.

[9] Edgeworth, F. Y. *Mathematical Psychics*, C. Kegan Paul & Co., London, 1881. See also Francis Ysidro Edgeworth, by Peter Newman, *The New Palgrave, Utility and Probability*, W. W. Norton & Co., New York, 1990.

[10] Ehtamo, H., Ruusunen, J., and Hämäläinen, R. P. Solution for a dynamic bargaining problem with application to resource management, *JOTA*, **59** (2):391–405, 1988.

[11] Ehtamo, H., Ruusunen, J., and Hämäläinen, R. P. A hierarchical approach to bargaining in power pool management, *IEEE Trans. on Auto. Cont.*, **Ac-34**, pp. 666–669, 1989.

[12] Friedman, A. *Differential Games*, Wiley-Interscience, New-York, 1971.

[13] Fudenberg, D. and Maskin, E. The Folk theorem in repeated games with discounting and with incomplete information, *Econometrica*, **54**, pp. 533–554, 1986.

[14] Gao, L., Jakubowski, A., Klompstra, M. B., and Olsder, G. J. Time-dependent cooperation in games, in T. Başar and P. Bernhard, eds., *Differential Games and Applications*, Lecture Notes in Control and Information Sciences, Vol. 119, Springer-Verlag, Berlin, 1989.

[15] Goffin, J. L. and Haurie A. Necessary conditions and sufficient conditions for Pareto optimality in a multicriterion perturbed system, *Proc. 5th IFIP Conference on Optimization Techniques*, Rome, Springer-Verlag, 1973.

[16] Hämäläinen, R. H., Kaitala, V., and Haurie, A. Bargaining on whales: A differential game model with Pareto optimal equilibria, *Oper. Res. Let.*, **3** (1):5–11, 1984.

[17] Hämäläinen, R. P., Kaitala., V., and Haurie, A. Equilibria and threats in a fishery management game, *Optimal Cont. Appl. Meth.*, **6** (1):315–333, 1985.

[18] Haurie, A. On Pareto optimal decision for a coalition of a subset of players, *IEEE Trans. Auto. Cont.*, **AC-18** (2), April 1973.

[19] Haurie, A. On some properties of the characteristic function and the Core of a multistage game of coalitions, *IEEE Trans. Auto. Cont.*, **AC-20** (2), April 1975.

[20] Haurie, A. A note on Nonzero-sum differential games with bargaining solutions, *JOTA*, **13** (1):31–39, 1976.

[21] Haurie, A. and Delfour, M. Individual and collective rationality in a dynamic pareto equilibrium, *JOTA*, **13** (3):290–302, 1974.

[22] Haurie, A. and Pohjola, M. Efficient equilibria in a differential game of capitalism, *J. Econ. Dyn. Cont.*, **11**, pp. 65–78, 1987.

[23] Haurie, A. and Tolwinski, B. Acceptable equilibria in dynamic bargaining games, *Large Scale Systems*, **6**, pp. 73–89, 1984.

[24] Haurie, A. and Tolwinski, B. Definition and properties of cooperative equilibria in a two-player game of infinite duration, *JOTA*, **46** (4):525–534, 1985.

[25] Ho, Y. C. Differential games, dynamic optimization, and generalized control theory, *JOTA*, **6** (3):179–207, 1970.

[26] Ho, Y. C. Comment on a paper by Medanic and Andjelic, *JOTA*, **10** (3):187–189, 1972.

[27] Hogan, H. L. Preferred coalitions in cooperative differential games, *JOTA*, **13** (2):186–202, 1974.

[28] Krassovski, N. N. and Subbotin, A. I. *Jeux Différentiels*, Nauka, Moscow, 1977.

[29] Klinger, A. Vector-valued performance criteria, *IEEE Trans. Auto. Cont.*, **AC-9** (1), 1964.

[30] Kwon, Y. K. and Yu, P. L. Stabilization through taxation in n-person games, *JOTA*, **23** (2):277–284, 1977.

[31] Leitmann, G., Rocklin, S., and Vincent, T. L. A note on control space properties of cooperative games, *JOTA*, **9** (6):379–390, 1972.

[32] Lin, J. G. Maximal vectors and multi-objective optimization, *JOTA*, **18** (1):41–64, 1976.

[33] Liu, P. T. Nonzero-sum differential games with bargaining solutions, *JOTA*, **11** (3):284–292, 1973.

[34] Medanic, J. and Andjelic, M. On a class of differential games without saddle-point solutions, *JOTA*, **8** (6):413–430, 1971.

[35] Nash, J. The bargaining problem, *Econometrica*, **18**, 155–162, 1950.

[36] Nash, J. Two-person cooperative games, *Econometrica*, **21**, 128–140, 1953.

[37] Pareto, V. *Cours d'Économie Politique*, Sapurious Rouge, Lausanne, 1896.

[38] Petrosjan, L. A. and Murzov, N. V. Game of overpulling with many participants (in Russian), *Viestnik Leningradskogo Univ.*, **13**, 1967.

[39] Ray, A. and Blaquière, A. Sufficient conditions for optimality of threat strategies in a differential game, *JOTA*, **30** (1):99–109, 1981.

[40] Reid, R. W. and Citron, S. J. On noninferior performance index vectors, *JOTA*, **7** (1):11–27, 1969.

[41] Salukvadze, M. On the existence of solutions in problems of optimization under vector-valued criteria, *JOTA*, **13** (2):203–217, 1974.

[42] Tolwinski, B. A concept of cooperative equilibrium for dynamic games, *Automatica*, **18**, (4):431–441, 1982.

[43] Scarf, H. The Core of a n-person game, *Econometrica*, **35**, 50–69, 1967.

[44] Schmitendorf, W. E. and Moriarty, G. A sufficiency condition for coalitive Pareto-optimal solutions, *JOTA*, **18** (1):93–102, 1976.

[45] Stadler, W. Sufficient conditions for preference optimality, *JOTA*, **18** (1):119–140, 1976.

[46] Starr, A. W. and Ho, Y. C. Nonzero-sum differential games, *JOTA*, **3** (4):207–219, 1969.

[47] Starr, A. W. and Ho, Y. C. Nonzero-sum differential games, *JOTA*, **3** (3):184–206, 1969.

[48] Tolwinski, B., Haurie, A., and Leitmann, G. Cooperative equilibria in differential games, *J. Math. Anal. Appl.*, **119**, pp. 182–202, 1986.

[49] Varaiya, P. N-player stochastic differential games, *SIAM J. Cont. and Opt.* **14** (3):538–545, 1976.

[50] Vincent, T. L. and Leitmann, G. Control space properties of cooperative games, *JOTA*, **6** (2):91–113, 1970.

[51] von Neumann, J. and Morgenstern, O. *Theory of Games and Economic Behavior*, Princeton University Press, 1944.

[52] Yu, P. L. and Leitmann, G. Compromise solutions, domination structures and salukvadze's solution, *JOTA*, **13** (3):362–378, 1974.

[53] Zadeh, L. A. Optimality of non-scalar-valued performance criteria, *IEEE Trans. Auto. Cont.*, **AC-8** (1), 1963.

Certainty Equivalence Principle and Minimax Team Problems

Pierre Bernhard
University of Nice
and
I3S CNRS-UNSA
Sophia Antipolis, France

Naira Hovakimyan
National Academy of Sciences
of Armenia
Institute of Mechanics
Yerevan, Armenia

Abstract

We consider a team problem, with two decision makers for simplicity, where the uncertainties are dealt with in a minimax fashion rather than in a stochastic framework. We do not assume that the players exchange information at any time. Thus, new ideas are necessary to investigate that situation. In contrast with the classical literature, we do not use necessary conditions, but investigate to what extent ideas from the (nonlinear) minimax certainty equivalence theory allow one to conclude here. We are led to the introduction of a "partial-team" problem, where one of the decision makers has perfect state information. We then investigate the full-team problem, but the main result concerning it is shown still to be rather weak. We nevertheless apply it to the linear quadratic case, where it yields an original result.

1 Introduction

The origin of this research was in nonlinear robust control, as explained in [4]. There we show that the natural nonlinear equivalent to the (linear) $\mathcal{H}_\infty$-optimal control approach is *not* the so-called "nonlinear $\mathcal{H}_\infty$" control problem, but a minimax team problem. By this we mean a team problem where the uncertainties are dealt with in a minimax fashion, looking for a guaranteed outcome, rather than in a stochastic fashion, looking for a mean outcome.

As far as we know, explicit results published for stochastic dynamic team problems all involve either a form of imbeddedness of the information available to the players, as in [10] and [7], or other very special features (see [1], for instance). In the case of minimax team problems, often investigated under the umbrella of

Nash equilibria, the results we are aware of almost always assume a dicrete-time model with a one-step delay information sharing pattern (see, e.g., [8] and [14], except for some asymptotic results for weak coupling, see [15]).

Here we attempt to deal with nonimbedded information, without any information sharing. While the result obtained is rather weak, as stressed by the section dealing with an abstract analysis of that result, it allows us to give an original result for the continuous-time linear quadratic problem. It would clearly be a simple matter to derive the equivalent result for the discrete-time equivalent.

As a matter of fact, our basic derivation applies to a nonlinear setting, since it uses the basically nonlinear minimax certainty equivalence principle. Practically, however, there is little hope to get tractable results other than to the LQ problem or to some simple low-dimensional nonlinear problems, as was done with the basic principle.

2 The System Considered

2.1 Dynamics and Cost Function

Consider a team of two decision makers, whom we call *players* for short, each controlling different actions and having access to different information. There is a common payoff for both players, which they have to minimize. (In [4] the case is considered, when their dynamics are separated with respect to all variables.) To be more precise, let $x(t)$ be the state variable of the system and let u_1, u_2, and w respectively denote the control variables for each of the players and the disturbance, in terms of which the dynamic equation of system's evolution over time in the nonlinear general setup can be presented as

$$\dot{x} = f(t, x, u_1, u_2, w), \qquad x(t_0) = x^0, \tag{1}$$

where $t \in [t_0, +\infty)$, $x(t) \in \mathbb{R}^n$, $u_i \in \mathbb{R}^{m_i}$, $i = 1, 2$, $w \in \mathbb{R}^l$. The disturbance variable w is dealt with by considering the "guaranteed performance level," leading to a formulation in terms of a minimax or dynamic game problem.

The players' controls and the disturbance obey the following restrictions:

$$\begin{pmatrix} u_1 \\ u_2 \end{pmatrix} \in \mathsf{U} := \mathsf{U}_1 \times \mathsf{U}_2, \qquad w \in \mathsf{W},$$

where the U_1, U_2, and W are compact convex sets in appropriate spaces. The sets $\mathcal{U}_i$ and $\mathcal{W}$ of admissible open-loop controls $u_i(\cdot)$ and $w(\cdot)$ will contain all measurable functions from $[t_0, \infty)$ into U_i and W, respectively.

Under the necessary regularity assumptions (specified below), we shall denote for a given initial time $t_0 \in \mathbb{R}$ by $x(.) = S(t_0, x^0, u_1(.), u_2(.), w(.))$ the unique (Cauchy) solution of system (1). (The first argument in S is often omitted.)

We shall consider the following performance index:

$$J = M(x(T)) + \int_{t_0}^{T} L(t, x, u_1, u_2, w)\, dt + N(x^0), \tag{2}$$

where L, M, and N are given differentiable functions from the appropriate spaces into $\mathbb{R}$.

Regularity assumptions. We shall assume that the functions f, L are of class C^1 and that a growth condition that holds on f guarantees the existence of a unique solution S to (1) over $[t_0, T]$ for any $(U, W) \in (\mathcal{U}, \mathcal{W})$.

The precise formulations of the problems depend on information structures and are given in the following sections.

2.2 The Classical Game Problem Formulation and Its Solution

Let us introduce the standard problem in perfect information, that is, with admissible strategies in state feedback, namely, of the form

$$u_i = \varphi_i(t, x), \qquad i = 1, 2,$$

and recall the classical Hamilton–Jacobi–Isaacs solution ([11] and [3]). Given the initial time and state (t_0, x^0), determine, if it exists, the Isaacs' upper-value function:

$$V(t_0, x^0) = \min_{u_1} \min_{u_2} \max_{w} \tilde{J}, \tag{3}$$

where $\tilde{J} = J - N(x^0)$.

Proposition 2.1. *If there exists a C^1-function $V : [t_0, T] \times \mathbb{R}^n \to \mathbb{R}$, solution of the partial differential equation*

$$-\frac{\partial V}{\partial t} = \min_{u_1 \in \mathsf{U}_1} \min_{u_2 \in \mathsf{U}_2} \max_{w \in \mathsf{W}} H\left(t, x, \frac{\partial V}{\partial x}, u_1, u_2, w\right) \tag{4}$$

with boundary condition

$$\forall x, \; V(T, x) = M(x),$$

where

$$H(t, x, \mu, u_1, u_2, w) = L + \langle \mu, f \rangle,$$

is the Hamiltonian of the system (the angled brackets $\langle \cdot, \cdot \rangle$ denote the scalar product in $\mathbb{R}^n$), then the upper value of the game (3) is $V(t_0, x^0)$. Moreover, if the Hamiltonian has a saddle point in (u, w) for all (x, μ), and if there exist admissible strategies

$$u(t) = \varphi^*(t, x(t)) = \begin{pmatrix} \varphi_1^*(t, x(t)) \\ \varphi_2^*(t, x(t)) \end{pmatrix}, \qquad w(t) = \psi^*(t, x(t)), \tag{5}$$

which are a saddle point of $H(t, x, \partial V/\partial x, u_1, u_2, w)$, then they are optimal.

Hereafter, φ^*, ψ^*, together with V, will be referred to as the *Isaacs solution.*

Standing assumption. In the sequel, we always assume that the Isaacs solution exists and is unique.

3 The State-Feedback Partial-Team Problem

3.1 Statement of the Problem

In the problem investigated in this section the players have different informations about the evolution of the system over time: we shall suppose that the first player (indicated by subindex 1) can measure only

$$y_1 = h_1(t, x, w), \tag{6}$$

while the second one (indicated by subindex 2) has access to exact and instantaneous state measurement.

For any function $a(\cdot) : t \to a(t)$, we shall use the notation a^τ for its restriction to $[t_0, \tau]$. Notice, that with a mild abuse of notation we may write causality of S as $S^\tau(x^0, u_1, u_2, w) = S^\tau(x^0, u_1^\tau, u_2^\tau, w^\tau)$.

The admissible strategies are thus of the form

$$u_1(t) = \varphi_1(t, y_1^t), \qquad u_2(t) = \varphi_2(t, x(t)). \tag{7}$$

For arbitrary initial conditions (t_0, x^0), call disturbances the pairs $\omega := (x^0, w) \in \Omega := \mathbb{R}^n \times \mathcal{W}$, and define $X^0 := \mathbb{R}^n$. We shall consider several information structures beyond (7), where $x^0 \in X^0$ is not known to the players. This is why we have added the "initial cost" $N(x^0)$ in (2).

The *state-feedback partial-team problem* is the following: under the information structure (7), find optimal controls for the minimizing players, guaranteeing

$$\min_{\varphi_1} \min_{\varphi_2} \max_{\omega} J(t_0, x^0, \varphi_1, \varphi_2, w),$$

where J is given by (2).

3.2 Solution via the Minimax Certainty Equivalence Principle

We shall consider the system driven by $u_2(t) = \varphi_2^\star(t, x)$. If this strategy is fixed, then the problem for player 1 is a classical partial information minimax control problem, as in [2] or [6]. Observe also that the corresponding full state information minimax control problem for player 1 has φ_1^* and V as its unique solution.

Let the trajectories of this system be denoted by $x(\cdot) = S_1(x_0, u_1, w) = S_1(u_1, \omega)$, and denote by

$$\Omega_1^\tau(u_1^\tau, y_1^\tau) = \{\omega \in \Omega \mid h_1^\tau(\cdot, S_1(u_1, \omega), w(\cdot)) = y_1^\tau\}$$

the set of ω's that are compatible with the past observations of the first player. It is clear that $\forall t,\ \Omega_1^t \in \Omega$.

We introduce, following [2] or [6], the *auxiliary problem* : for every $\tau \in [t_0, T]$ and for any fixed u_1^τ and y_1^τ, we define a maximization problem in ω as follows. Let

$$\dot{x} = f(t, x, u_1^\tau(t), \varphi_2^\star(t, x), w), \qquad x(t_0) = x^0.$$

For $t \in [t_0, \tau]$, the state x is now a function of ω (and more specifically of its restriction ω^τ to $[t_0, \tau]$). Let

$$G(\tau, u_1^\tau, \omega) = V(\tau, x(\tau)) + \int_{t_0}^{\tau} L(t, x(t), u_1(t), \varphi_2^*(t, x(t)), w(t))dt + N(x^0),$$

The auxiliary problem is to characterize

$$\widehat{\Omega}_1^t = \arg \max_{\omega \in \Omega_1^t(u_1^\tau, y_1^\tau)} G(t, u_1, \omega)$$

and

$$\widehat{X}_1(t) = \{\hat{x}_1(t) \mid \hat{x}_1(\cdot) = S_1(u_1^t, \hat{\omega}), \text{ and } \hat{\omega} \in \widehat{\Omega}_1^t\}.$$

Following [2, 6], we assume

Crucial Assumption. For all pairs (ω, u_1, u_2) and for all $t \in [t_0, T]$, suppose that $\widehat{X}_1$ is a singleton [6].

Therefore, let $\widehat{X}_1 = \{\hat{x}_1(t)\}$. (This does not necessarily imply that $\widehat{\Omega}_1^t$ is a singleton.)

Theorem 3.1. *Under the crucial assumption above, the pair of optimal controls* $\varphi_1^*(t, \hat{x}_1(t))$, $\varphi_2^*(t, x(t))$ [$\varphi_i^*(t, x(t)$ *being defined by* (5)] *solves the partial team problem. Moreover,*

$$\max_{\omega \in \Omega} J(t_0, \varphi_1^*(t, \hat{x}_1(t)), \varphi_2^*(t, x(t)), \omega) = \max_{x^0 \in X^0} [V(t_0, x^0) + N(x^0)]. \tag{8}$$

Proof. Apply the certainty equivalence principle of [2] to the problem obtained replacing everywhere u_2 by $\varphi_2^*(t, x)$ in the original problem (1)(2)(6). This yields (8). Now, notice that the value (8) is equal to the Isaacs value, corresponding to the full information case, and [on the basis of uniqueness of $\hat{x}_1(t)$] can be presented as

$$\max_{\omega \in \Omega} J(t_0, \varphi_1^*(t, \hat{x}_1), \varphi_2^*(t, x), \omega) = V(t_0, \hat{x}_1^0) + N(\hat{x}_1^0). \tag{9}$$

Hereafter we shall call the strategy $\hat{u}_1(t) = \varphi_1^*(t, \hat{x}_1(t))$ the partial-team strategy of the first player.

It is clear that a similar result can be proved for the second player, supposing that the first player has complete knowledge of the system's evolution in time, whereas the second one has access only to an output map, other than (6):

$$y_2 = h_2(t, x, w). \tag{10}$$

The partial-team strategy of the second player, solving his or her partial team problem, will be

$$\hat{u}_2(t) = \varphi_2^*(t, \hat{x}_2(t)), \tag{11}$$

$\hat{x}_2$ being generated by a similar auxiliary problem, corresponding to measurement equation (10), well posed for player 2. The value of the game in this case is also equal to Isaacs' value and can be presented as

$$\max_{\omega \in \Omega} J(t_0, \varphi_1^*(t, x), \varphi_2^*(t, \hat{x}_2), \omega) = V(t_0, \hat{x}_2^0) + N(\hat{x}_2^0). \tag{12}$$

Denote the worst disturbance in this case by $\widehat{\Omega}_2 = (\hat{x}_2^0, \hat{w}_2)$ and by $\widehat{\Omega}_2^t$ the set of worst disturbances compatible with the observations of the second player.

Notice that $\hat{x}_1(t)$ and $\hat{x}_2(t)$ are the estimates of the *same* state variable $x(t)$ from the viewpoint of two player-minimizers 1 and 2, depending on their information, given by (6) and (10), respectively. In addition, $\widehat{\Omega}_1$ and $\widehat{\Omega}_2$ are the worst disturbances in the corresponding contexts.

One can conclude on the basis of (8), (9), and (12), that

$$\hat{x}_1^0 = \hat{x}_2^0 = \arg\max_x [V(t_0, x) + N(x)]. \qquad \square$$

4 The Full-Team Problem

4.1 A Direct Analysis

Consider now the case when both players have different imperfect measurements of the state variable (6), (10), that is, the admissible strategies are of the form

$$u_1(t) = \varphi_1(t, y_1^t), \qquad u_2(t) = \varphi_2(t, y_2^t). \tag{13}$$

Write, for short, $\hat{\varphi}_1$ for $\varphi_1^*(t, \hat{x}_1)$ and $\hat{\varphi}_2$ for $\varphi_2^*(t, \hat{x}_2)$.

The *full-team problem* is the following: under the information structure (6),(10),(13), find optimal controls for player-minimizers, guaranteeing

$$\min_{\varphi_1} \min_{\varphi_2} \max_{\omega} J(t_0, x_0, \varphi_1, \varphi_2, w),$$

where J is defined by (2).

We need the following assumption.

Crucial Assumption. The crucial assumption of the partial-team problem for player 1 and the corresponding one for player 2 hold, and in addition, the functional $\omega \mapsto J(\hat{\varphi}_1, \hat{\varphi}_2, \omega)$ has a unique local and global maximum (e.g., it is quasiconcave).

Technical Assumptions.

- The value function in (4) and the optimal strategies φ^* are continuously differentiable with respect to x.
- The optimal controls $u_i = \varphi_i^*(t, x(t))$ are interior to the sets U_i along an optimal trajectory.
- Under the the partial-team strategies, the “worst-case” estimates $\hat{x}_1, \hat{x}_2$ satisfy filterlike equations as in [9]:

$$\dot{\hat{x}}_i = \hat{f}_i(t, \hat{x}_i, y_i) \tag{14}$$

with $\hat{f}_i$ continuously differentiable in $\hat{x}_i$.

Remark 4.1. Such filterlike equations were derived in [9] for the case where the disturbance variable is split into a dynamics disturbance, say w, and additive measurement disturbance(s), say $y_i = h_i(t,x) + v_i$, and the integrand in the performance index is a sum $L = L_0(t,x,u,w) + K(t,x,u,v_1,v_2)$. A similar approach can be used to derive filter equations in the more general setting of this paper, but they would lack the appealing character of Didinsky's filter of [9].

Theorem 4.1. *Under the above two assumptions, the pair of control laws* $\hat{\varphi}_1 = \varphi_1^*(t,\hat{x}_1)$, $\hat{\varphi}_2 = \varphi_2^*(t,\hat{x}_2)$ *solves the full-team problem. Moreover,*

$$\max_{\omega\in\Omega} J(t_0,\hat{\varphi}_1,\hat{\varphi}_2,\omega) = \max_{x^0\in X^0}[V(t_0,x^0) + N(x^0)].$$

Proof. Recall first that the "filter equations" (14) for $i = 1, 2$ have the same initial state:

$$\hat{x}^0 = \arg\max_x [V(t_0,x) + N(x)].$$

Consider the following system of differential equations:

$$\begin{cases} \dot{x} = f(t,x,\varphi_1^*(t,\hat{x}_1),\varphi_2^*(t,\hat{x}_2),w), & x(t_0) = x^0, \\ \dot{\hat{x}}_1 = \hat{f}_1(t,\hat{x}_1,y_1), & \hat{x}_1(t_0) = \hat{x}^0, \\ \dot{\hat{x}}_2 = \hat{f}_2(t,\hat{x}_2,y_2), & \hat{x}_2(t_0) = \hat{x}^0. \end{cases} \tag{15}$$

the following performance index:

$$J(t_0,\omega) = M(T,x(T)) + \int_{t_0}^{T} L(t,x,\hat{\varphi}_1,\hat{\varphi}_2,w)dt + N(x^0), \tag{16}$$

and the goal

$$\max_{\omega} J(t_0,\omega). \tag{17}$$

The system (15), (16), (17) presents a classical optimal control problem with respect to the control variable w and the initial state x^0. Under the technical assumption, it can be investigated via Pontryagin's maximum principle [12].

The following notations are accepted below:

$$\frac{\partial\varphi_i^*}{\partial x} = \frac{\partial\varphi_i^*}{\partial x}(t,x), \qquad \frac{\partial\hat{\varphi}_i}{\partial\hat{x}_i} = \frac{\partial\varphi_i^*}{\partial x}(t,\hat{x}_i).$$

Let us write the Hamiltonian and necessary conditions:

$$H(t,x,\hat{x}_1,\hat{x}_2,\lambda,\hat{\lambda}_1,\hat{\lambda}_2,w) = L + \lambda' f + \hat{\lambda}_1^t \hat{f}_1 + \hat{\lambda}_2^t \hat{f}_2,$$

where $\lambda, \hat{\lambda}_1, \hat{\lambda}_2$ are the corresponding adjoint variables for $x, \hat{x}_1, \hat{x}_2$.

We write the adjoint equations for the maximization problem (17):

$$\begin{cases} -\dot{\lambda}^t = \dfrac{\partial L}{\partial x} + \lambda^t \dfrac{\partial f}{\partial x} + \hat{\lambda}_1^t \dfrac{\partial \hat{f}_1}{\partial y_1} \dfrac{\partial h_1}{\partial x} + \hat{\lambda}_2^t \dfrac{\partial \hat{f}_2}{\partial y_2} \dfrac{\partial h_2}{\partial x}, \\ -\dot{\hat{\lambda}}_1^t = \left(\dfrac{\partial L}{\partial u_1} + \lambda^t \dfrac{\partial f}{\partial u_1} \right) \dfrac{\partial \hat{\varphi}_1}{\partial \hat{x}_1} + \hat{\lambda}_1^t \dfrac{\partial \hat{f}_1}{\partial \hat{x}_1}, \\ -\dot{\hat{\lambda}}_2^t = \left(\dfrac{\partial L}{\partial u_2} + \lambda^t \dfrac{\partial f}{\partial u_2} \right) \dfrac{\partial \hat{\varphi}_2}{\partial \hat{x}_2} + \hat{\lambda}_2^t \dfrac{\partial \hat{f}_2}{\partial \hat{x}_2}, \end{cases} \tag{18}$$

and the following boundary conditions:

$$\begin{cases} \lambda(t_0) + \dfrac{\partial N}{\partial x}(x^0) = 0, \\ \lambda(T) - \dfrac{\partial M}{\partial x}(T, x(T)) = 0, \\ \hat{\lambda}_i(T) = 0. \end{cases} \tag{19}$$

Pontryagin's principle [12] claims that the optimal control $\hat{w}$, solving the problem (15), (16), and (17), yields:

$$\max_w H(x, \hat{x}_1, \hat{x}_2, \lambda, \hat{\lambda}_1, \hat{\lambda}_2, w) = H(x, \hat{x}_1, \hat{x}_2, \lambda, \hat{\lambda}_1, \hat{\lambda}_2, \hat{w}). \tag{20}$$

Denote by $x^*(\cdot)$ the optimal trajectory, that is, the trajectory generated from x_0 by the strategies φ_i^* and ψ^*. Then we claim:

Proposition 4.1. *The following solves Pontryagin's necessary conditions* (18) *to* (20)*:*

$$x(t) = \hat{x}_1(t) = \hat{x}_2(t) = x^*(t), \tag{21}$$

$$\lambda(t) = \frac{\partial V}{\partial x}(t, x^*(t)), \tag{22}$$

$$\hat{\lambda}_1(t) = \hat{\lambda}_2(t) = 0, \tag{23}$$

$$\hat{w}(t) = \psi^*(t, x^*(t)). \tag{24}$$

Proof of the proposition. Observe that, if indeed the $\hat{\lambda}_i$'s are zero, then the first equations in (18) and (19) are the standard adjoint equations associated with the full-information Isaacs' equation, and because we have assumed V to be C^1 in x, its gradient solves them along an optimal trajectory. Also, because we have assumed that the optimal strategies φ_i^* take their values in the interior of the control sets U_i, with the proposed solution,

$$\frac{\partial L}{\partial u_1} + \lambda^t \frac{\partial f}{\partial u_1} = 0, \qquad \frac{\partial L}{\partial u_2} + \lambda^t \frac{\partial f}{\partial u_2} = 0,$$

so that the differential equations for $\hat{\lambda}_1$ and $\hat{\lambda}_2$ are homogeneous. Also, as their boundary condition are 0, $\hat{\lambda}_i \equiv 0$ is indeed a solution. The rest follows easily: the proposed w indeed maximizes the Hamiltonian.

As a consequence of the proposition, the disturbance generated by ψ^* solves the optimization problem (15), (16), (17). Therefore, the worst possible performance index under the proposed strategies for the team players is again Isaacs' value, and no causal controller can do better. □

4.2 An Abstract Analysis

An abstract view of the last result is as follows: Let $J : \mathcal{U} \times \mathcal{W} \to \mathbb{R}$ be a performance index from two Banach spaces $\mathcal{U}$ (here $\mathcal{U}_1 \times \mathcal{U}_2$) and $\mathcal{W}$ into the real line, with Fréchet derivatives $\partial J/\partial u$ and $\partial J/\partial w$, the first one jointly continuous in (u, w). Let $\Phi : \mathcal{W} \to \mathcal{U}$ be a set of admissible strategies for player 1 (here, causal strategies or state feedbacks) and $\widetilde{\Phi}$ a subset of Φ [here partial information decentralized strategies of the form (13)].

We make the following hypotheses on Φ:

- Φ is a vector space;
- Φ contains the constant maps, i.e., $\forall \bar{u} \in \mathcal{U}$, the map $w \mapsto \bar{u}, \forall w \in \mathcal{W}$ is in Φ.

Assume there exist $\varphi^* \in \Phi$ solution of the problem

$$\min_{\varphi \in \Phi} \sup_{w \in \mathcal{W}} J(\varphi(w), w)$$

and a *unique* w^* solution of the related maximization problem:

$$J(\varphi^*(w^*), w^*) = \max_{w \in \mathcal{W}} J(\varphi^*(w), w).$$

Lemma 4.1. *We have the following fact:*

$$\frac{\partial J}{\partial u}(\varphi^*(w^*), w^*) = 0.$$

Proof. Let $G(\varphi, w) = J(\varphi(w), w)$ and $\bar{G}(\varphi) = \sup_{w \in \mathcal{W}} G(\varphi, w)$. Place on Φ the topology of the uniform convergence. It is clear that G has a partial Fréchet derivative in φ, given by

$$\forall \psi \in \Phi, \qquad \frac{\partial G}{\partial \varphi}(\varphi, w) \cdot \psi = \frac{\partial J}{\partial u}(\varphi(w), w) \cdot \psi(w).$$

This partial derivative is continuous in (φ, w) (still for the topology of the uniform convergence on Φ). Hence, by Danskin's theorem [5], everytime the max in w is reached at a unique point $\hat{w}$, there exists

$$\frac{d\bar{G}}{d\varphi}(\varphi) = \frac{\partial G}{\partial \varphi}(\varphi, \hat{w}).$$

Since φ^* minimizes $\bar{G}$ over Φ, and we assumed the maximizing w^* to exist and be unique, we get

$$0 = \frac{d\bar{G}}{d\varphi}(\varphi^*) = \frac{\partial G}{\partial \varphi}(\varphi^*, w^*).$$

Hence, $\forall\psi \in \Phi$,

$$\frac{\partial J}{\partial u}(\varphi^*(w^*), w^*) \cdot \psi(w^*) = 0.$$

But we have also assumed that constant maps were in Φ. Hence,

$$\forall u \in \mathcal{U}, \qquad \frac{\partial J}{\partial u}(\varphi^*(w^*), w^*) \cdot u = 0,$$

whence the claim of the lemma. □

We add now a technical assumption: φ^* is differentiable.

Differentiating $J(\varphi^*(w), w)$ at w^*, we easily deduce that in addition to the above, $(\partial J/\partial w)(\varphi^*(w^*), w^*) = 0$.

Now let $\hat{\varphi} \in \widetilde{\Phi}$ (in our case our set of two partial-team strategies together) be differentiable and such that

- $\hat{\varphi}(w^*) = \varphi^*(w^*)$,
- $w \mapsto J(\hat{\varphi}(w), w)$ has a unique local and global maximum (say, is quasiconcave).

Then we have the following proposition.

Proposition 4.2. $\hat{\varphi}$ *solves the problem*

$$\min_{\varphi\in\widetilde{\Phi}} \sup_{w\in\mathcal{W}} J(\varphi(w), w).$$

Proof. Clearly, $J(\hat{\varphi}(w), w)$ is stationary at w^*, which suffices to show that its max in w is $J(\hat{\varphi}(w^*), w^*) = J(\varphi^*(w^*), w^*)$, which is the smallest possible $\sup_w J(\varphi(w), w)$ for φ ranging over Φ, and a fortiori for φ ranging over the subset $\widetilde{\Phi}$. □

This abstract version of our last result only stresses that it is indeed weak. One could *assume* that the payoff J with the *open-loop* controls $\hat{u}_i$ has only one local and global maximum in ω, and then these open-loop controls would constitute an optimal decentralized strategy. Hence, our result needs to be complemented by some kind of reciprocal, showing that if our "partial-team" strategies do not succeed, none will, in some sense. Investigation of this matter is in progress.

5 The Linear Quadratic Case

5.1 The System

Let the system with its observed outputs be

$$\begin{aligned} \dot{x} &= Ax + B_1u_1 + B_2u_2 + Dw, \\ y_1 &= C_1x + E_1w, \\ y_2 &= C_2x + E_2w. \end{aligned}$$

We shall assume that the pair (A, D) is stabilizable, that

$$\begin{pmatrix} D \\ E_1 \\ E_2 \end{pmatrix} \begin{pmatrix} D^t & E_1^t & E_2^t \end{pmatrix} = \begin{pmatrix} M & 0 & 0 \\ 0 & N_1 & 0 \\ 0 & 0 & N_2 \end{pmatrix},$$

with N_1 and N_2 positive definite.

Let the performance index be of the form

$$J = \|x(T)\|_X^2 + \int_0^T (\|x\|_Q^2 + \|u_1\|_{R_1}^2 + \|u_2\|_{R_2}^2 - \gamma^2 \|w\|^2)\, dt - \gamma^2 \|x^0\|_Y^2.$$

Finally, assume that $Q \geq 0$ and that the pair $(Q^{1/2}, A)$ is completely reconstructible.

Using, for instance, the theory of [2], it is a simple matter to apply the above theory to that minimax team problem.

5.2 The Isaacs Solution

The above system admits a state-feedback minimax control of the form

$$\varphi^*(t, x) = \begin{pmatrix} -F_1(t)x \\ -F_2(t)x \end{pmatrix},$$

with $F_i = R_i^{-1} B_i^t P(t)$, $i = 1, 2$, if, first, the Riccati equation

$$\dot{P} + PA + A^t P - F_1^t R_1 F_1 - F_2^t R_2 F_2 + \gamma^{-2} PMP + Q = 0, \qquad P(T) = X, \quad (25)$$

has no conjugate point, defining the symmetric matrix function $P(\cdot)$ over $[0, T]$.

Then, the worst disturbance is given by $w(t) = \psi^*(t, x(t))$ with $\psi^*(t, x) = \gamma^{-2} D^t P(t) x$. The value function for the game $\tilde{J}$ (i.e., without the initial term) is $V(x^0) = \|x^0\|_{P(0)}^2$, so that the minimax of the criterion exists if, second, $P(0) - \gamma^2 Y < 0$. It is known that this last condition can also be written in terms of $Z := Y^{-1}$ as $\rho(ZP(0)) < \gamma^2$.

In accordance with the above theory, the standing assumption will be that these conditions are fulfilled.

5.3 The Partial Team Problem

Let us now examine our partial-team problem. We set $u_2(t) = \varphi_2^*(t, x(t))$, so that the dynamics of the problem at hand are now

$$\dot{x} = A_1 x + B_1 u_1 + Dw,$$

with $A_1 := A - B_2 F_2$.

Also, the relevant criterion is

$$J_1 = \|x(T)\|_X^2 + \int_0^T (\|x\|_{Q_1}^2 + \|u_1\|_{R_1} - \gamma^2 \|w\|^2)\, dt + \|x^0\|_Y^2,$$

with $Q_1 := Q + F_2^t R_2 F_2$.

To apply the theory of [2] to this problem, we should first define a matrix $P_1(t)$-solution of a game Riccati equation written with the sole control u_1, and A and Q replaced by A_1 and Q_1. We notice, however, that under the standing assumption that P exists, this equation has a solution, namely, $P(\cdot)$ itself. Therefore, the two partial-team problems and the full state-feedback control problem share the same P-matrix (denoted Z in [2]).

Let us further write the dual Riccati equation of that problem:

$$\dot{\Sigma}_1 = A_1\Sigma_1 + \Sigma_1 A_1^t - \Sigma_1 C_1^t N_1^{-1} C_1 \Sigma_1 + \gamma^{-2}\Sigma_1 Q_1 \Sigma_1 + M, \qquad \Sigma_1(0) = Z. \tag{26}$$

The crucial assumptions for this partial-team problem are that

1. the Riccati equation (26) has no conjugate point over $[0, T]$,
2. $\forall t \in [0, T], \quad \rho(\Sigma_1(t)P(t)) < \gamma^2$.

Then, the "worst" state $\hat{x}_1(t)$ is given as the solution of the filter equation

$$\dot{\hat{x}}_1 = A_P\hat{x}_1 + K_1(y_1 - C_1\hat{x}_1), \quad \hat{x}_1(0) = 0, \tag{27}$$

with $A_P := A - B_1F_1 - B_2F_2 + \gamma^{-2}MP$ and $K_1 = (I - \gamma^{-2}\Sigma_1 P)^{-1} C_1^t N_1^{-1}$. The partial-team minimax strategy is

$$u_1(t) = -F_1(t)\hat{x}_1(t). \tag{28}$$

5.4 The Full-Team Problem

Let us now turn to the full-team problem. The maximization problem to investigate has its state in $\mathbb{R}^{3n}$ and is as follows:

$$\begin{pmatrix} \dot{x} \\ \dot{\hat{x}}_1 \\ \dot{\hat{x}}_2 \end{pmatrix} = \mathcal{A}\begin{pmatrix} x \\ \hat{x}_1 \\ \hat{x}_2 \end{pmatrix} + \mathcal{D}w, \qquad \begin{pmatrix} x(0) \\ \hat{x}_1(0) \\ \hat{x}_2(0) \end{pmatrix} = \begin{pmatrix} x^0 \\ 0 \\ 0 \end{pmatrix},$$

with

$$\mathcal{A} = \begin{pmatrix} A & -B_1F_1 & -B_2F_2 \\ K_1C_1 & A_P - K_1C_1 & 0 \\ K_2C_2 & 0 & A_P - K_2C_2 \end{pmatrix}, \qquad \mathcal{D} = \begin{pmatrix} D \\ K_1E_1 \\ K_2E_2 \end{pmatrix},$$

and the performance index to be maximized by w is

$$G = \|x(T)\|^2 + \int_0^T (\|x\|_Q^2 + \|\hat{x}_1\|^2_{F_1^t R_1 F_1} + \|\hat{x}_2\|^2_{F_2^t R_2 F_2} - \gamma^2\|w\|^2)\,dt - \gamma^2\|x^0\|_Y^2.$$

This maximization problem has a (unique) solution if, first, the Riccati equation for the $3n \times 3n$ block matrix $\Pi = (\Pi_{ij})$, $i, j = 0, 1, 2$,

$$\dot{\Pi} + \Pi\mathcal{A} + \mathcal{A}^t\Pi + \gamma^{-2}\Pi\mathcal{D}\mathcal{D}^t\Pi + \mathcal{Q} = 0, \qquad \Pi(T) = \mathrm{diag}\,(X, 0, 0), \tag{29}$$

with $\mathcal{Q} = \operatorname{diag}(Q, F_1^t R_1 F_1, F_2^t R_2 F_2)$, has no conjugate point over $[0, T]$ and, second $\Pi_{00} - \gamma^2 Y < 0$ or, equivalently, $\rho(Z\Pi_{00}(0)) < \gamma^2$.

Let us summarize:

Theorem 5.1. *If*

1. *the Riccati equation* (25) *has no conjugate point over* $[0, T]$,
2. *the Riccati equations* (26) *and the similar one reversing the indices 1 and 2 have no conjugate points over* $[0, T]$,
3. $\forall t \in [0, T], \quad \rho(\Sigma_i(t)P(t)) < \gamma^2,\ i = 1, 2,$
4. *the Riccati equation* (29) *has no conjugate point over* $[0, T]$,
5. $\rho(Z\Pi_{00}(0)) < \gamma^2$,

then the strategies (28) *where* $\hat{x}_1(\cdot)$ *is given by* (27), *and similarly mutatis mutandis for* u_2, *are team minimax optimal.*

[The condition on $\rho(ZP(0))$ has been omitted because it is implied by conditions (3) at time zero.]

Explicitly writing the six blocks of Riccati equation (29) is not very nice. It is nevertheless nothing more than an exercise in elementary matrix calculus. The noteworthy point about it is that it is not difficult to check on those equations that

$$\Pi(t)\begin{pmatrix} I \\ I \\ I \end{pmatrix} = \begin{pmatrix} P(t) \\ 0 \\ 0 \end{pmatrix}.$$

As a consequence, one can check, as should be expected, that if w is set to the worst disturbance, all three vectors x, $\hat{x}_1$ and $\hat{x}_2$ agree. [Also, the absence of a conjugate point to (29) implies the absence of a conjugate point to (25).]

REFERENCES

[1] Bansal, R. and Başar T. Solutions to a class of linear-quadratic-Gaussian (LQG) stochastic team problems with nonclassical information, *Sys. & Cont. Lett.*, **9**, pp. 125–130, 1987.

[2] Başar, T. and Bernhard P. *$\mathcal{H}_\infty$-optimal Control and Related Minimax Control Problems: A Dynamic Game Approach*, Birkhauser, Boston, 1991 (second edition: 1995).

[3] Bernhard, P. Differential Games: Isaacs' equation, in *Encyclopedia of Systems and Control*, M. Singh, ed., Pergamon Press, Pergamon, 1987.

[4] Bernhard, P. and Hovakimyan, N. Nonlinear robust control and minimax team problems, *Int. J. Robust & Nonlinear Cont.* **9**, pp. 239–257, 1999.

[5] Bernhard, P. and Rapaport, A. On a theorem of Danskin with an application to a theorem of Von Neumann-Sion, *Nonlinear Anal., Theory, Meth. & Appl.*, **24**, pp. 1163–1181, 1995.

[6] Bernhard, P. and Rapaport, A. Min-max certainty equivalence principle and differential games, *Int. J. Robust & Nonlinear Cont.*, **6**, pp. 825–842, 1996.

[7] Chu, K-C. Team decision theory and information structures in optimal control problems II, *IEEE-AC*, **17**, pp. 22–28, 1972.

[8] Didinsky, G. and Başar, T. Minimax decentralized controllers for discrete time linear systems, *Proc. 31st IEEE-CDC*, Tucson, AZ., December 1992.

[9] Didinsky, G., Başar, T., and Bernhard, P. Structural properties of minimax controllers for a class of differential games arising in nonlinear H^{∞} control, *Sys. & Cont. Lett.* **21**, pp. 433–441, 1993.

[10] Ho, Y-C. and Chu, K-C. Team decision theory and information structures in optimal control problems I, *IEEE-AC*, **17**, pp. 15–22, 1972.

[11] Isaacs, R. *Differential Games.* Wiley and Sons, New York, 1965.

[12] Pontryagin, L. S., Boltyanskii, V. G., Gamkrelidze, R. V., and Mischenko, E. F. *The Mathematical Theory of Optimal Processes*, John Wiley & Sons, New York, London, 1962.

[13] Radner R. Team decision problems, *Ann. Math. Stat.*, **33**, pp. 857–881, 1962.

[14] Srikant, R. Relationship between decentralized controller design using H_∞ and stochastic risk averse criteria, *IEEE-AC*, **39**, pp. 861–864, 1994.

[15] Srikant, R. and Başar, T. Approximate Nash strategies and suboptimal team solutions in Markov chains with weakly coupled decision makers, *Int. Yearbook on Game Theory & Appl.* **1**, pp. 65–82, 1993.

Evolutionary Processes in Signaling Games: The Impact of Different Learning Schemes

Gisèle Umbhauer
Bureau d'Economie Théorique et Appliquée
Strasbourg Cedex, France

Abstract

This paper deals with an evolutionary approach to two-player signaling games. The sender can be of two types, H and L. H, in contrast to L, is supposed to be better off being recognized. We stick to the case in which the H-sender's preferred equilibrium path is the Hellwig pooling equilibrium one. The studied games satisfy a single-crossing property, according to which an H-sender moves more easily to high messages than an L-sender does. Within this context, we bring to light the idea that different learning schemes can lead to different long-term behaviors. Fast learning seems to favor the Hellwig equilibrium, whereas smooth learning favors the perfect-revelation Riley equilibrium. Namely, we show that these results derive from a different use of the strategic asymmetries of the game. Fast learning exploits imitation but does not make heavy use of the single crossing assumption. By contrast, smooth learning perfectly exploits the single crossing property of the game.

1 Introduction

This paper deals with an evolutionary approach to extensive form signaling games. We aim to study the information transmission made possible by specific evolutionary models. We show that long-term behavior differs according to some specifications of the learning process, such as the speed of learning. We claim that fast learning processes lead to little information revelation in the studied games, whereas smooth learning processes rather favor information revelation. The reason for these contrasted results lies in a different exploitation of the properties of the games.

The games are two-player, two-type signaling games. The first player, the sender, has private information, called his *type*, which is either L or H. He sends a message to the second player, the receiver, who observes the message, but not the type, and chooses a response. We study the context in which the sender aims to be recognized by the receiver only if he is of type H. This context gives rise to different incentives about the revelation of information, which often result in the existence of quite different equilibria. In particular, we focus on signaling games with many perfect Bayesian equilibrium (PBE hereafter) paths, among them the

Hellwig [1] path, with no information revelation, and the traditional Riley [10] path, with complete information revelation. We stick to the case in which the player 1 of type H (H-player hereafter) prefers the Hellwig path to any other PBE path. So *we focus on the case that, a priori, is the least favorable to information revelation*, given that the H-player himself prefers an equilibrium outcome with no information transmission. This case is the most interesting from a strategic point of view: classical Nash refinements, and especially those based on forward induction, lead in this case to rather contrasted results with quite different information content (see, for example, [14]).

An immediate application is the market of experience goods: the sender is the producer, selling a product which can be of high (H) or low (L) quality. He starts the game by charging a price for the good. The receiver is the consumer, who observes the price but not the quality and buys a certain amount of the good. So this paper, which is partly from [15], displays links with the papers by Nöldeke & Samuelson ([8] and [9]), which deal with a screening-experience good market model, where each seller has only one unit to sell. The game begins with a buyer, who associates to each signal a reservation price depending on the expected quality. Then a seller, who knows the quality of his product, chooses a signal and sells the unit at the price the buyer charges. This approach gives rise to an evolutionary result, which is that many different equilibria (and even nonequilibria), with quite different information content, are observed over the long run.

Our paper starts with a similar result. The purpose is to go beyond this first result by examining the learning behavior of the agents. We show that, according to the specifications of the learning process, the actions played in the long run will not all be observed with the same probability.

The paper is organized as follows. Section 2 describes the model. Section 3 yields first insights into evolutionary results. Sections 4 and 5 establish that if many equilibrium outcomes appear in the long run, fast learning favors the Hellwig outcome, that is, no information transmission. Section 6 turns to a binomial learning process that makes both fast and slow learning possible. We show that the results on long-term behavior are more moderated than the ones obtained with only fast learning. Section 7 comments on the reasons for these contrasted results and some of their consequences. We show that a binomial learning process exploits structural asymmetries that cannot be exploited by fast learning. In particular, single crossing proves to have a stronger dynamic impact in the binomial process. These facts make it possible to give, in the experience good market model, some insights into the way in which prices can get a bad reputation (i.e., become assigned to low quality). Finally, Section 8 concludes with the consequences of smooth learning processes and gives additional comments. In smooth learning processes, single crossing has a main impact, leading us to conjecture great information revelation in long-run behavior.

2 Description of the Game and Forward-Induction Results

The sender's type t, with t belonging to $T = \{L, H\}$, is drawn according to a prior probability distribution $\rho(.)$, which is common knowledge of both players. We write H- (resp. L-) player, or H- (resp. L-) sender, for a player 1 of type H (resp. L).

Player 1 sends a message m, with m belonging to a set M, which can depend on t, in which case we write $M(t)$. Player 2, after observing the message, chooses a response r belonging to a set R, which can depend on m, in which case we denote it by $R(m)$. $M(.)$ and $R(.)$ are intervals of real numbers. Player 2's beliefs on the type, after observing m, are given by a posterior probability distribution $\phi(./m)$ on T. $\phi(./m)$ belongs to Δ_T, the set of probability distributions on T. The receiver's best response to m, when the beliefs are given by ϕ, is denoted by $r(m, \phi)$; when the beliefs concentrate on H (resp. L), we simply write $r(m, H)$ (resp. $r(m, L)$). $u(t, m, r)$ denotes the sender's payoff when he is of type t and sends m and player 2 responds with r.

Let's introduce some additional assumptions.

(i) $\forall t \in T,\ \exists\, \underline{m}_t \in M(t)$ such that $\forall \phi \in \Delta_T$: $u(t, m, r(m, \phi)) = 0$ for $m \leq \underline{m}_t$;
$\forall \phi \in \Delta_T\ \exists\, \widetilde{m}_\phi \in M(L) \cap M(H)$, $\widetilde{m}_\phi > \underline{m}_L$, $\widetilde{m}_\phi > \underline{m}_H$ such that
$\forall t \in T,\ u(t, m, r(m, \phi)) = 0$ for $m \geq \widetilde{m}_\phi$;
$u(t, m, r(m, \phi))$ is strictly concave in m, for m between $\underline{m}_t$ and $\widetilde{m}_\phi$,
and continuous in m, for fixed t and ϕ.

(ii) $\forall t \in T,\ \forall\, \phi$ and $\phi' \in \Delta_T,\ \forall\, m \in]\underline{m}_t, \widetilde{m}_{\phi'}[$ such that $\phi'(H/m) > \phi(H/m)$:
$u(t, m, r(m, \phi)) < u(t, m, r(m, \phi'))$;
$\forall t \in T,\ \forall\, \phi$ and $\phi' \in \Delta_T,\ \forall\, m \in]\underline{m}_t, \widetilde{m}_{\phi'}[$ such that $\phi'(H/m) = \phi(H/m)$:
$u(t, m, r(m, \phi)) = u(t, m, r(m, \phi'))$;

(iii) $\forall \phi \in \Delta_T,\ \forall\, m \in]\underline{m}_L, \widetilde{m}_\phi[$, $u(L, m, r(m, \phi)) > u(H, m, r(m, \phi))$.

(iv) $\forall m, m'$ with $m' > m$, $\forall r \in R(m)$, $\forall r' \in R(m')$ with $u(H, m', r') > 0$:
$u(L, m', r') \geq u(L, m, r) \Rightarrow u(H, m', r') > u(H, m, r)$.

Assumption (i) ensures that the function $u(t, m, r(m, \phi))$ is well shaped, being strictly positive and concave between $\underline{m}_t$ and $\widetilde{m}_\phi$. $\underline{m}_t$, respectively $\widetilde{m}_\phi$, denotes the lowest, respectively the highest, message that yields a positive payoff to the t-player. The assumptions on $\underline{m}_t$ and $\widetilde{m}_\phi$ ensure that the sender can achieve a strictly positive payoff regardless of player 2's beliefs. By assumption (ii), the payoff is increasing in the probability assigned to H. This assumption is behind *the incentive of an L-player to imitate an H-player*. Assumption (iii) means that, ceteris paribus, an L-sender gets a higher payoff than an H one. Assumption (iv) is the *single-crossing property*. It ensures that if an L-player gets a higher payoff by sending a message m' higher than m, then the H-player gets a strictly higher payoff by doing so.

These assumptions find an immediate application in the experience good market (see [14]). In particular, the single crossing assumption captures the idea that higher prices are "easier" to charge by high-quality producers than by low-quality ones.

These games have generically an infinite number of PBE paths, among them the Riley equilibrium path and the Hellwig equilibrium path. We recall their characteristics hereby:

Definition 2.1. The Riley path, E_R, is characterized by the following:

The L- (resp. H-) player sends the message m_L (m_H), where m_L and m_H satisfy the following: m_L is the message sent by an L-player in a context of perfect and complete information; m_H is the revelating message that maximizes the H-player's payoff, under the constraint that the L-player has no incentive to mimic the message. Player 2 responds with $r(m_L, L)$ (resp. $r(m_H, H)$) after observing m_L (resp. m_H).

The Hellwig path, E_ρ, is characterized by the following:

Player 1 sends m_ρ regardless of type, and player 2 responds with $r(m_\rho, \rho)$, where m_ρ is the message that maximizes the H-player's payoff, when player 2 assigns each message to both types with the prior probabilities.

By definition, E_R is the H-player's preferred separating PBE path, whereas E_ρ is the H-player's preferred pooling path. In this paper, we focus on the case in which the H-player prefers the Hellwig path to the Riley one. In this case, the Hellwig path is the H-sender's preferred path among all PBE paths (see [14] for a justification). Within this framework, traditional Nash refinements do not converge to a same equilibrium selection. While most forward-induction criteria (among others [3]) select only the Riley outcome, hence claim for complete information revelation, some forward induction criteria select only the Hellwig outcome ([13]) or a subset of outcomes including the Hellwig one ([6]), hence advocate for much less information transmission (for the reasons for this contrasted selection, see, for example, [14]).

The studied case is depicted in Figure 1. Assumptions (i) to (iv), in particular single crossing, ensure that the interval $]m', m''[$ (where m' and m'' are defined as in Figure 1) exists.[1]

3 Learning–Mutation Processes

We stick to the class of *discrete evolutionary models in which the learning dynamic is perturbed by mutations*, pioneered by Kandori, Mailath & Rob [2], Nöldeke & Samuelson [7], Samuelson [11] and Young [17]. Few evolutionary processes work with extensive form games; that is why we turn, for a start, to the process in Nöldeke & Samuelson [7] (NS hereafter).

---

[1]The only assumptions specific to Figure 1, that do not follow from the above assumptions are $\forall m \in]m', m''[\ u(L, m, r(m, \rho)) < u(L, m_L, r(m_L, L))$ and $u(H, m, r(m, \rho)) < u(H, m_H, r(m_H, H))$.

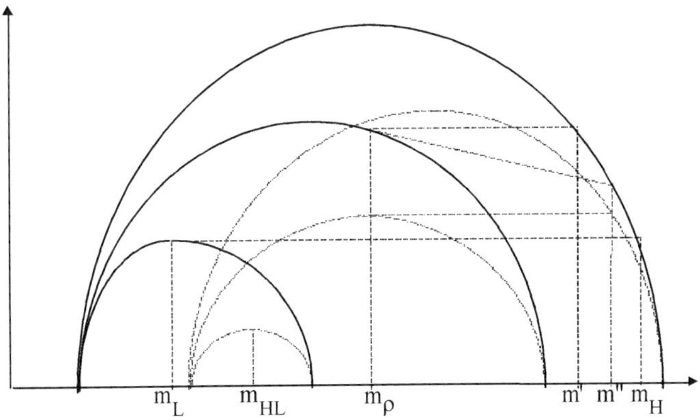

Figure 1: The continuous curves represent the L-sender's payoff when player 2 assigns the messages to L (curve 1), to both types with the prior probabilities (curve 2), and to H (curve 3); curve 1 is below curve 2, which is below curve 3. The broken curves represent the H-sender's payoff when player 2 assigns the messages to L (curve 1), to both types with the prior probabilities (curve 2), and to H (curve 3); curve 1 is below curve 2, which is below curve 3.

As usual in an evolutionary framework, each player i becomes a population of agents. The number of agents, Δ, is supposed to be finite. The game is repeated over and over, an infinite number of times. At each time t, each agent is randomly matched with a randomly drawn agent of the opposing population. Each agent, at time t, is described by a characteristic, which consists of his (pure) strategy, coupled with a conjecture, which specifies his beliefs (at each of his information sets) about the actions taken by the other agents. Agent 1's strategy has to specify an action for each possible type. A state of the system at time t is a description of how many agents in each population have each possible characteristic.

At each time t, each agent, when called on to play, learns with positive probability μ; if so, he updates his beliefs to match the observed actions and plays accordingly. Thus, he repeats his past action whenever it is a best response to the updated conjectures, and he switches to a best response—chosen according to a probability distribution that puts positive weight on all best responses—in the other case. If a player does not learn, his actions and beliefs are the preceding period's ones.

Besides having the possibility to learn, each agent may mutate. He mutates with positive probability λ, in which case he changes to any possible characteristic with strictly positive probability.

Coupling the learning and mutation processes leads to a Markov process, where γ_{ij}, the probability to switch from state i to state j, is strictly positive for all i and j.

NS are interested in the stationary distributions of these processes, namely, in the profiles of strategies and beliefs observed over the long run. Due to the support of the mutation mechanism, a unique stationary distribution exists for each λ. This stationary distribution has a limit, called the *limit distribution*, when λ goes to 0. NS establish that the support of the limit distribution restricts to the absorbing sets that are easiest to reach, in terms of mutations required when starting from the other absorbing sets. Consequently, as soon as two sets are at one mutation from the other (in both directions), both or neither appears in the long run (see NS for more details).

NS's process is defined for finite strategy sets. Hence, we discretize the strategy sets $M(t)$, $t \in T$, and $R(m)$, $m \in M(t)$, in order to keep them finite. We introduce a smallest and a highest message (response) for each $t \in T$ [after observing each $m \in M(t)$] and define a message (response) grid in the following way: the message (resp. response) closest to a given message m (resp. response r) is $m + \varepsilon$, $r + \varepsilon'$, where ε and ε' are the uniform step sizes of the message and response grids. ε and ε' are very small in order to allow (a) best replies to different actions to differ sufficiently and, (b) interesting values, such as m_ρ, not to be excluded from the strategy set. Different message and response step sizes for different types, and after observing different messages, could be introduced, but their impact on the study is not worth the additional notations.

To compare some of our results to the ones obtained in [9], we also arbitrarily restrict $M(.)$ and $R(.)$ to best responses to at least one system of conjectures (resp. on the agents 2's responses and on the types of the agents 1). Consequently, the set of PBEs is the same as the set of Nash equilibria. This introduced restriction is linked to the restriction in NS [9], who only allowed exploration in the beliefs space. It has only limited impact (we mention the impact in what follows each time it exists).

4 Fast-Learning Evolutionary Processes: First Results

In this section we stick to the learning process with $\mu = 1$, an assumption adopted in the papers by Nöldeke & Samuelson ([8],[9]) and Spence [12]. This assumption is very strong, but it makes it possible to shed light on some phenomena worth of study.[2]

The first result, similar to the one obtained in [8] and [9], is that NS's process (with $\mu = 1$) has little cutting power in the model, a fact derived from Propositions 4.1 and 4.2 and the proofs of those propositions.

Proposition 4.1. *The model has only two kinds of absorbing sets, "enlarged" PBE and cycles characterized by the H-agents send $\widetilde{m}$, all L-agents play $\widetilde{m}$ or*

[2]When μ equals 1, the process, due to the support of the mutations, is still an irreducible and aperiodic Markov process. Hence, it still has a unique stationary probability distribution. The proofs that establish the existence of a limit distribution when λ goes to 0 and the fact that only absorbing sets can appear in the limit distribution do not change.

m_L, the agents 2 respond with $r(m_L, L)$ after m_L and play $r(\widetilde{m}, H)$ or $r(\widetilde{m}, \rho)$ after $\widetilde{m}$, according to the play of the L-agents.[3]

If E_ρ or E_R is in the limit distribution, then both are in the limit distribution.

Proof. For the first part of the proposition, see [15].

For the second part of the proposition, it is enough to show that one mutation allows a switch from E_ρ to E_R and vice versa.

To switch from E_ρ to E_R, start in an E_ρ-state in which every message different from m_ρ is assigned to L, except for m_H assigned to H, and introduce a mutation from an H-agent to a message m^* between m' and m'' (where m' and m'' are defined as in Figure 1). It follows that the agents 2 learn and switch to $r(m^*, H)$ after this message. Then the H-agents switch to m^*, which induces the agents 2 to switch to $r(m_\rho, L)$ after m_ρ. This switch causes the L-agents to play m^*, too, which leads the agents 2 to respond with $r(m^*, \rho)$. Consequently, the L-agents switch to m_L and the H-agents switch to m_H.[4]

Conversely, start from an E_R-state and introduce a mutation from an H-agent to m_ρ. The agents 2 are induced to respond with $r(m_\rho, H)$ after this message; it follows that all agents 1 switch to m_ρ regardless of type. Then the agents 2 play $r(m_\rho, \rho)$ after m_ρ, which stops the system in $E\rho$. □

Proposition 4.2. *$E\rho$ appears in the limit distribution.*

Proof. By Proposition 4.1 we know that each absorbing set is either an equilibrium state or a cycle of two states. We show that each set is at one mutation from E_ρ, which ensures that E_ρ is in the support of the limit distribution.

- For one, if the absorbing set is a cycle with messages $\widetilde{m}$ and m_L, m_L being played only by the L-agents, then $\widetilde{m}$ is different from m_ρ; otherwise the L-agents would never play m_L (by single crossing, m_L followed by $r(m_L, L)$ yields a lower payoff than m_ρ assigned to both types with the prior probabilities). Hence, consider the state in this cycle where $\widetilde{m}$ is followed by $r(\widetilde{m}, \rho)$; in this state, by definition of m_ρ, the H-agents are worse off than in E_ρ. So suppose that an H-agent mutates to m_ρ. This sets off the following learning sequence: player 2 responds with $r(m_\rho, H)$ after m_ρ, which prompts all H- and L-agents to switch to m_ρ. Then, the agents 2 play $r(m_\rho, \rho)$ after m_ρ, causing the agents 1 to stick to this message, in that both types are better off

[3]By "enlarged" PBE, we mean a state in which the played actions correspond to a PBE path. Such a state is not necessarily an equilibrium, in that beliefs can be wrong. In the sequel of the paper, we always mean "enlarged PBE" when speaking about PBE, but, for the sake of easier notations, we omit the term "enlarged." If the actions of player 2 were not restricted to best responses to at least one system of beliefs, then we would obtain the possibility of cycles in which the L-agents, instead of the two messages $\widetilde{m}$ and m_L, would play in turn $\widetilde{m}$ and another message $\widetilde{m}'$, with $\widetilde{m}'$ assigned to L.

[4]If, by contrast to Figure 1, $u(H, m'', r(m'', \rho)) > u(H, m_H, r(m_H, H))$, it is still possible to switch from E_ρ to E_R by starting from an E_ρ-state in which all messages m belonging to $[m'', m_H]$ are assigned to H.

with m_ρ followed by $r(m_\rho, \rho)$ than with $\widetilde{m}$ followed by $r(\widetilde{m}, \rho)$ and with any message assigned to L.

- For another, consider a PBE-path E-state, with E different from E_ρ. The H-agents are better off in E_ρ than in E (this stems from the fact that the Hellwig equilibrium is preferred to the Riley equilibrium and from assumptions (i) to (iv); see [14] for more precision). There are two cases to examine: m_ρ either is or isn't played in E.
- If m_ρ is not played in E, it is enough to introduce a mutation from an H-agent toward m_ρ. Then m_ρ is assigned to H, causing all H-agents to switch to m_ρ: the L-agents switch to this message, too, immediately or after an adjustment of the agents 2's responses after the messages sent by the L-agents. It follows that m_ρ is assigned to both types, causing the H-agents to stay on this message. The L-agents stay on this message, too, or switch back to it, once the agents 2 have adjusted their responses after the possible messages sent by the L-agents. Hence, the learning sequence ends in E_ρ.
- If m_ρ is played, then the L-agents necessarily play this message. Either they play only m_ρ, or they play it together with another message. Assume first that the L-agents play only m_ρ. Here we use the restriction according to which player 2's strategy set reduces to best responses to at least one system of beliefs. Given this restriction, the H-agents necessarily play m_ρ with positive probability [otherwise m_ρ would be followed by $r(m_\rho, L)$ and the L-agents would not play m_ρ]. If all H-agents play m_ρ, then the system is already in E_ρ. If not, suppose that an additional H-agent switches to m_ρ; this mutation changes the agents 2's response after m_ρ (provided the response grid is sufficiently fine, as we suppose), causing all H-agents to play m_ρ; finally, the agents 2 play $r(m_\rho, \rho)$ after m_ρ, and the system is in E_ρ.

  Assume now that the L-agents mix on m_ρ and another message. Necessarily, by the best-response restriction on the receiver's strategy sets, the H-agents mix on m_ρ and another message (note that m_ρ is necessarily assigned to the H-type with a probability lower than the prior one). If one additional H-agent switches to m_ρ, the agents 2 adjust their response in a way that induces both the H- and L-agents to switch to m_ρ, causing the agents 2 to respond with $r(m_\rho, \rho)$ after m_ρ. This leads both types of sender agents to stay on m_ρ. □

Propositions 4.1 and 4.2 and the proof of Proposition 4.1 show that NS's process has no cutting edge, in that the proof of Proposition 4.1, together with the fact that E_ρ is in the limit distribution, clearly establish that the support of this distribution is very large. In fact, we studied the one-mutation switch from E_ρ to E_R, but we could have established one-mutation switches from E_ρ to many other equilibria or cycles. All these states appear in the long run. This result, close to some results of the screening model developed by Nöldeke & Samuelson [8], [9] and similar to some results obtained by Laffond [4] and Lesourne [5], is not astonishing, in that limit-distribution supports in extensive form games are often large when the strategy sets are large.

5 Fast Learning Processes: Second Insights

Up to now, the results have lent only partial support to the different forward-induction criteria and back both complete and null information revelation in the long run.

However, the dynamics of the process clearly highlight the different strategic positions H- and L-agents hold. For example, the *imitation incentives* of the L-agents show through the dynamics. So the system switches from E_R to E_ρ because, after an H-agent switches to m_ρ, the L-agents *hurry up to mimic this message* to benefit from the payoff associated to m_ρ followed by $r(m_\rho, H)$. In a similar way, when starting from E_ρ, the switch from an H-agent to a message m^* between m' and m'', by inducing the response $r(m^*, H)$ after m^* and the switch of all H-agents to m^*, leads the agents 2 to play $r(m_\rho, L)$ after m_ρ, which compels the L-agents to mimic m^*. The dynamics also clearly highlight that the imitation goes only one way (from the L-agents to the messages sent by the H-agents).

We now show that these *asymmetries can be exploited further*. We claim that, despite the fact that many states appear in the long run, some of them appear with a *higher probability*. So we aim to go beyond NS's results by looking at the values of the probabilities in the long run (see [16] for a way to address the probabilities in the limit distributions).

To put it more precisely, we conjecture that fast learning should favor the Hellwig equilibrium path:

Conjecture 5.1. E_ρ is assigned with a high (the highest?) probability when $\mu = 1$.

In support of this conjecture, we state Propositions 5.1 and 5.2.

Proposition 5.1. *Each absorbing set is at one mutation from E_ρ, regardless of the starting state in this set.*

Proof. The proof is given in the proof of Proposition 4.2, where we established that one mutation is enough to switch from any equilibrium-absorbing state to E_ρ. We also showed that each cycle is at one mutation from E_ρ. Given that no mutation is necessary to switch from any state in a cycle to any other state in the same cycle, it follows that any state in an absorbing cycle is at one mutation from E_ρ. □

Proposition 5.1 highlights the fact that E_ρ is very easy to reach, in that, regardless of the starting state, one mutation is enough to join it. Moreover, it illustrates how the strategic asymmetries in the game are used in the dynamic evolution. In fact, the dynamic reasoning in the proof of Proposition 5.1 makes heavy use of the strategic revelation incentive advantage of an H-agent, which consists of the fact that, in contrast to an L-agent, he does not fear being recognized. That is why all the moves toward E_ρ are set off by an H-agent's switch toward m_ρ. In turn, the learning sequence that follows this mutation rests heavily on the will or the necessity for the L-agents to *copy* the H-agents' behavior.

Proposition 5.2. *For each absorbing set, there exists an E_ρ-state that is more than one mutation from this absorbing set.*

Proof. See Appendix A at the end of this chapter. □

Proposition 5.1 establishes that E_ρ is easy to reach; Proposition 5.2 shows that E_ρ may be difficult to leave. The proof of Proposition 5.2 again rests heavily on the mimetic incentive. It is because an L-agent is always incited to mimic an H-agent that it is often not possible to leave E_ρ with one mutation. In particular, though the H-agent could be better off with messages different from m_ρ, he cannot ensure the evolution toward such messages, because of the mimesis of the L-agents. In other terms, messages between m' and m'' cannot get, within the dynamic framework, the *reputation* of being sent only by H-agents. The information they convey cannot weight H with a probability higher than the prior probability.

Propositions 5.1 and 5.2, when restricted to two absorbing sets, are illustrated in Figure 2.

By Propositions 5.1 and 5.2 it seems easier to join E_ρ than to leave it. In terms of probabilities, this leads us to conjecture that E_ρ should be observed with a high (the highest?) probability. In other words, fast learning seems to favor the H-agent, in that it favors the evolution toward the H-player's preferred equilibrium path.

This is of course only a conjecture, because, even in the restricted Figure 2, we did not represent all the one-mutation paths. For example, we looked only for the existence of at least one one-mutation path leading from a state to the other set, without caring about the possible existence of many such paths. Yet, despite this fact, the strong asymmetry put into light in this section is worth interest as it helps us discriminate among the many states in the support of the limit distribution.

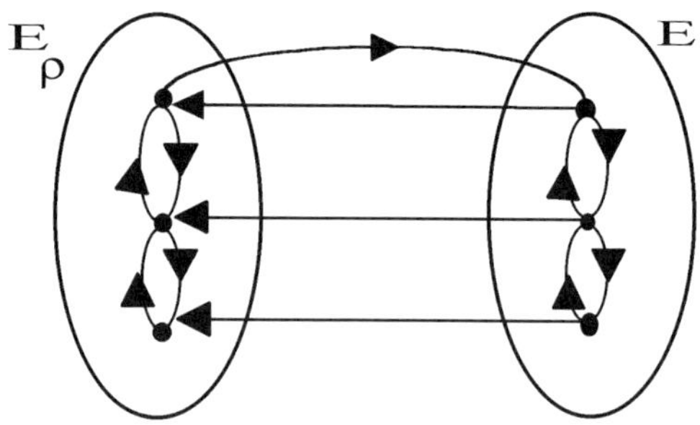

Figure 2: Each Venn diagram is the set of absorbing states corresponding to a given PBE path. Each point is one of these states. The first diagram figures the E_ρ-set, whereas E is a set of absorbing states of another PBE path. Each arrow figures a move that requires exactly one mutation.

6 What About Information Transmission When the Probability of Learning is Lower than 1?

Let us now turn to the true NS process, with $\mu < 1$. A first observation is about the absorbing sets. When they reduce to singletons, they can only be ("enlarged") PBEs, given our best-response restriction on $R(.)$ and $M(.)$. But the absorbing sets different from singletons can be richer than in Proposition 4.1, in that an agent no longer automatically switches to a best response at any moment of time. For example, the absorbing cycles in Proposition 4.1 have to be changed as follows:

> The following cycles constitute absorbing sets: the H-agents send $\widetilde{m}$, the L-agents are spread over $\widetilde{m}$ and m_L, and the agents 2 responds with $r(m_L, L)$ after m_L and are spread over $r(\widetilde{m}, H)$ and $r(\widetilde{m}, \rho)$ after $\widetilde{m}$.

This is due to the fact that not everybody is learning. Hence, starting from a state where $\widetilde{m}$ is assigned to H and followed by $r(\widetilde{m}, H)$, the L-agents may only smoothly switch to $\widetilde{m}$, and the agents 2 may only smoothly update beliefs and actions. In other terms, different speeds of learning lead to noncoordinated moves that result in more states in a cycle.

Despite this fact, Proposition 5.1 still holds, a result we state in form of another proposition.

Proposition 6.1. *Even for $\mu < 1$, each state in any absorbing set is one mutation from E_ρ.*

Proof. See Appendix B. □

The important point behind the proof is that, despite μ's being strictly lower than 1, a binomial learning process still makes fast learning possible. Hence, the evolution obtained with fast learning is still possible, which ensures that it is still possible to reach E_ρ with one mutation.

However, given that a binomial learning process also makes it possible that at a given time only few agents learn, one can expect the existence of different evolution trajectories. It therefore follows that Proposition 5.2 is no longer true.

Proposition 6.2. *It is still possible to find E_ρ-states that are more than one mutation from any separating equilibrium path. But it is possible to find some pooling equilibrium paths such that any E_ρ-state is one mutation from these paths.*

Proof. Start from the E_ρ-state in which every message different from m_ρ, is assigned to L. It immediately follows that it is not possible to switch to a separating PBE with only one mutation. The reason is that at least one H-agent would have to mutate to a separating message m higher or equal to m_H. However, such a mutation can lead only the mutant agent back to m_ρ.

By contrast, consider, for example, the pooling-equilibrium path $E_{m_\rho+\varepsilon}$, in which both agents play $m_\rho+\varepsilon$. It is impossible to prevent the system from evolving to this path when starting from any E_ρ-state. The main steps of the reasoning are illustrated in Figure 3. The proof runs as follows.

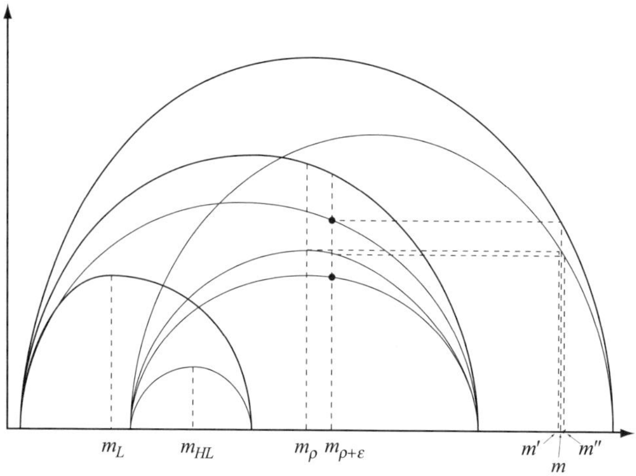

Figure 3: The additional curves (in comparison to Figure 1) are the H- and L-sender's payoff when the receiver agents assign the messages to L with a probability a little higher than the prior one.

One mutation from an H-agent toward $m_\rho + \varepsilon$ can set off the following adjustments. The agents 2 adjust in a way that induces only the H-agents to switch to $m_\rho + \varepsilon$ (such a reaction exists because of single crossing). The H-agents switch to this message, hence causing the agents 2 to play $r(m_\rho, L)$ after m_ρ. It ensues that the L-agents switch to $m_\rho + \varepsilon$, and the agents 2 switch to $r(m_\rho + \varepsilon, \rho)$. Now there are two possibilities. Either the system stays in $E_{m_\rho+\varepsilon}$ or it does not. In the latter case, if only the L-agents are incited to deviate, the response after $m_\rho + \varepsilon$ and the deviating message can lead only the system back to $E_{m_\rho+\varepsilon}$. If only the H-agents are induced to deviate to a given message m, then m has to be in the interval $]m', m''[$ (where m' and m'' are fixed as in Figure 3) and it has to be assigned to H with high probability (for the sake of easier notation, we now set this probability to 1). It follows that the system can evolve as follows: all H-agents prompt to m, but the agents 2 only *smoothly* change their response after $m_\rho + \varepsilon$; it may be that the L-agents switch to m *as soon* as the payoff they get with m followed by $r(m, H)$ is higher than the one obtained with $m_\rho + \varepsilon$, that is, as soon as this payoff is the highest black point in Figure 3. Now, if the agents 2 quickly adapt their behavior after m, by playing $r(m, \rho)$, both H and L-agents can switch back to $m_\rho + \varepsilon$, in that they still get a high payoff (figured by the two black points in Figure 3). The agents 2 then quickly adapt their response, which stabilizes the system in the $E_{m_\rho+\varepsilon}$ path. Of course, it may be that, in the preceding step, the H-agents, rather then switch back to $m_\rho + \varepsilon$, deviate to a new other message; if so, however, one applies the previous reasoning as before until the H- and L-agents go back to $m_\rho + \varepsilon$.

Finally, one checks that if both H- and L-agents are induced to switch to a same message m, then a *fast* response updating by the receiver agents can lead both H- and L-agents back to $m_\rho + \varepsilon$. □

7 Comments

The two interesting points in the proof of Proposition 6.2 are (1) the alternative use of fast and slow updating and (2) a heavy use of the single-crossing assumption.

In particular, the heavy use of the single-crossing assumption becomes possible because of the chance to work with slow adjustments. To see this point, observe that with fast learning it is not possible to get a one-mutation switch toward $E_{m_\rho+\varepsilon}$ when starting from an E_ρ-state in which every out-of-equilibrium message is assigned to L; this impossibility stems from the fact that a mutation from an H-agent toward $m_\rho + \varepsilon$ is *immediately* followed by $r(m_\rho + \varepsilon, H)$, causing the H- *and* L-agents to immediately switch to this message. In other words, the fact that an H-agent is more likely to play $m_\rho + \varepsilon$ than an L-agent (single crossing) cannot be exploited, that is, fast updating does not make it possible for single crossing to have an impact.

It is worth observing that this fact can result in a different *reputation building*. We hereafter say that a message m has a *bad reputation* if it is assigned to L. To illustrate this point, we again study the previous switch. When only fast updating is possible, then a switch from an H-agent to $m_\rho + \varepsilon$, by prompting the H- *and* L-agents to $m_\rho + \varepsilon$, does *not* make it possible for m_ρ to get a bad reputation, because *nobody* keeps on playing this message. So the receiver agents do not change their beliefs after this message, which ensures the possibility to come back to it. By contrast, by slow updating, only the H-agents switch to $m_\rho + \varepsilon$, the L-agents keep on playing m_ρ. This makes it possible for m_ρ to get a bad reputation, which further prevents both the H- and L-agents from again sending this message.

In economic terms, fast learning leads the consumer agents to quickly assign high prices to high quality; it follows that all types of producers send the products at these high prices, thereby allowing unused lower prices such as m_ρ. It follows that a rational consumer has no reason to attribute these prices to low quality, which always allows the producer, especially if he is of high quality, to come back to them. By contrast, with slow adjustments in prices and quantities, high prices may for a while be charged only by high-quality producer agents. Hence, lower prices, charged by low-quality agents, become assigned, as time goes on, to low quality, which prevents the high-quality producer agents from switching back to these prices in the future (namely, once the low-quality producers, due to the consumer behavior's updating, have switched to high prices).

It follows that, in this model, the H-agents are more penalized by slow updating than by fast updating, given that m_ρ gets a bad reputation with slow updating and slow imitation more easily than with fast updating and immediate imitation. Of course, in our context, people play in a nonstrategic way, by just exploring and adapting. But it can be worth studying these phenomena with more strategic purposes. So, if the agents 2 were strategic players, and if they preferred E_ρ to $E_{m_\rho+\varepsilon}$, they should choose to update their behavior after $m_\rho + \varepsilon$ quickly [by playing $r(m_\rho + \varepsilon, H)$] in order to lead the L-agents to play $m_\rho + \varepsilon$, too, so that finally, after the new updated response $(r(m_\rho + \varepsilon, \rho))$, both L- and H-agents go

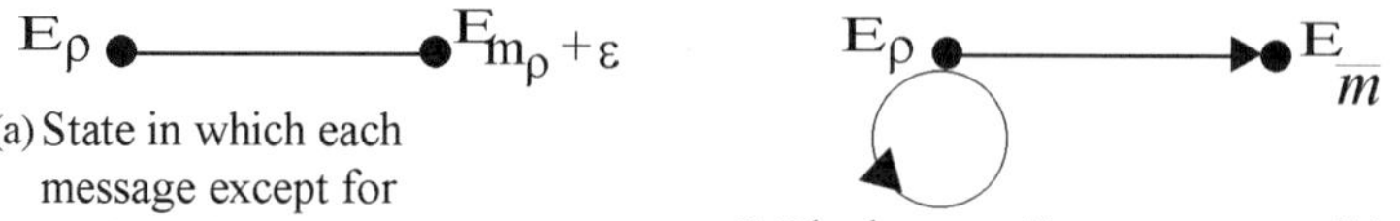

(a) State in which each message except for m_ρ is assigned to L

(b) The loop on E_ρ was not possible with fast moving.

Figure 4.

back to m_ρ. In other words, the above study advocates using in a more strategic way notions such as waiting and fast or slow updating. The strategic asymmetries, here single crossing and mimesis, especially have a different impact according to the speed with which people react.

Figure 4(a) symbolizes this new one-mutation switch to which values of μ different from 1 can give rise.

In addition, by fast learning, in the general case, each one-mutation move induces only one possible evolution. This is no longer true with binomial learning. A mutation, which necessarily leaves the system away from E_ρ with fast learning, may keep the system in E_ρ with binomial learning. For example, consider the E_ρ-state in which every out-of-equilibrium message except $\overline{m}$ [as defined in Figure 5(a)] is assigned to L, $\overline{m}$ being assigned to both types with the prior probabilities. Suppose that an H-agent mutates to m^* between m' and m'', as defined in Figure 1. By fast learning, the system necessarily ends in the pooling state $E_{\overline{m}}$ with $\overline{m}$ as the sent message. This is no longer true with binomial learning because the receiver agents may smoothly change their behavior after m_ρ, whereas all L-agents may switch to m^* as soon as they get more with m^* followed by $r(m^*, H)$ than with m_ρ. Then the agents 2 may play $r(m^*, \rho)$ after m^*, leading both H- and L-agents to switch back to m_ρ, which sticks the system back in E_ρ. This fact is symbolized in Figure 4(b).

A one-mutation switch, by binomial learning, can also lead to several new different equilibrium paths, thanks to a systematic exploitation of single crossing. For example, start from an E_ρ-state in which $\overline{m}$, $\overline{\overline{m}}$ [as defined in Figure 5(a)] and m_ρ are assigned to both types with the prior probabilities—all other out-of-equilibrium messages being assigned to L—and suppose that an H-agent switches to m^* [as depicted in Figure 5(a)]. In that case, fast learning induces the agents 2 to respond with $r(m^*, H)$ after m^*, causing the H-agents to switch to m^* and the agents 2 to respond with $r(m_\rho, L)$ after m_ρ; the L-agents switch to $\overline{\overline{m}}$, followed by $r(\overline{\overline{m}}, L)$ after behavior updating. So the L-agents switch to m^*, leading the agents 2 to respond with $r(m^*, \rho)$, causing all agents 1 to end in the pooling-equilibrium path $E_{\overline{m}}$.

Though this evolution is still possible with a probability of learning different from 1, it is no longer the only possible one. For example, it may be that once the L-agents switched to $\overline{\overline{m}}$, the agents 2 smoothly adjust behavior so that the L-agents are just slightly worse off than with m^* followed by $r(m^*, H)$ [their utility level is the highest black point in Figure 5(a), the lowest black point being the H-agents'

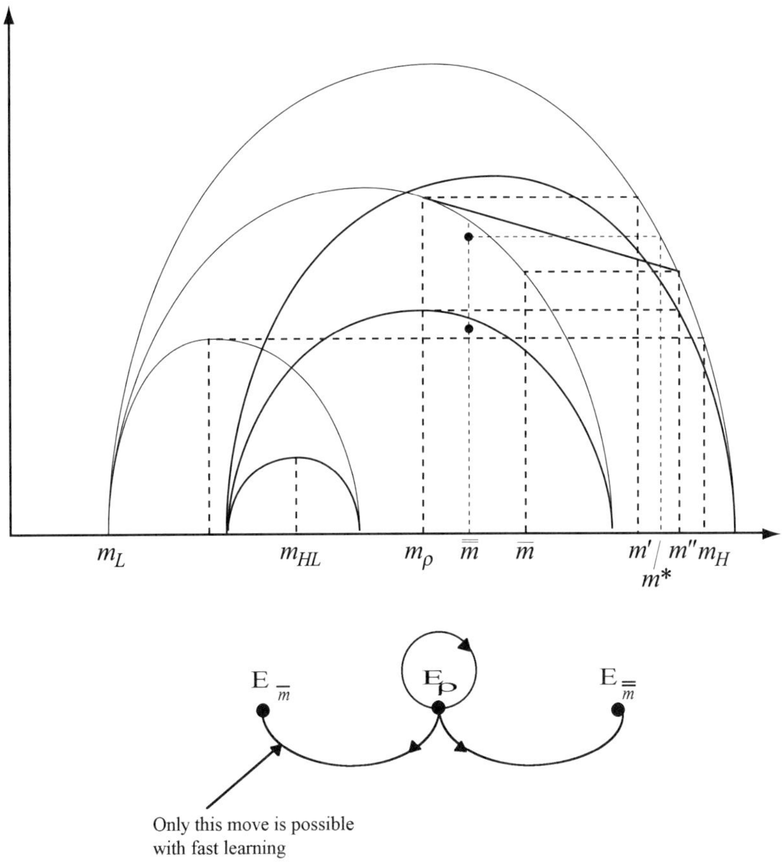

Figure 5: (a) and (b).

payoff for the same message and the same response]. When all H- and L-agents send m^*, it follows that a fast reaction by the agents 2 can lead both the H- and L-agents back to $\overline{\overline{m}}$, which leads the system to the pooling-equilibrium path $E_{\overline{\overline{m}}}$, with $\overline{\overline{m}}$ as the sent message. Similarly, it is possible to show that the same one-mutation switch can lead back to E_ρ. These results are symbolized in Figure 5(b).

To summarize, for $\mu < 1$, it is less likely that E_ρ will be observed with the highest probability. But, given Propositions 6.1 and 6.2, however, low-information revelation still seems to be a likely result in the long run.

8 A Smooth Nonbinomial Learning Process: Some Insights and Additional Comments

Section 5 allows only fast learning. Section 6 allows all speeds of learning. We now again turn to a unique speed of learning, but a smooth and regular one.

Stated informally, we give up the binomial nature of the learning process to allow agents only to smoothly adjust their behavior in the direction of a best response (on this point, we come closer to models based on traditional adjustment schemes).

Not surprisingly, such smooth learning processes best exploit the single-crossing property.

A first consequence is that Proposition 5.1 no longer holds, a fact we state in the form of a proposition.

Proposition 8.1. *There are absorbing states which are at more than one mutation from an E_ρ-state. In particular, each separating equilibrium path includes such states.*

Proof. Let us give the proof for E_R. Start from an E_R-state, where m_ρ is assigned to H with a probability that deters any H-agent from playing this action. For a switch to E_ρ to be possible, it is necessary that at least one H-agent switches to m_ρ. But a smooth learning process, coupled with the *single-crossing property*, now ensures that *only* the L-agents will be induced to switch to m_ρ. So the receiver agents switch to a response that finally leads the system back to E_R. A similar proof holds for any other separating path. □

Proposition 8.1 strengthens the phenomenon already observed when turning from $\mu = 1$ to $\mu < 1$. The smoother the learning process, the more impact the single-crossing property has.

One can even reverse the conjectures about the state most often observed in the long run, provided one adds the restriction that beliefs are right (a restriction we do not want to add).

Proposition 8.2. *If beliefs are supposed to be right, then it is impossible to leave E_R with only one mutation.*

Proof. To leave E_R with one mutation, it is necessary that an H-agent mutates. If he switches to a message higher than m_H, then he is better off by switching back to m_H. If he switches to a message lower than m_H, then the agents 2, by smoothly updating their behavior, only lead the L-agents toward this message (due to single crossing). It follows that new updating can only move the system back to the starting point. □

Hence, this kind of learning process leads to a radically different conjecture, which is that the outcome most observed in the long run is the Riley outcome.

We conclude with three comments.

The first is about *forward induction*. Obviously, given the above developments, it is possible to back contrasted traditional Nash refinement results by referring to different learning processes. Some forward-induction criteria, in particular those that eliminate equilibria by other equilibria and select the Hellwig outcome, make limited use of the single-crossing property: it is therefore intuitive to see their results backed by fast learning processes or more generally by binomial learning processes. The more classical forward-induction criteria, which select the Riley outcome, make heavy use of the single-crossing property: it is therefore logical to see their results backed by smooth, nonbinomial learning processes.

The second comment is about *generalization*. It is not obvious to generalize the conjectures of the paper to a model with more than two types H and L. Working with many types increases the number of absorbing sets and may also raise the number of one-mutation switches between these sets. So it is complicated to make precise the values of the limit-distribution probabilities without introducing additional assumptions in the model. Yet the comments about the different speeds of learning and some properties like single crossing and imitation incentives are general and consequently still hold: only their impact on the limit-distribution probabilities may be less important.

The last comment is about *information transmission*. Perhaps one of the most interesting observations in the paper is the fact that information diffusion and reputation building strongly depend on the speed of learning. Perhaps surprisingly, fast learning and therefore fast imitation can favor the H-agents, by leading more surely in the long term to their preferred equilibrium. It is because the L-agents quickly imitate the H-agents that few messages (generally, only one) are played at a given moment in time, causing the unplayed messages *not* to get a bad reputation. The selected equilibrium path conveys no information, as the L-agents systematically mimic the H-agents. By contrast, smooth learning limits imitation in the model at a given moment in time and hence makes it possible that several messages are played at a same time. It follows that many messages can get a bad reputation, which may prevent H-agents from switching back to some of them; in some ways, these messages are "locked out." It results in a long-term behavior with differentiated messages and much information revelation. However, given that some messages are "locked-out," the end state may be appreciated neither by the H-agents nor by the L-agents. In other words, the selected equilibrium path is neither the H-agents' nor the L-agents' preferred one.

Appendix A Proof of Proposition 5.2

We start in an E_ρ-state in which each message except m_ρ is assigned to L, and beliefs match actions. Is one mutation enough to leave E_ρ?

To induce actions that lead away from E_ρ, a mutation has to be generated by an H-agent (or an agent that mutates regardless of type). If the switch is toward a message lower than m_ρ, then, by single crossing, the L-agents switch to this message before or together with the H-agents and the system finally goes back to E_ρ. If the switch is from an H-agent toward a message higher than m_ρ and lower than m' (as defined in Figure 1), then the receiver agents' adjustments induce both types to switch to this message. So the new behavior updating induces both types to go back to m_ρ. A switch to a message higher than m'' (as defined in Figure 1) cannot make the system stray from E_ρ. Finally, if an H-agent switches to a message m^* between m' and m'', then we already know that, after some learning steps, m_ρ is assigned to L, all L- and H-agents play m^*, and the agents 2 assign this message to both types with the prior probabilities. If the H-agents prefer m^* followed by $r(m^*, \rho)$ to m_{HL} followed by $r(m_{HL}, L)$ (by contrast to what happens in Figure 1), then the process ends in a pooling-equilibrium path state E_{m^*}, with m^* as the sent message. If not, the H-agents switch to m_{HL}, causing the L-agents to switch to m_L once the agents 2 have adjusted their behavior after m^*; then the agents 2 respond with $r(m_{HL}, H)$ after m_{HL}, which leads all agents 1 to switch to m_{HL} (because $m_{HL} < m_H$). In turn, this message is finally followed by $r(m_{HL}, \rho)$, and the system ends either in an pooling-equilibrium path state $E_{m_{HL}}$, with m_{HL} as the sent message, or in a cycle with the messages m_L and m_{HL}.

It follows that a unique mutation, when it leads away from the E_ρ-state, leads to E_{m^*} if this state exists or to $E_{m_{HL}}$ or to a cycle with m_L and m_{HL}. To avoid the evolution toward the first equilibrium state, it is enough to start from a state where $m^* - \varepsilon$ is assigned to both types with the prior probabilities. To avoid the evolution to the two other absorbing sets, it is enough to suppose that m_H is assigned to H.

Appendix B Proof of Proposition 6.1

It is obvious for a PBE state.

Let us consider an absorbing set that does not reduce to a singleton. We first show that there exists at least one state in the cycle where an H-agent is worse off than in E_ρ, and then we start from this state to switch to E_ρ.

To show that in at least one state of the cycle some H-agents are worse off than in E_ρ, first observe that learning can lead both H- and L-agents to play only the messages that yield the highest payoff. Suppose that all agents 2 adapt their responses and that all agents 1 adapt their messages. Call I the set of messages played both by H- and L-agents. Necessarily, by single crossing, either I is empty or it is a singleton.

If I is empty, the agents 2, if they learn fast, can lead all H- (resp. L-) agents to focus on a same message m (resp. m'), with m' different from m. Hence, either the L-agents switch to m, in which case the agents 2 can respond with $r(m, \rho)$, causing the H-agents to be worse off than in E_ρ (unless $m = m_\rho$) or the L-agents stay on m', which means that $m > m_H$, namely, that the H-agents are worse off than in E_ρ. In both cases, one mutation from an H-agent toward m_ρ allows the system to evolve to E_ρ.

If $I = \{m\}$ and if all L-agents play m, then, if m is different from m_ρ, the H-agents are worse off with m than in E_ρ and switching one H-agent to m_ρ can lead the system to E_ρ; if m is equal to m_ρ, either the system is in E_ρ or the same mutation as before leads to E_ρ. If not all L-agents play m, then, given that an L-agent is indifferent between m_L followed by $r(m_L, L)$ and m_H followed by $r(m_H, H)$, m can only be lower than m_H. This ensures that the H-agents are strictly worse off with m than with m_H followed by $r(m_H, H)$ (by single crossing). It follows that if m is different from m_ρ, one mutation from H to m_ρ can lead the system to E_ρ.

If $m = m_\rho$, then not all H-agents play this message: hence, one mutation from an H-agent toward m_ρ can again lead the system to E_ρ.

REFERENCES

[1] Hellwig, M. Some recent developments in the theory of competition in markets with adverse selection, *Eur. Econ. Rev.*, 1, pp. 319–325, 1987.

[2] Kandori, M., Mailath, G., and Rob, R. Learning, mutation, and long-run equilibria in games. *Econometrica*, **61**, pp. 29–56, 1993.

[3] Kohlberg, E. and Mertens, J. F. On the strategic stability of equilibria, *Econometrica*, **54**, pp. 1003–1038, 1986.

[4] Laffond, G. La révélation de la qualité par les prix sur un marché en auto-organisation, *Econ. Appl.* **XLII**, pp. 129–153, 1989.

[5] Lesourne, J. *The Economics of Order and Disorder*, Clarendon Press, Oxford, 1992.

[6] Mailath, G. J., Okuno-Fujiwara, M., and Postlewaite, A. On belief-based refinements in signaling games, *J. Econ. Th.* **60**, pp. 241–276, 1993.

[7] Nöldeke, G. and Samuelson, L. An evolutionary analysis of backward and forward induction, *Games and Econ. Beh.*, **5**, pp. 425–454, 1993.

[8] Nöldeke, G. and Samuelson, L. Learning to signal in markets, *Working Paper 9409*, University of Wisconsin Madison, 1994.

[9] Nöldeke, G. and Samuelson, L. A dynamic model of equilibrium selection in signaling markets, *J. Econ. Th.* **73**, pp. 118–156, 1997.

[10] Riley, J. Informational equilibrium, *Econometrica*, **47**, pp. 331–359, 1979.

[11] Samuelson, L. Stochastic stability in games with alternative best replies, *J. Econ. Th.* **64**, pp. 35–65, 1994.

[12] Spence, A. M. *Market Signaling*, Harvard University Press, Cambridge MA, 1974.

[13] Umbhauer, G. Forward induction, consistency, and rationality of ε perturbations, *Working Paper BETA 9104*, Univ. Louis Pasteur, Strasbourg, France, 1991.

[14] Umbhauer, G. Information transmission in signaling games: Confrontation of different forward induction criteria, in *Models and Experiments in Risk and Rationality*, B. Munier and M. Machina, eds., Kluwer Academic Publishers, New York, 1994.

[15] Umbhauer, G. On the stationary probability distribution of evolutionary processes, *Working Paper BETA 9505*, Univ. Louis Pasteur, Strasbourg, France, 1995.

[16] Umbhauer, G. Forward induction, evolutionary processes and behavior evolution, *J. Evol. Econ.* **7**, pp. 415–433, 1997.

[17] Young, P. The evolution of conventions, *Econometrica*, **61**, pp. 57–84, 1993.

Mixed Strategies for Hierarchical Zero-Sum Games

Lina Mallozzi
and
Jacqueline Morgan
Università degli Studi di Napoli "Federico II"
Dipartimento di Matematica e Applicazioni
Napoli, Italy

Abstract

New concepts of approximate mixed solutions for hierarchical saddle-point problems are introduced in the situation in which the leader cannot influence the followers and minimize the worst, under the lack of convexity assumptions. Then sufficient conditions for the existence of such approximate solutions and convergence of the corresponding values are presented.

1 Introduction

A great number of papers have been devoted to the Stackelberg problem, relating a two-player game with a leader–follower behavior in which a solution is an optimal action chosen by the leader when the follower reacts by playing optimally. When the follower's optimal reaction set (set of the solutions of the lower-level problem) is always a singleton, we obtain the concept introduced by von Stackelberg [19] in the context of static economic competitions and by Chen and Cruz [5] and Simaan and Cruz [17] in a control theory framework. When the lower-level problem does not always have a unique solution, two concepts of generalized solutions have been introduced by Leitmann [7] (called "strong" and "weak solutions" by Breton, et al. [4]). Moreover, this leader–follower behavior can be extended to a situation with two or more followers playing a noncooperative game (see, for example, Basar and Olsder [2]). An economic example is the Stackelberg–Nash–Cournot equilibria, in which a market leader seeks to maximize his utility subject to the equilibrium conditions of a set of followers. For existence results in the case where the reaction set of the followers is a singleton, see, for example, [16] and [18], which models the supply side of an oligopolistic market noncooperatively supplying a homogeneous product. Note that this problem is a special case of the general problem of optimization subject to variational inequality constraints. For a general survey about variational inequalities and their applications, see, for example, Harker and Pang [6] and Lou et al.'s book about optimization problems with equilibrium constraints [8].

In this paper we consider the case where the followers' reaction set is not always a singleton. More precisely, let X, Y_1, and Y_2 be nonempty, compact subsets of three Euclidean spaces and let $l,\ f$ be two real-valued functions defined on $X \times Y_1 \times Y_2$. We investigate the following problem in this paper:

$$(w - HP) = \begin{cases} \text{find } \overline{x} \in X \text{ such that} \\ \inf_{x \in X} \sup_{(y_1, y_2) \in S(x)} l(x, y_1, y_2) = \sup_{(y_1, y_2) \in S(\overline{x})} l(\overline{x}, y_1, y_2), \end{cases}$$

where $S(x)$, for all $x \in X$, is the set of solutions to the parametric saddle-point problem (see, for example, Aubin [1] and Owen [20]):

$$S(x) = \begin{cases} \text{find } (\overline{y}_1, \overline{y}_2) \in Y_1 \times Y_2 \text{ such that} \\ f(x, \overline{y}_1, \overline{y}_2) = \sup_{y_2 \in Y_2} f(x, \overline{y}_1, y_2) = \inf_{y_1 \in Y_1} f(x, y_1, \overline{y}_2). \end{cases}$$

$v = \inf_{x \in X} \sup_{(y_1, y_2) \in S(x)} l(x, y_1, y_2)$ will be called the "value" of the problem.

In this case $l,\ f,\ -f$ can represent the cost functions of a leader and of two followers, respectively; $S(x)$ is the reaction set of the followers, and the problem $(w - HP)$ is related to the situation in which the leader cannot influence the followers and minimizes the worst. Such a problem will be called a *weak hierarchical saddle-point problem* as in [10] and in [12]. When the leader can force the followers to play the best for himself, the problem is called a *strong* hierarchical saddle-point problem as in [11] and in [13].

For the weak hierarchical saddle-point problem, the existence of a solution is not guaranteed even for nice cost functions on compact sets. Nevertheless, existence results for a regularized version have been given in [12] under some convexity-like assumptions on the follower cost function f. In order to remove these convexity assumptions, in this paper we will consider an approximate mixed extension for the lower-level saddle-point problem. Then we introduce different concepts of ε-mixed solutions for the problem $(w - HP)$ for $\varepsilon > 0$, by using mixed extensions first only for the followers (in Section 3) and, then for all the players (in Section 4). Then, in both cases, we give sufficient conditions on the data for the existence of such approximate solutions, together with the convergence of the corresponding values as $\varepsilon \to 0^+$. Moreover, the different approximate average values considered are compared and illustrated by an example.

Let us note that we extend to the zero-sum game the concepts of approximate mixed solutions we introduced in previous papers ([9] and [10]) for Stackelberg problems where the lower-level problem is only an optimization problem. The results obtained in the two different situations (zero-sum game and lower-level optimization problem) are not exactly the same. In fact, for zero-sum games we have convergence of the approximate average value to a value greater than or equal to the value of the problem, differently from the lower-level optimization problem in which the approximate average value always converges to the value of the problem.

2 Mixed Extension

As observed in the introduction, the existence of a solution to the weak hierarchical saddle-point problem $(w - HP)$ is not guaranteed even for nice cost functions, as the following example shows:

Example 2.1. Let $X = Y_1 = Y_2 = [0, 1]$ and

$$l(x, y_1, y_2) = x - (y_1 + y_2),$$
$$f(x, y_1, y_2) = -y_1(x + y_2).$$

The set of the lower-level saddle points is

$$S(x) = \begin{cases} [0, 1] \times \{0\} & \text{if } x = 0, \\ \{(1, 0)\} & \text{if } x \neq 0, \end{cases}$$

and we can compute

$$w(x) = \sup_{(y_1, y_2) \in S(x)} l(x, y_1, y_2) = \begin{cases} 0 & \text{if } x = 0, \\ x - 1 & \text{if } x \neq 0. \end{cases}$$

So the value of the problem is

$$v = \inf_{x \in [0,1]} \sup_{(y_1, y_2) \in S(x)} l(x, y_1, y_2) = -1,$$

and there exists no solution to the weak problem $(w - HP)$.

Nevertheless, in some good situations (under convexity assumptions), by using the following regularized problem (Morgan and Raucci [12]), we can obtain approximate solutions. Let $\varepsilon > 0$ and for all $x \in X$, let $S(x, \varepsilon)$ be the set of the ε-saddle points to the lower level, that is, the set of solutions to the problem:

$$\mathcal{S}(x, \varepsilon) = \begin{cases} \text{find } (\overline{y}_1, \overline{y}_2) \in Y_1 \times Y_2 \text{ such that} \\ \sup\limits_{y_2 \in Y_2} f(x, \overline{y}_1, y_2) - \inf\limits_{y_1 \in Y_1} f(x, y_1, \overline{y}_2) \text{ is well defined and} \\ \sup\limits_{y_2 \in Y_2} f(x, \overline{y}_1, y_2) - \inf\limits_{y_1 \in Y_1} f(x, y_1, \overline{y}_2) \leq \varepsilon, \end{cases}$$

where $a - b$ is well defined if it is not $(+\infty) - (+\infty)$ or $(-\infty) - (-\infty)$.

The regularized version of $(w - HP)$ is

$$(w - HP)(\varepsilon) = \begin{cases} \text{find } \overline{x}_\varepsilon \in X \text{ such that} \\ \inf\limits_{x \in X} \sup\limits_{(y_1, y_2) \in S(x, \varepsilon)} l(x, y_1, y_2) = \sup\limits_{(y_1, y_2) \in S(\overline{x}_\varepsilon, \varepsilon)} l(\overline{x}_\varepsilon, y_1, y_2). \end{cases}$$

Let $w(x, \varepsilon) = \sup_{(y_1, y_2) \in S(x, \varepsilon)} l(x, y_1, y_2)$; $v(\varepsilon) = \inf_{x \in X} w(x, \varepsilon)$ be called the "approximate value" of the problem.

An existence result in [12] requires the following convexity assumption on f in order to obtain the existence of a solution to the problem $(w - HP)(\varepsilon)$ and the convergence of the corresponding approximate value:

(A) the function $f(x, \cdot, y_2)$ is convex on Y_1 for all $y_2 \in Y_2$, and the function $f(x, y_1, \cdot)$ is concave on Y_2 for all $y_1 \in Y_1$ and for all $x \in X$.

In fact, in Example 2.1, we can compute the approximate solutions. More precisely, let $\varepsilon > 0$, $Y = Y_1 \times Y_2$ and $y = (y_1, y_2)$; the set of the ε-saddle points is

$$S(x,\varepsilon) = \begin{cases} \{y \in Y \text{ s.t. } y_1 \in [0,1],\, y_2 \in [0, \varepsilon - x + y_1 x]\} & \text{if } x \le \varepsilon, \\ \{y \in Y \text{ s.t. } y_1 \in [1-\varepsilon/x, 1],\, y_2 \in [0, \varepsilon - x + y_1 x]\} & \text{if } x > \varepsilon. \end{cases}$$

We can compute

$$w(x,\varepsilon) = \begin{cases} x & \text{if } x \in [0,\varepsilon], \\ x - 1 + \varepsilon/x & \text{if } x \in (\varepsilon, 1]. \end{cases}$$

So the approximate value is

$$\inf_{x\in[0,1]} \sup_{(y_1,y_2)\in S(x,\varepsilon)} x - (y_1 + y_2) = -1 + 2\sqrt{\varepsilon},$$

and an approximate solution to the problem $(w - HP)$ is $\overline{x}_\varepsilon = \sqrt{\varepsilon}$ for all $\varepsilon < 1/4$.

Unfortunately, if the convexity assumption (A) is not satisfied, the existence of the solutions is not guaranteed, as the following example shows.

Example 2.2. Let $X = Y_1 = Y_2 = [0,1]$ and

$$l(x, y_1, y_2) = x - (y_1 + y_2),$$
$$f(x, y_1, y_2) = y_1 y_2 - (1+x) y_1^2.$$

The set of the lower-level saddle points is

$$S(x) = \begin{cases} \{(0,1),(1,1)\} & \text{if } x = 0, \\ \{(1,1)\} & \text{if } x \neq 0, \end{cases}$$

and we can compute

$$w(x) = \begin{cases} -1 & \text{if } x = 0, \\ x - 2 & \text{if } x \neq 0. \end{cases}$$

So the value of the problem is

$$v = \inf_{x\in[0,1]} \sup_{(y_1,y_2)\in S(x)} x - (y_1 + y_2) = -2,$$

and there exists no solution to the weak problem.

Let $\varepsilon > 0$, $Y = Y_1 \times Y_2$, and $y = (y_1, y_2)$; the set of the ε-saddle points is

$$S(x,\varepsilon) = \begin{cases} \{y \in Y \text{ s.t. } y_2 \in [1-\varepsilon, x+1-\varepsilon],\, y_1 \in [\alpha_2(y_2), 1]\} \cup \\ \{y \in Y \text{ s.t. } y_2 \in [x+1-\varepsilon, 1],\, y_1 \in [0, \alpha_1(y_2)] \cup [\alpha_2(y_2), 1]\} \\ & \text{if } x \le \varepsilon, \\ \{y \in Y \text{ s.t. } y_2 \in [1-\varepsilon, 1],\, y_1 \in [\alpha_2(y_2), 1]\} & \text{if } x > \varepsilon, \end{cases}$$

where

$$\alpha_1(y_2), \alpha_2(y_2) = \frac{1}{[2(1+x)]} \mp \sqrt{1/[4(1+x)^2] + 1 - (\varepsilon + y_2)/(1+x)}.$$

We can compute

$$w(x, \varepsilon) = \begin{cases} -1 + \varepsilon & \text{if } x \in [0, \varepsilon], \\ x - 2 + \varepsilon & \text{if } x \in (\varepsilon, 1]. \end{cases}$$

Thus, the approximate value is

$$v(\varepsilon) = \inf_{x \in [0,1]} \sup_{(y_1, y_2) \in S(x,\varepsilon)} x - (y_1 + y_2) = -2 + 2\varepsilon,$$

and there exists no approximate solution to the weak problem $(w - HP)$ even if the approximate value $v(\varepsilon)$ converges to v for $\varepsilon \to 0^+$. In fact, the function f does not satisfy the sufficient conditions for the existence of an approximate solution given in Theorem 5.1 by Morgan and Raucci [12].

When convexity assumptions on the cost functions are not satisfied, a classical approach in game theory is to consider mixed strategies. When X is a nonempty subset of a reflexive Banach space, we recall the concept of the mixed extension of X (Borel [3] and von Neumann and Morgenstern [21]).

Definition 2.1. The mixed extension of X is the set of the Radon probability measures on X, that is,

$$\overline{\mathcal{M}}(X) = \{\mu \in C^*(X) \text{ s.t. } \langle \mu, f \rangle \geq 0 \, \forall f \in C_+(X) \text{ and } \langle \mu, \theta \rangle = 1 \text{ with } \theta = 1\},$$

where $C_+(X)$ is the cone of positive continuous functions defined on X, $C^*(X)$ is the dual of the space of continuous functions on X, $C(X)$ is supplied with the sup norm, and $\langle \cdot, \cdot \rangle$ is the duality product.

If X is a compact set, then $\overline{\mathcal{M}}(X)$ is compact and convex, and it is the closure of the set of the discrete probability measures on X with respect to the weak topology (Prop. 11, p. 82, in [1]). The convergence of the measures defined by the weak topology in $C^*(X)$ is characterized by

$$\lim_{n \to +\infty} \mu_n = \mu \overset{\text{def}}{\Longleftrightarrow} \lim_{n \to +\infty} \langle \mu_n, f \rangle = \langle \mu, f \rangle, \qquad \forall f \in C(X),$$

and called "vague" in Schwartz [15]. We note that, X being compact, the topology of the vague convergence on $\overline{\mathcal{M}}(X)$ is metrizable (Th. 5.13.34 in [15]).

If δ is the Dirac measure, the correspondence $x \in X \mapsto \delta(x) \in \overline{\mathcal{M}}(X)$ is an immersion of X into $\overline{\mathcal{M}}(X)$.

For a mixed extension of a game that uses the mixed extensions of the strategy spaces, see, for example, [21], [20], [1], [14], and [2].

For what concerns the hierarchical saddle-point problem, we can consider the mixed extension of the parametric zero-sum game in the lower level of the problem $(w - HP)$ when $l(x, \cdot, \cdot)$, $f(x, \cdot, \cdot)$ are continuous functions on $Y_1 \times Y_2$ for all $x \in X$ and Y_1 and Y_2 are supposed to be compact subsets. More precisely, let $\overline{\mathcal{M}}(Y_1)$, $\overline{\mathcal{M}}(Y_2)$ be the sets of the Radon probability measures on Y_1, Y_2, respectively, and let $\hat{l}(x, \mu_1, \mu_2)$, $\hat{f}(x, \mu_1, \mu_2)$ be the linearized objective functions, defined on

$X \times \overline{\mathcal{M}}(Y_1) \times \overline{\mathcal{M}}(Y_2)$ as follows:

$$\hat{l}(x, \mu_1, \mu_2) = \langle \mu_2, \langle \mu_1, l(x, \cdot, \cdot)\rangle\rangle,$$
$$\hat{f}(x, \mu_1, \mu_2) = \langle \mu_2, \langle \mu_1, f(x, \cdot, \cdot)\rangle\rangle.$$

Note that for all $x \in X, \hat{l}(x, \mu_1, \mu_2)$ and $\hat{f}(x, \mu_1, \mu_2)$ are continuous functions on $\overline{\mathcal{M}}(Y_1) \times \overline{\mathcal{M}}(Y_2)$ (such results can be proved by using Theorem 5.13.30 in [15]).

The extended lower-level problem is as follows:

$$\hat{\mathcal{S}}(x) = \begin{cases} \text{find } (\overline{\mu}_1, \overline{\mu}_2) \in \overline{\mathcal{M}}(Y_1) \times \overline{\mathcal{M}}(Y_2) \text{ such that} \\ \hat{f}(x, \overline{\mu}_1, \overline{\mu}_2) = \sup\limits_{\mu_2 \in \overline{\mathcal{M}}(Y_2)} \hat{f}(x, \overline{\mu}_1, \mu_2) = \inf\limits_{\mu_1 \in \overline{\mathcal{M}}(Y_1)} \hat{f}(x, \mu_1, \overline{\mu}_2), \end{cases}$$

and we denote by $\hat{S}(x)$ the set of solutions to $\hat{\mathcal{S}}(x)$, for all $x \in X$, that is, the set of parametric, mixed, lower-level saddle points.

A natural extension for the hierarchical problem $(w - HP)$ is the following:

$$(\hat{w} - HP) = \begin{cases} \text{find } \overline{x} \in X \text{ such that} \\ \inf\limits_{x \in X} \sup\limits_{(\mu_1, \mu_2) \in \hat{S}(x)} \hat{l}(x, \mu_1, \mu_2) = \sup\limits_{(\mu_1, \mu_2) \in \hat{S}(\overline{x})} \hat{l}(\overline{x}, \mu_1, \mu_2). \end{cases}$$

Let $\hat{w}(x) = \sup_{(\mu_1, \mu_2) \in \hat{S}(x)} \hat{l}(x, \mu_1, \mu_2)$; $\hat{v} = \inf_{x \in X} \hat{w}(x)$ will be called the mixed average value.

Unfortunately, again, the existence of a solution to the problem $(\hat{w} - HP)$ is not guaranteed, as shown by Example 2.2.

Let $X = Y_1 = Y_2 = [0, 1]$ and

$$l(x, y_1, y_2) = x - (y_1 + y_2),$$
$$f(x, y_1, y_2) = y_1 y_2 - (1 + x) y_1^2.$$

In this case,

$$\hat{w}(x) = w(x) = \begin{cases} -1 & \text{if } x = 0, \\ x - 2 & \text{if } x \neq 0, \end{cases}$$

and there exists no solution to the mixed weak problem $(\hat{w} - HP)$.

Since $S(x) = S_1(x) \times S_2(x)$ for all $x \in X$, with $S_1(x) \subseteq Y_1$ and $S_2(x) \subseteq Y_2$, we can obtain the following inclusion:

$$\overline{\mathcal{M}}(S_1(x)) \times \overline{\mathcal{M}}(S_2(x)) \subseteq \hat{S}(x) \qquad \text{for all } x \in X, \tag{1}$$

which is a consequence of the following proposition.

Proposition 2.1. *Let g be a continuous, real-valued function defined on $Y_1 \times Y_2$, and let S, assumed to be nonempty, be the set of saddle points of g on $Y_1 \times Y_2$. Since $S = S_1 \times S_2$ with $S_1 \subseteq Y_1$ and $S_2 \subseteq Y_2$, we have*

$$\overline{\mathcal{M}}(S_1) \times \overline{\mathcal{M}}(S_2) \subseteq \hat{S},$$

where $\hat{S}$ is the set of the mixed saddle points of g, that is, the solutions of the problem

$$\hat{\mathcal{S}} = \begin{cases} \textit{find } (\overline{\mu}_1, \overline{\mu}_2) \in \overline{\mathcal{M}}(Y_1) \times \overline{\mathcal{M}}(Y_2) \textit{ such that} \\ \hat{g}(\overline{\mu}_1, \overline{\mu}_2) = \sup\limits_{\mu_2 \in \overline{\mathcal{M}}(Y_2)} \hat{g}(\overline{\mu}_1, \mu_2) = \inf\limits_{\mu_1 \in \overline{\mathcal{M}}(Y_1)} \hat{g}(\mu_1, \overline{\mu}_2). \end{cases}$$

Proof. Let $\overline{\mu} \in \overline{\mathcal{M}}(S_1) \times \overline{\mathcal{M}}(S_2)$; then $\overline{\mu} = (\overline{\mu}_1, \overline{\mu}_2)$ with $\overline{\mu}_i \in \overline{\mathcal{M}}(S_i)$ for $i = 1, 2$. Then there exist two sequences $(\overline{\mu}_{1_n})_n$ and $(\overline{\mu}_{2_n})_n$ converging to $\overline{\mu}_1$ and $\overline{\mu}_2$, respectively, with $\overline{\mu}_{i_n} \in \mathcal{M}(S_i)$ for $i = 1, 2$ and for all $n \in \mathcal{N}$, that is for $i = 1, 2$ and for $j = 1, \dots, \overline{l}_i(n)$ there exist $\overline{y}^j_{i_n} \in S_i$ and positive real numbers $\overline{\lambda}^j_{i_n}$ such that $\overline{\mu}_{i_n} = \sum_{j=1}^{\overline{l}_i(n)} \overline{\lambda}^j_{i_n} \delta(\overline{y}^j_{i_n})$ and $\sum_{j=1}^{\overline{l}_i(n)} \overline{\lambda}^j_{i_n} = 1$. Then for all $(y_1, y_2) \in Y_1 \times Y_2$,

$$g(\overline{y}^j_{1_n}, y_2) \le g(\overline{y}^j_{1_n}, \overline{y}^k_{2_n}) \le g(y_1, \overline{y}^k_{2_n}). \tag{2}$$

Let us show that $\overline{\mu} = (\overline{\mu}_1, \overline{\mu}_2) \in \hat{S}$, that is, for all $(\mu_1, \mu_2) \in \overline{\mathcal{M}}(Y_1) \times \overline{\mathcal{M}}(Y_2)$,

$$\hat{g}(\overline{\mu}_1, \mu_2) \le \hat{g}(\overline{\mu}_1, \overline{\mu}_2) \le \hat{g}(\mu_1, \overline{\mu}_2), \tag{3}$$

where $\hat{g}(\mu_1, \mu_2) = \langle \mu_2, \langle \mu_1, g(\cdot, \cdot) \rangle \rangle$. Let $\mu_i \in \overline{\mathcal{M}}(Y_i)$ for $i = 1, 2$. Then there exist two sequences $(\mu_{1_n})_n$ and $(\mu_{2_n})_n$ converging to μ_1 and μ_2, respectively, with $\mu_{i_n} \in \mathcal{M}(Y_i)$ for $i = 1, 2$ and for all $n \in \mathcal{N}$, that is, for $i = 1, 2$ and for $j = 1, \dots, l_i(n)$, there exist $y^j_{i_n} \in Y_i$ and positive real numbers $\lambda^j_{i_n}$ such that $\mu_{i_n} = \sum_{j=1}^{l_i(n)} \lambda^j_{i_n} \delta(y^j_{i_n})$ and $\sum_{j=1}^{l_i(n)} \lambda^j_{i_n} = 1$. Then, by using (2), we obtain

$$\begin{aligned} \hat{g}(\overline{\mu}_{1_n}, \overline{\mu}_{2_n}) &= \sum_{j=1}^{\overline{l}_1(n)} \sum_{k=1}^{\overline{l}_2(n)} \overline{\lambda}^j_{1_n} \overline{\lambda}^k_{2_n} g(\overline{y}^j_{1_n}, \overline{y}^k_{2_n}) \le \sum_{j=1}^{\overline{l}_1(n)} \sum_{k=1}^{\overline{l}_2(n)} \overline{\lambda}^j_{1_n} \overline{\lambda}^k_{2_n} g(y_1, \overline{y}^k_{2_n}) \\ &= \sum_{k=1}^{\overline{l}_2(n)} \overline{\lambda}^k_{2_n} g(y_1, \overline{y}^k_{2_n}) \end{aligned}$$

for all $y_1 \in Y_1$.

Particularly, we have

$$\hat{g}(\overline{\mu}_{1_n}, \overline{\mu}_{2_n}) \le \sum_{k=1}^{\overline{l}_2(n)} \overline{\lambda}^k_{2_n} g(y^j_{1_n}, \overline{y}^k_{2_n}) \qquad \text{for all } j = 1, \dots, l_1(n),$$

and

$$\hat{g}(\overline{\mu}_{1_n}, \overline{\mu}_{2_n}) \le \sum_{j=1}^{l_1(n)} \sum_{k=1}^{\overline{l}_2(n)} \lambda^j_{1_n} \overline{\lambda}^k_{2_n} g(y^j_{1_n}, \overline{y}^k_{2_n}) = \hat{g}(\mu_{1_n}, \overline{\mu}_{2_n}). \qquad \square$$

In an analogous way, it can be proved that $\hat{g}(\overline{\mu}_{1_n}, \mu_{2_n}) \le \hat{g}(\overline{\mu}_{1_n}, \overline{\mu}_{2_n})$ for all $n \in \mathcal{N}$. Since $\hat{g}(\cdot, \cdot)$ is continuous with respect to the vague convergence, we obtain the result in (3).

We can prove the following proposition.

Proposition 2.2. *Assume that f is a continuous function on $X \times Y_1 \times Y_2$. We have*

$$w(x) \leq \hat{w}(x), \qquad \text{for all } x \in X,$$
$$v \leq \hat{v},$$

where $w(x) = \sup_{(y_1,y_2)\in S(x)} l(x, y_1, y_2)$, $\hat{w}(x) = \sup_{(\mu_1,\mu_2)\in\hat{S}(x)} \hat{l}(x, \mu_1, \mu_2)$ and $v = \inf_{x\in X} w(x)$, $\hat{v} = \inf_{x\in X} \hat{w}(x)$.

Proof. By using (1), we have

$$\sup_{(y_1,y_2)\in S(x)} l(x, y_1, y_2) = \sup_{(\mu_1,\mu_2)\in\overline{\mathcal{M}}(S_1(x))\times\overline{\mathcal{M}}(S_2(x))} \hat{l}(x, \mu_1, \mu_2)$$
$$\leq \sup_{(\mu_1,\mu_2)\in\hat{S}(x)} \hat{l}(x, \mu_1, \mu_2).$$
□

Remark 2.1. Let us note that differently from the weak Stackelberg problem (in which the lower-level problem is a minimum problem with nonuniqueness of the solution), we do not have $v = \hat{v}$ in general, but only $v \leq \hat{v}$.

The aim of the following sections will be to introduce more suitable concepts of approximate solutions when the followers' objective functions do not satisfy convexity assumptions.

3 (Lower) ε-Mixed Extension

First we introduce approximate mixed strategies for the lower-level $S(x)$ of the hierarchical problem $(w - HP)$.

Definition 3.1. Let $x \in X$ and $\varepsilon > 0$; an ε-mixed saddle point to the parametric saddle-point problem $S(x)$, defined in the Introduction, is a solution to the problem

$$\hat{\mathcal{S}}(x, \varepsilon) = \begin{cases} \text{find } (\overline{\mu}_1, \overline{\mu}_2) \in \overline{\mathcal{M}}(Y_1) \times \overline{\mathcal{M}}(Y_2) \text{ such that} \\ \sup_{\mu_2\in\overline{\mathcal{M}}(Y_2)} \hat{f}(x, \overline{\mu}_1, \mu_2) - \inf_{\mu_1\in\overline{\mathcal{M}}(Y_1)} \hat{f}(x, \mu_1, \overline{\mu}_2) \leq \varepsilon. \end{cases}$$

We denote by $\hat{S}(x, \varepsilon)$ the set of the ε-mixed saddle points to the problem $\hat{\mathcal{S}}(x, \varepsilon)$.

Now we can define a regularized mixed version of the hierarchical problem $(w - HP)$ by using ε-mixed saddle points for the followers.

Definition 3.2. Let $\varepsilon > 0$; a (lower) ε-mixed solution to the problem $(w - HP)$ is a solution to the following problem:

$$(\hat{w} - HP)(\varepsilon) = \begin{cases} \text{find } \hat{x}_\varepsilon \in X \text{ such that} \\ \inf_{x\in X} \sup_{(\mu_1,\mu_2)\in\hat{S}(x,\varepsilon)} \hat{l}(x, \mu_1, \mu_2) = \sup_{(\mu_1,\mu_2)\in\hat{S}(\hat{x}_\varepsilon,\varepsilon)} \hat{l}(\hat{x}_\varepsilon, \mu_1, \mu_2). \end{cases}$$

Let $\hat{w}(x,\varepsilon) = \sup_{(\mu_1,\mu_2)\in\hat{S}(x,\varepsilon)} \hat{l}(x,\mu_1,\mu_2)$; $\hat{v}(\varepsilon) = \inf_{x\in X} \hat{w}(x,\varepsilon)$ will be called the (lower) ε-mixed average value of the problem $(w-HP)$. For the problem $(\hat{w}-HP)(\varepsilon)$, the following existence result holds.

Theorem 3.1. *Suppose that l and f are continuous functions on $X \times Y_1 \times Y_2$.*

Then, for all $\varepsilon > 0$, there exists at least a solution $\hat{x}_\varepsilon$ to the (lower) mixed regularized problem $(\hat{w}-HP)(\varepsilon)$ and

$$\lim_{\varepsilon\to 0^+} \hat{w}(\hat{x}_\varepsilon,\varepsilon) = \hat{v}.$$

Proof. $\hat{l}$ and $\hat{f}$ satisfy the following assumptions:

- $\hat{l}$ and $\hat{f}$ are continuous functions on $X \times \overline{\mathcal{M}}(Y_1) \times \overline{\mathcal{M}}(Y_2)$;
- $\hat{f}(x,\cdot,\mu_2)$ is a linear function on $\overline{\mathcal{M}}(Y_1)$ for all all $\mu_2 \in \overline{\mathcal{M}}(Y_2)$ and $\hat{f}(x,\mu_1,\cdot)$ is a linear function on $\overline{\mathcal{M}}(Y_2)$ for all $\mu_1 \in \overline{\mathcal{M}}(Y_1)$.

Then, by Theorem 5.1 in [12], there exists at least a solution to the problem $(\hat{w}-HP)(\varepsilon)$ for all $\varepsilon > 0$, and by Theorem 5.2 in [12], we have

$$\lim_{\varepsilon\to 0^+} \hat{v}(\varepsilon) = \hat{v}.$$ □

Remark 3.1. If we consider mixed strategies for players 2 and 3 in Example 2.2, the linearized objective functions, in terms of the mathematical expectations, are

$$\hat{l}(x,\mu_1,\mu_2) = x - (E(y_1) + E(y_2)),$$
$$\hat{f}(x,\mu_1,\mu_2) = E(y_1)E(y_2) - (1+x)E(y_1^2).$$

For all $\varepsilon > 0$, we have

$$\hat{w}(x,\varepsilon) = \begin{cases} \varepsilon - 1 & \text{if } 0 \le x \le \varepsilon, \\ x - 2 + \varepsilon/x & \text{if } x > \varepsilon, \end{cases}$$

and $\hat{x}_\varepsilon = \sqrt{\varepsilon}$ is a (lower) ε-mixed solution to the approximate hierarchical problem $(\hat{w}-HP)(\varepsilon)$. Moreover, the (lower) ε-mixed average value $\hat{v}(\varepsilon) = -2 + 2\sqrt{\varepsilon}$ converges to $\hat{v} = v = -2$ as $\varepsilon \to 0^+$.

4 ε-Mixed Extension

In order to get a lower average value of the approximate problem $(w-HP)(\varepsilon)$, in this section we consider a mixed extension for the upper-level problem as well.

Let l and f be continuous functions on $X \times Y_1 \times Y_2$; we can consider the sets of Radon probability measures $\overline{\mathcal{M}}(X)$, $\overline{\mathcal{M}}(Y_1)$, $\overline{\mathcal{M}}(Y_2)$ on X, Y_1, Y_2, respectively, and the linearized objective functions, defined as follows on $\overline{\mathcal{M}}(X) \times \overline{\mathcal{M}}(Y_1) \times \overline{\mathcal{M}}(Y_2)$:

$$\bar{l}(m,\mu_1,\mu_2) = \langle m, \hat{l}(\cdot,\mu_1,\mu_2)\rangle,$$
$$\bar{f}(m,\mu_1,\mu_2) = \langle m, \hat{f}(\cdot,\mu_1,\mu_2)\rangle.$$

By considering a mixed solution also for player 1, we have the following extension for the hierarchical problem $(w - HP)$:

$$(\bar{w} - HP) = \begin{cases} \text{find } \bar{m} \in \overline{\mathcal{M}}(X) \text{ such that} \\ \inf_{m \in \overline{\mathcal{M}}(X)} \sup_{(\mu_1, \mu_2) \in \bar{S}(m)} \bar{l}(m, \mu_1, \mu_2) = \sup_{(\mu_1, \mu_2) \in \bar{S}(\bar{m})} \bar{l}(\bar{m}, \mu_1, \mu_2), \end{cases}$$

where $\bar{S}(m)$, for all $m \in \overline{\mathcal{M}}(X)$, is the set of the solutions to

$$\bar{S}(m) = \begin{cases} \text{find } (\overline{\mu}_1, \overline{\mu}_2) \in \overline{\mathcal{M}}(Y_1) \times \overline{\mathcal{M}}(Y_2) \text{ such that} \\ \bar{f}(m, \overline{\mu}_1, \overline{\mu}_2) = \sup_{\mu_2 \in \overline{\mathcal{M}}(Y_2)} \bar{f}(m, \overline{\mu}_1, \mu_2) = \inf_{\mu_1 \in \overline{\mathcal{M}}(Y_1)} \bar{f}(m, \mu_1, \overline{\mu}_2). \end{cases}$$

Let $\bar{w}(m) = \sup_{(\mu_1, \mu_2) \in \bar{S}(m)} \bar{l}(m, \mu_1, \mu_2)$ and $\bar{v} = \inf_{m \in \overline{\mathcal{M}}(X)} \bar{w}(m)$; $\bar{v}$ will be called the "ε-mixed average value" of the problem $(w - HP)$.

As in the previous section, in order to obtain existence results, we have to deal with approximate mixed solutions.

Definition 4.1. Let $\varepsilon > 0$; an ε-mixed solution to the problem $(w - HP)$ is a solution to the following problem:

$$(\bar{w} - HP)(\varepsilon) = \begin{cases} \text{find } \bar{m}_\varepsilon \in \overline{\mathcal{M}}(X) \text{ such that} \\ \inf_{m \in \overline{\mathcal{M}}(X)} \sup_{(\mu_1, \mu_2) \in \bar{S}(m, \varepsilon)} \bar{l}(m, \mu_1, \mu_2) = \sup_{(\mu_1, \mu_2) \in \bar{S}(\bar{m}_\varepsilon, \varepsilon)} \bar{l}(\bar{m}_\varepsilon, \mu_1, \mu_2), \end{cases}$$

where $\bar{S}(m, \varepsilon)$, for all $m \in \overline{\mathcal{M}}(X)$ and any $\varepsilon > 0$, is the set of the solutions to the problem

$$\bar{S}(m, \varepsilon) = \begin{cases} \text{find } (\overline{\mu}_1, \overline{\mu}_2) \in \overline{\mathcal{M}}(Y_1) \times \overline{\mathcal{M}}(Y_2) \text{ such that} \\ \sup_{\mu_2 \in \overline{\mathcal{M}}(Y_2)} \bar{f}(m, \overline{\mu}_1, \mu_2) - \inf_{\mu_1 \in \overline{\mathcal{M}}(Y_1)} \bar{f}(m, \mu_1, \overline{\mu}_2) \leq \varepsilon. \end{cases}$$

Let $\bar{w}(m, \varepsilon) = \sup_{(\mu_1, \mu_2) \in \bar{S}(m, \varepsilon)} \bar{l}(m, \mu_1, \mu_2)$; $\bar{v}(\varepsilon) = \inf_{m \in \overline{\mathcal{M}}(X)} \bar{w}(m, \varepsilon)$ will be called the ε-mixed average value.

The following result holds.

Theorem 4.1. *Suppose that l and f are continuous functions on $X \times Y_1 \times Y_2$.*

Then, for all $\varepsilon > 0$, there exists at least a solution $\bar{m}_\varepsilon$ to the mixed regularized problem $(\bar{w} - HP)(\varepsilon)$ and

$$\lim_{\varepsilon \to 0^+} \bar{w}(\bar{m}_\varepsilon, \varepsilon) = \bar{v}.$$

Moreover, for $\varepsilon > 0$ sufficiently small,

$$\bar{v}(\varepsilon) \leq \hat{v}(\varepsilon). \tag{4}$$

Proof. Again, as in Theorem 3.1, it is sufficient to apply Theorem 5.1 in [12] in order to obtain the existence result and Theorem 5.2 in [12] in order to have the convergence result.

Moreover, for all $x \in X$ we have

$$\hat{l}(x, \mu_1, \mu_2) = \bar{l}(x, \mu_1, \mu_2) \qquad \forall(\mu_1, \mu_2) \in \overline{\mathcal{M}}(Y_1) \times \overline{\mathcal{M}}(Y_2),$$
$$\hat{S}(x, \varepsilon) = \bar{S}(x, \varepsilon).$$

Then

$$\bar{v}(\varepsilon) = \inf_{m \in \overline{\mathcal{M}}(X)} \Big(\sup_{(\mu_1, \mu_2) \in \bar{S}(m, \varepsilon)} \bar{l}(m, \mu_1, \mu_2) \Big)$$
$$\leq \inf_{x \in X} \Big(\sup_{(\mu_1, \mu_2) \in \bar{S}(x, \varepsilon)} \bar{l}(x, \mu_1, \mu_2) \Big) = \hat{v}(\varepsilon),$$

and we obtain the result in (4). □

Remark 4.1. If we take the limit for $\varepsilon \to 0^+$ in (4), we have $\bar{v} \leq \hat{v}$. It could be possible to have $\bar{v}(\varepsilon) < \hat{v}(\varepsilon)$, as the following example shows.

Example 4.1. Let $X = Y_1 = Y_2 = [0, 1]$ and

$$l(x, y_1, y_2) = \sqrt{x} - (y_1 + y_2),$$
$$f(x, y_1, y_2) = y_1 y_2 - (1 + x) y_1^2.$$

In this case, the problems $(w - HP)$, $(w - HP)(\varepsilon)$, and $(\hat{w} - HP)$ have no solution. The lower ε-mixed problem $(\hat{w} - HP)(\varepsilon)$ has the solution $\hat{x}_\varepsilon = \sqrt[3]{(2\varepsilon)^2}$ for all $\varepsilon > 0$.

If we consider the following probability distribution on X:

$$\bar{m}_\varepsilon = \begin{cases} \sqrt[3]{(2\varepsilon)^2} - \varepsilon & \text{w.p. } 1/2, \\ \sqrt[3]{(2\varepsilon)^2} + \varepsilon & \text{w.p. } 1/2, \end{cases}$$

we have:

$$\bar{v}(\varepsilon) \leq \bar{w}(\bar{m}_\varepsilon, \varepsilon) = \frac{1}{2}\left[\sqrt{\sqrt[3]{(2\varepsilon)^2} - \varepsilon} + \sqrt{\sqrt[3]{(2\varepsilon)^2} + \varepsilon}\,\right] - \left(2 - \varepsilon / \sqrt[3]{(2\varepsilon)^2}\right) < \hat{v}(\varepsilon)$$

for ε sufficiently small.

5 Conclusion

In this paper we considered hierarchical games with two followers, playing a zero-sum game and without convexity assumptions on their cost functions, and we extended the concepts of approximate mixed solutions introduced in previous papers in the case of a unique follower. We emphasize that the results obtained in the two different situations for the lower-level problem are not exactly the same. So, a question naturally arises: what can we obtain in the case of nonzero-sum games?

Let us note that the results given in this paper could be adapted for general constraints or constraints defined by a finite number of equalities and/or inequalities.

Moreover, the case of linear or quadratic cost functions, interesting for applications, could be investigated together with the case of matrix games.

REFERENCES

[1] Aubin, J. P. *Mathematical Methods of Game and Economic Theory*, North-Holland, Amsterdam, 1982.

[2] Basar, T. and Olsder, G. J. *Dynamic Noncooperative Games*, Academic Press, New York, 1995.

[3] Borel, E. La theorie du jeu et les equations integrales a noyeau symmetrique, *Comptes Rendus Aca. Sc.*, **173**, pp. 1304–1308, 1921.

[4] Breton, M., Alj, A., and Haurie, A. Sequential Stackelberg equilibria in two-person games, *J. Opt. Th. Appl.*, **59**, pp. 71–97, 1988.

[5] Chen, C. and Cruz, J. Stackelberg solution for two-person games with biased information patterns, *IEEE Trans. Auto. Cont.*, **AC-17**, pp. 791–798, 1972.

[6] Harker, P. T. and Pang, J. S. Finite-dimensional variational inequality and nonlinear complementarity problems: A survey of theory, algorithms and applications, *Math. Program.*, **48**, pp. 161–220, 1990.

[7] Leitmann, G. On Generalized Stackelberg Strategies, *J. Opt. Th. Appl.*, **26**, pp. 637–643, 1978.

[8] Luo, Z. Q., Pang, J. S., and Ralph, D. *Mathematical Programs with Equilibrium Constraints*, Cambridge Univ. Press, New York, 1996.

[9] Mallozzi, L. and Morgan, J. ε-mixed strategies for static continuous kernel stackelberg problem, *J. Opt. Th. Appl.*, **78**, pp. 303–316, 1993.

[10] Mallozzi, L. and Morgan, J. Weak Stackelberg problem and mixed solutions under data perturbations, *Optimization*, **32**, pp. 269–290, 1995.

[11] Mallozzi, L. and Morgan, J. Hierarchical systems with weighted reaction set, in *Nonlinear Optimization and Applications*, D. Pillo and F. Giannessi, eds., Plenum Press, New York, pp. 271–283, 1996.

[12] Morgan, J. and Raucci, R. Continuity properties of ε-solutions for generalized parametric saddle point problems and application to hierarchical games, *J. Math. Anal. Appl.*, **211**, pp. 30–48, 1997a.

[13] Morgan, J. and Raucci, R. Approximate solutions for strong hierarchical saddle point problems, *Atti Sem. Mat. Fis. Univ. Modena*, **XLV**, pp. 395–409, 1997b.

[14] Moulin, H. *Game Theory for the Social Sciences*, NYU Press, New York, 1986.

[15] Owen, G. *Game Theory*, Academic Press, New York, 1982.

[16] Schwartz, L. *Analyse III Calcul Integral*, Hermann, Paris, 1993.

[17] Sheraly, H. D., Soyster, A. L., and Murphy, F. H. Stackelberg-Nash-Cournot equilibria: Characterizations and computations, *Oper. Res.*, **31**, pp. 253–276, 1983.

[18] Simaan, M. and Cruz, J. On the Stackelberg strategies in nonzero sum games, *J. Opt. Th. Appl.*, **11**, pp. 533–555, 1973.

[19] Tobin, R. L. Uniqueness results and algorithm for Stackelberg-Cournot-Nash equilibria, *Ann. Oper. Res.*, **34**, pp. 21–36, 1992.

[20] von Neumann, J. and Morgenstern, O. Theory of Games and Economic Behavior, Princeton Univ. Press, Princeton, 1934.

[21] von Stackelberg, H. *Marktform und Gleichgewicht*, Julius Springer, Vienna, 1934.

The Existence and Uniqueness of Equilibria in Convex Games with Strategies in Hilbert Spaces

Dean A. Carlson
University of Toledo
Department of Mathematics
Toledo, Ohio, USA

Abstract

In this paper we extend the approach of Rosen [7] for existence and uniqueness of Nash equilibria for finite-dimensional convex games to an abstract setting in which the strategies of each player are in separable Hilbert spaces. Through the use of an extension of the Kakutani fixed-point theorem, we are able to extend Rosen's existence result to this setting. Our uniqueness results are obtained by extending Rosen's notion of *strict diagonal convexity* to this setting. Several examples, in the context of open-loop dynamic games, to which our results may be applied are presented.

1 Introduction

This paper represents an effort to extend to infinite dimensions the theory initiated in Rosen [7] concerning the existence and uniqueness of Nash equilibria for convex games in $\mathbb{R}^n$. While the existence question has been addressed before, apparently the question of uniqueness has not been extended. While these conditions are stringent, nevertheless, in the absence of some additional restrictions that require the players to seek a more restrictive equilibrium, there are occasions in which it is of paramount importance to know *a priori* that a unique equilibrium exists. With regards to the approach generally taken in the development of this theory, it is customary to define a set-valued mapping that represents each player's best reply to a fixed equilibrium. The existence of a feasible point is now established by applying an appropriate fixed-point theorem. This necessitates that an optimization problem must, in principle at least, be solved for each player in order to define the above-mentioned set-valued mapping. In Rosen's approach, a combined weighted cost function is defined and an analog of the best reply mapping is defined which requires that only one optimization problem be solved. This approach has apparently not been considered in the infinite-dimensional setting. As we shall see, perhaps one reason for this is that in the abstract setting the analog of Rosen's existence results requires stronger continuity conditions than are typically required in more recent results. Moreover, these continuity conditions are not typically satisfied in dynamic games formulated in terms of control theory.

In light of the above remarks, our primary goal is to focus on the uniqueness, rather than the existence, of a Nash equilibrium. Nevertheless, what makes Rosen's approach attractive is that it leads to conditions for uniqueness that are formulated in terms of an extension of standard convexity assumptions. Also, these conditions are independent from the fixed-point argument used in establishing existence. This avoids the use of highly restrictive fixed-point theorems (e.g., the contraction mapping theorem) and permits the investigation of uniqueness when existence cannot be directly established. Finally, Rosen's approach permits the introduction of *normalized equilibria* in which the corresponding Lagrange multipliers for the corresponding players are related by the weighting structure used to define the combined costs. These normalized Lagrange multipliers may provide useful schemes for imposing penalties on individual players in a consistent manner.

With these brief remarks, the plan of our paper is as follows. In Section 2 we introduce the abstract game in a Hilbert space, define notation, and present the extension of Rosen's theory to this infinite-dimensional setting. In Section 3 we present several examples arising in dynamic games to which we can give specific conditions for which uniqueness holds.

2 The Game in Hilbert Space

We begin by formulating a general abstract convex game in Hilbert space. To this end we let $P \doteq \{1, 2, \ldots, p\}$, where $p \in \mathbb{N}$ denotes the total number of players. For each $j \in P$, we let E_j be a real Hilbert space with inner product $\langle \cdot, \cdot \rangle_j$ and we let $E = \Pi_{j \in P} E_j$ be considered as a Hilbert space with inner product defined for $\mathbf{x} = (x_1, x_2, \ldots, x_p) \in E$ (here $x_j \in E_j$ for each $j \in P$) and $\mathbf{y} = (y_1, y_2, \ldots, y_p) \in E$ as

$$\langle \mathbf{x}, \mathbf{y} \rangle = \sum_{j \in P} \langle x_j, y_j \rangle_j.$$

The cost associated with each player is described by a function $F_j : E \to \mathbb{R}$, $j \in P$, which is assumed to be convex in the jth component and continuous with respect to the weak topology on E. Additionally, we impose constraints on the players through coupled linear equality and convex inequality constraints. We denote the linear constraints by a single vector-valued function $H : E \to \mathbb{R}^n$ and the inequality constraints by convex functions $G_l : E \to \mathbb{R}$, for $l = 1, 2, \ldots, k$. That is, a vector $\mathbf{x} \in E$ is feasible for the convex game if

$$H(\mathbf{x}) = v, \tag{1a}$$

$$G_l(\mathbf{x}) \leq 0 \text{ for } l = 1, 2, \ldots, k, \tag{1b}$$

in which $v \in R^n$ is a fixed vector. Finally, we say that a feasible point $\mathbf{x}^* \in E$ is a *Nash equilibrium* if for each $j \in P$ we have

$$F_j(\mathbf{x}^*) \leq F_j([\mathbf{x}^{*j}, y]) \tag{2}$$

holds for all $y \in E_j$ for which the vector $[\mathbf{x}^{*j}, y] \doteq (x_1, x_2, \ldots, x_{j-1}, y, x_{j+1}, \ldots, x_p)$ is feasible. For brevity we let Ω denote the set of all feasible points and observe that Ω is a convex subset of E. To report our results for each vector $\mathbf{r} = (r_1, r_2, \ldots, r_p)$ with $r_j > 0$, for $j \in P$, we define the combined weighted cost function $\rho_r : \Omega \times \Omega \to \mathbb{R}$ by the formula

$$\rho_r(\mathbf{x}, \mathbf{y}) \doteq \sum_{j \in P} r_j F_j([\mathbf{x}^j, y_j]). \tag{3}$$

We observe that this combined cost functional is weakly continuous on $E \times E$ and that for each fixed $\mathbf{x} \in E$ the map $\mathbf{y} \to \rho_r(\mathbf{x}, \mathbf{y})$ is convex.

2.1 The Existence of a Nash Equilibrium

Our first result is well known but is repeated here for completeness as well as for illustrative purposes. We begin by defining the set-valued mapping $\Gamma : E \to 2^E \setminus \emptyset$ by the formula

$$\Gamma(\mathbf{x}) : \{\mathbf{y} \in E : \quad \mathbf{y} \text{ is feasible and } \rho_r(\mathbf{x}, \mathbf{y}) \leq \rho_r(\mathbf{x}, \mathbf{z}) \quad \text{for all } \mathbf{z} \in \Omega\}. \tag{4}$$

We first state the following extension of the classical Kakutani fixed-point theorem to infinite dimensions.

Theorem 2.1. *Let H be a weakly separable Hilbert space with S a convex, weakly closed set in E. Let $B : S \to 2^S \setminus \{\emptyset\}$ be a set-valued mapping satisfying the following:*

1. *$B(\mathbf{x})$ is convex for each $\mathbf{x} \in S$.*
2. *The graph of B, $\{(\mathbf{x}, \mathbf{y}) \in S \times S : \mathbf{y} \in B(\mathbf{x})\}$, is weakly closed in $H \times H$. That is, if $\{\mathbf{x}_n\}$ and $\{\mathbf{y}_n\}$ are two sequences in S such that $\mathbf{x}_n \to \mathbf{x}$, $\mathbf{y}_n \to \mathbf{y}$, weakly in H with $\mathbf{x}_n \in B(\mathbf{y_n})$, then necessarily we have $\mathbf{x} \in B(\mathbf{y})$.*
3. *$\bigcup_{\mathbf{x} \in S} B(\mathbf{x})$ is contained in a sequentially weakly compact set T.*

Then there exists $\mathbf{x}^* \in S$ such that $\mathbf{x}^* \in B(\mathbf{x}^*)$.

Proof. See Theorem 5 [2]. □

With this result we can now give a proof of the following well-known theorem.

Theorem 2.2. *If, in addition to the above hypotheses, we assume that the set of feasible points Ω is closed, convex, and bounded, then there exists a Nash equilibrium.*

Proof. We first observe that for each $\mathbf{x} \in \Omega$ the function $\rho_r(\mathbf{x}, \cdot) : \Omega \to \mathbb{R}$ is weakly continuous and that Ω is convex and weakly compact. Therefore, there exists $\mathbf{y} \in \Omega$ such that $\mathbf{y} \in \Gamma(\mathbf{x})$, giving us that $\Gamma(\mathbf{x})$ is nonempty for each $\mathbf{x} \in \Omega$. Further, since $\rho_r(\mathbf{x}, \cdot)$ is a convex function, it follows that $\Gamma(\mathbf{x})$ is a convex set for each $\mathbf{x} \in \Omega$. Finally, we observe that if $\{\mathbf{x}_n\}$ and $\{\mathbf{y}_n\}$ are two sequences in Ω that converge weakly to $\mathbf{x}$ and $\mathbf{y}$ and additionally satisfy $\mathbf{y}_n \in \Gamma(\mathbf{x}_n)$ for each $n \in \mathbb{N}$, we have for each $\mathbf{z} \in \Omega$ that

$$\begin{aligned}\rho_r(\mathbf{x}, \mathbf{y}) &= \lim_{n \to +\infty} \rho_r(\mathbf{x}_n, \mathbf{y}_n) \\ &\leq \lim_{n \to +\infty} \rho_r(\mathbf{x}_n, \mathbf{z}) \\ &= \rho_r(\mathbf{x}, \mathbf{z}).\end{aligned} \tag{5}$$

This of course implies that $\mathbf{y} \in \Gamma(\mathbf{x})$, so that we can conclude that by Kakutani's fixed-point theorem (as extended above) there exists $\mathbf{x}^* \in \Omega$ such that $\mathbf{x}^* \in \Gamma(\mathbf{x}^*)$.

It remains to show that $\mathbf{x}^*$ is a Nash equilibrium. To this end, suppose that there exists an index $j \in P$ and $y_j \in E_j$ such that $[\mathbf{x}^{*j}, y_j] \in \Omega$ and

$$F_j([\mathbf{x}^{*j}, y_j]) < F_j(\mathbf{x}^*).$$

This implies that

$$\rho_r(\mathbf{x}^*, [\mathbf{x}^{*j}, y_j]) < \rho_r(\mathbf{x}^*, \mathbf{x}^*),$$

which contradicts the fact that $\mathbf{x}^*$ is a fixed point of $\Gamma(\cdot)$. Thus, we have that $\mathbf{x}^*$ is a Nash equilibrium, as desired. □

Remark 2.1. The above existence theorem is a well-known result that is typically proven through the use of the Ky–Fan inequality (see, e.g., Aubin [1]). The above proof is modeled after Rosen [7], given for finite-dimensional convex games. We further observe that the above result is valid for any constraint set Ω that is closed, bounded, and convex and does not rely on the structure of Ω assumed here.

2.2 Necessary Conditions for Nash Equilibria

We now briefly outline the classic necessary conditions for a Nash equilibrium. To do this we introduce the following Slater-type constraint qualification.

Definition 2.1. We say that the Slater constraint qualification holds at $\mathbf{x}$ for player j if there exists $\tilde{x}_j \in E_j$ such that

$$H([\mathbf{x}^j, \tilde{x}_j]) = v \quad \text{and} \quad G_l([\mathbf{x}^j, \tilde{x}_j]) < 0 \quad \text{for} \quad l = 1, 2, \ldots, k,$$

and if $v \in \operatorname{int}\left(\left\{z \in \mathbb{R}^n : z = H([\mathbf{x}^j, y]) \text{ for } y \in E_j\right\}\right)$ (here $\operatorname{int}(A)$ denotes the interior of A).

We also observe that if $\mathbf{x}^*$ is a Nash equilibrium, then for each $j \in P$ we have that x_j^* solves the mathematical programming problem, briefly described by

$$\min_{y \in E_j} F_j([\mathbf{x}^{*j}, y])$$

subject to

$$\begin{aligned} H([\mathbf{x}^{*j}, y]) &= v, \\ G_l([\mathbf{x}^{*j}, y]) &\leq 0 \quad \text{for} \quad l = 1, 2, \ldots, k. \end{aligned}$$

With this information we have the following result.

Theorem 2.3. *Let* $\mathbf{x}^*$ *be a Nash equilibrium and assume that the Slater constraint qualification holds at* $\mathbf{x}^*$ *for each player* $j \in P$. *Then for each* $j \in P$, *there exist scalars* $p_{jl} \geq 0$ *for* $l = 1, 2, \ldots, k$ *and scalars* q_{js}, $s = 1, 2, \ldots, n$, *such that*

$$F_j([\mathbf{x}^{*j}, y]) + \sum_{l=1}^{k} p_{jl} G_l([\mathbf{x}^{*j}, y]) + \sum_{s=1}^{n} q_{js}(H_s([\mathbf{x}^{*j}, y]) - v_s) \geq F_j(\mathbf{x}^*) \quad \textit{for all} \quad y \in E_j,$$

$$\sum_{l=1}^{k} p_{jl} G_l(\mathbf{x}^*) = 0. \tag{6}$$

Proof. The proof of the above result follows immediately from the classical results from optimization theory. See, for example, [1], Theorem 1, Sect. 6. □

Remark 2.2. In [7] the above result is stated to be true under the weaker constraint qualification given below in Definition 2.3. Without further investigation, we remark that it is possible for this constraint qualification to hold without the constraint qualification given by Definition 2.1 holding for at least one $j \in P$. One such case where the two conditions are equivalent is when the constraints are decoupled (see Definition 2.5). Therefore, there is apparently a minor error in Rosen's paper.

2.3 Normalized Nash Equilibrium

An observation to be made regarding the above necessary conditions is that between separate players, the multipliers p_{jl} and q_{js} are not related. Rosen, aware of this difficulty, introduced the notion of a *normalized equilibrium*, defined below.

Definition 2.2. A Nash equilibrium $\mathbf{x}^* \in E$ is called a *normalized Nash Equilibrium* with respect to the weights $\mathbf{r} = (r_1, r_2, \ldots, r_p)$ if there exist two sets of constants $p_l \geq 0$ for $l = 1, 2, \ldots, k$ and $q_s \in \mathbb{R}$ for $s = 1, 2, \ldots, n$ such that the constants

$$p_{jl} = \frac{1}{r_j} p_l \qquad \text{and} \qquad q_{js} = \frac{1}{r_j} q_s, \tag{7}$$

for $l = 1, 2, \ldots, k$ and $s = 1, 2, \ldots, n$, form a set of Kuhn–Tucker multipliers for the convex game. That is, the equations described by (6) hold.

To discuss the existence of a normalized Nash equilibrium, we require the following constraint qualification.

Definition 2.3. We say that Slater constraint qualification holds if there exists $\tilde{\mathbf{x}} \in E$ such that

$$H(\tilde{\mathbf{x}}) = v \qquad \text{and} \qquad G_l(\tilde{\mathbf{x}}) < 0, \quad \text{for} \quad l = 1, 2, \ldots, k, \tag{8}$$

and $v \in \operatorname{int}(H(E))$.

Theorem 2.4. *If the set of feasible points Ω is closed, convex, and bounded, and if the Slater constraint qualification,* (2.3), *holds, then there exists a normalized Nash equilibrium for each set of positive weights* $\mathbf{r}$.

Proof. Under the conditions given above for each set of positive weights, $\mathbf{r} = (r_1, r_2, \ldots, r_p)$, there exists a fixed point $\mathbf{x}^*$ of the set-valued mapping Γ defined by Eq. (4); as we have seen in Theorem 2.2, such a point is a Nash equilibrium for the game. Moreover, we know that $\mathbf{x}^* \in E$ is also a minimizer of the combined cost function $\rho_r(\mathbf{x}^*, \cdot)$ over all feasible points. Therefore, since the Slater constraint qualification holds, there exist nonnegative numbers $p_l \geq 0$, for $l = 1, 2, \ldots, k$, and numbers q_s, for $s = 1, 2, \ldots, n$, such that

$$\rho_r(\mathbf{x}^*, \mathbf{y}) + \sum_{l=1}^{k} p_l G_l(\mathbf{y}) + \sum_{s=1}^{n} q_s (H_s(\mathbf{y}) - v) \geq \rho_r(\mathbf{x}^*, \mathbf{x}^*),$$

for all $\mathbf{y} \in E$, and

$$p_l G_l(\mathbf{x}) = 0, \quad \text{for} \quad l = 1, 2, \ldots, k.$$

To finish our proof, we fix $j \in P$ and let $p_{j,l}$ and q_{js} be as defined by Eq. (7). Then for each $y \in E_j$, we have $[\mathbf{x}^{*j}, y] \in E$ and so,

$$\begin{aligned}
\rho_r(\mathbf{x}^*, [\mathbf{x}^{*j}, y]) &+ \sum_{l=1}^{k} p_l G_l([\mathbf{x}^{*j}, y]) + \sum_{s=1}^{n} (H_s([\mathbf{x}^{*j}, y]) - v) - \rho_r(\mathbf{x}^*, \mathbf{x}^*) \\
&= r_j F_j([\mathbf{x}^{*j}, y]) + \sum_{l=1}^{k} p_l G_l([\mathbf{x}^{*j}, y]) \\
&\quad + \sum_{s=1}^{n} (H_s([\mathbf{x}^{*j}, y]) - v) - r_j F(\mathbf{x}^*) > 0.
\end{aligned}$$

Dividing by $r_j > 0$, we see that with these choices of p_{jl} and q_{js} Eq. (6) holds. Thus, we see that $\mathbf{x}^*$ is a normalized Nash equilibrium. □

2.4 Uniqueness of Nash Equilibria

On the one hand, the conditions for unique equilibria in convex games are generally quite stringent. On the other hand, having a unique equilibria is often of paramount importance. In finite dimensions this idea was investigated by Rosen in 1965, where he introduced the notion of strict diagonal convexity. Recently, Carlson and Haurie [3], [4] extended this idea to dynamic games with infinite horizons. We now extend Rosen's result to the abstract case considered here.

Definition 2.4. Let $\mathbf{r} = (r_1, r_2, \ldots, r_p)$ be fixed. We say that the combined cost function $\rho_r(\cdot, \cdot) : E \times E \to \mathbb{R}$ is strictly diagonally convex if

$$\sum_{j \in P} \left\{ r_j \langle (\pi_j^0 - \pi_j^1), (x_j^0 - x_j^1) \rangle_j \right\} < 0 \tag{9}$$

for all $x_j^0, x_j^1 \in E_j$ and all subgradients

$$\pi_j^0 \in \partial_{x_j} F_j(\mathbf{x}^0) \qquad \text{and} \qquad \pi_j^1 \in \partial_{x_j} F_j(\mathbf{x}^1).$$

With this definition we first consider the case when the constraints are decoupled, beginning with the following definition.

Definition 2.5. We say that the constraints described by (1) are *decoupled* if there exist integers $n_j \geq 0$ for $j \in P$ with $\sum_{j \in P} n_j = n$ so that $H(\cdot) = (H_1(\cdot), \ldots, H_p(\cdot))$, with $H_j(\cdot) : E_j \to \mathbb{R}^{n_j}$ and $\mathbf{v} = (v_1, v_2, \ldots, v_p)$, with $v_j \in \mathbb{R}^{n_j}$, and if there exists integers $k_j \geq 0$ with $\sum_{j \in P} k_j = k$ so that the convex inequality constraints take the form

$$G_{jl}(x_j) \leq 0 \quad \text{for} \quad l = 1, 2, \ldots, k_j,$$

for $j \in P$.

Remark 2.3. In the case of decoupled constraints, the constraint qualification given by Definition 2.3 is equivalent to having the constraint qualification given in Definition 2.1 holding for all $j \in P$.

With this definition, we have the following uniqueness result.

Theorem 2.5. *If the Slater constraint qualification holds and if there exists a positive vector* $\mathbf{r} \in \mathbb{R}^p$ *for which the combined cost function* $\rho_r(\cdot, \cdot)$ *is strictly diagonally convex, then there exists at most one Nash equilibrium.*

Proof. The proof of this result is by contradiction. Therefore, we let $\mathbf{x}^0$ and $\mathbf{x}^1$ be two Nash equilibrium. Since the constraints are decoupled and since the constraint qualification holds there exist multipliers $p_{ijl} \geq 0$ and $q_{ij} \in \mathbb{R}$ for $i = 0, 1$, $l = 1, 2, \ldots, k_j$, and $j \in P$ such that Eq. (6) holds for each $j \in P$ and $i = 0, 1$. Under our convexity assumptions, this is equivalent to saying that

$$0 \in \partial_{x_j} \left(F_j(\mathbf{x}^i) + \sum_{l=1}^{k_j} p_{ijl} G_{jl}(x_j^i) + q_{ij}(H_j(x_j^i) - v_j) \right)$$

and

$$p_{ijl} G_{jl}(x_j^i) = 0, \qquad l = 1, 2, \ldots, k_j,$$

hold for all $j \in P$ and $i = 0, 1$. From properties of subgradients, we have that there exist points π_j^i, η_j^i, and γ_{jl}^i in E_j such that

$$\pi_j^i + \eta_j^i + \sum_{l=1}^{k_j} \gamma_{jl}^i = 0$$

$$\pi_j^i \in \partial_{x_j} F(\mathbf{x}^i), \quad \gamma_{jl}^i \in p_{ijl} \partial G_{jl}(x_j^i), \quad \text{and} \quad \eta_j^i \in q_{ij}\, \partial H_j(x_j^i).$$

This implies that for each $j \in P$,

$$p_{0jl} G_{jl}(x_j^1) = p_{0jl} G_{jl}(x_j^1) - p_{0jl} G_{jl}(x_j^0) \geq \langle \gamma_{jl}^0, (x_j^1 - x_j^0) \rangle_j,$$

which gives us (since $\mathbf{x}^1$ is feasible and p_{0jl} is nonnegative)

$$\langle \gamma_{jl}^0, (x_j^1 - x_j^0)\rangle_j \leq 0, \qquad \text{for each } l = 1, 2, \ldots, k_j \text{ and } j \in P.$$

Similarly, we get

$$\langle \gamma_{jl}^1, (x_j^0 - x_j^1)\rangle_j \leq 0, \qquad \text{for each } j \in P.$$

Further, the linearity of H_j gives us

$$0 = \eta_j^0(x_j^1 - x_j^0)$$

and

$$0 = \eta_j^1(x_j^0 - x_j^1),$$

for all $j \in P$ (recall that η_j^i is a linear mapping from E_j into $\mathbb{R}^{n_j}$). Combining these, we have

$$\begin{aligned}
0 &= \sum_{j\in P} r_j \left[\left\langle \left(\pi_j^0 + \sum_{l=1}^{k_j} \gamma_{jl}^0 \right), (x_j^1 - x_j^0) \right\rangle_j + \eta_j^0(x_j^1 - x_j^0) \right.\\
&\quad \left. + \left\langle \left(\pi_j^1 + \sum_{l=1}^{k_j} \gamma_{jl}^1 \right), (x_j^0 - x_j^1) \right\rangle_j + \eta_j^1(x_j^0 - x_j^1) \right]\\
&= \sum_{j\in P} r_j \left[\langle (\pi_j^0 - \pi_j^1), (x_j^1 - x_j^0)\rangle_j \right]\\
&\quad + \left\{ \sum_{j\in P} r_j \left[\sum_{l=1}^{k_j} \langle \gamma_{jl}^0, (x_j^1 - x_j^0)\rangle \right] \right.\\
&\quad \left. + \sum_{j\in P} r_j \left[\sum_{l=1}^{k_j} \langle \gamma_{jl}^1, (x_j^0 - x_j^1)\rangle \right] \right\}\\
&= \alpha + \beta\\
&< 0,
\end{aligned}$$

where the strict inequality follows since $\alpha < 0$ by our strict diagonal convexity assumption and $\beta \leq 0$ by the convexity of the inequality constraints. Clearly, this is a contradiction, and we conclude that there exists at most one Nash equilibrium. □

We now direct our attention to the case when the constraints are not decoupled. In this case we impose additional smoothness conditions on the constraints. Specifically, we now assume that the functions $H_s : E \to \mathbb{R}$, for $s = 1, 2, \ldots, n$, and $G_l : E \to \mathbb{R}$, for $1, 2, \ldots, k$, are all Gateaux differentiable functions. In this situation, we have the following result.

Remark 2.4. The additional smoothness conditions are imposed in order to avoid the technical difficulties involved in combining partial subgradients. Clearly, weaker conditions could be imposed to handle a more general case. On the other hand, as we shall see in the examples below, this level of generality is often not required and it is our feeling that these technical difficulties actually add little to understanding the main ideas presented here.

Theorem 2.6. *Suppose that for some set of positive weights* $\mathbf{r} = (r_1, r_2, \ldots, r_p)$, *the combined cost function* $\rho_r : E \times E \to \mathbb{R}$ *is strictly diagonally convex. Then there exists at most one Nash equilibrium that is normalized with respect to the weights* $\mathbf{r}$.

Proof. Assume that for some fixed set of weights $\mathbf{r}$, there exist two normalized Nash equilibria, say $\mathbf{x}^0$ and $\mathbf{x}^1$. Then, from the definition of such equilibrium, for $i = 1, 2$, we have that there exist multipliers $p_l^i \geq 0$, for $l = 1, 2, \ldots, k$, and $q_s^i \in \mathbb{R}$, for $s = 1, 2, \ldots, n$, such that

$$F_j([\mathbf{x}^{ij}, y_j]) + \sum_{l=1}^{k} \frac{p_l^i}{r_j} G_l([\mathbf{x}^{ij}, y_j]) + \sum_{s=1}^{n} \frac{q_s^i}{r_j} H_s([\mathbf{x}^{ij}, y_j]) \geq F_j(\mathbf{x}^i),$$

$$p_l^i G_l(\mathbf{x}^i) = 0,$$

for all $j \in P$, $i = 0, 1$, $l = 1, 2, \ldots, k$, and $y_j \in E_j$. Thus, for $i = 0, 1$ and $j \in P$, we have

$$r_j \langle D_j F_j(\mathbf{x}^i), \tilde{v}_j \rangle_j + \sum_{l=1}^{k} p_l^i \langle D_j G_l(\mathbf{x}^i), \tilde{v}_j \rangle_j + \sum_{s=1}^{n} q_s^i H_s([\mathbf{0}^j, \tilde{v}_j]) = 0,$$

which holds for each $\tilde{v}_j \in E_j$. Here the notation $D_j F(\mathbf{x}^i)$ is the unique element in E_j representing the partial Gateaux derivative of F_j with respect to x_j at the point x_j^i. Further, we note that the last term on the left-hand side above follows since $H(\cdot)$ is a linear mapping from E into $\mathbb{R}^n$. Now for $i = 0$ and $\tilde{v}_j = x_j^1 - x_j^0$, the above becomes

$$r_j \langle D_j F_j(\mathbf{x}^0), x_j^1 - x_j^0 \rangle_j + \sum_{l=1}^{k} p_l^0 \langle D_j G_l(\mathbf{x}^0), x_j^1 - x_j^0 \rangle_j + \sum_{s=1}^{n} q_s^0 H_s([\mathbf{0}^j, x_j^1 - x_j^0]) = 0. \tag{10}$$

Similarly, for $i = 1$ and $\tilde{v}_j = x_j^0 - x_j^1$, we have

$$r_j \langle D_j F_j(\mathbf{x}^1), x_j^1 - x_j^0 \rangle_j + \sum_{l=1}^{k} p_l^1 \langle D_j G_l(\mathbf{x}^1), x_j^1 - x_j^0 \rangle_j + \sum_{s=1}^{n} q_s^1 H_s([\mathbf{0}^j, x_j^1 - x_j^0]) = 0. \tag{11}$$

Adding (11) to (10) and summing over $j \in P$, we have (after rearranging the double sums)

$$\begin{aligned} 0 = &\sum_{j \in P} \left\{ r_j \langle D_j F_j(\mathbf{x}^0) - D_j F_j(\mathbf{x}^1), x_j^1 - x_j^0 \rangle_j \right\} \\ &+ \sum_{l=1}^{k} p_l^0 \left\{ \sum_{j \in P} \langle D_j G_j(\mathbf{x}^0), x_j^1 - x_j^0 \rangle_j \right\} + \sum_{l=1}^{k} p_l^1 \left\{ \sum_{j \in P} \langle D_j G_j(\mathbf{x}^1), x_j^0 - x_j^1 \rangle_j \right\} \end{aligned}$$

$$
\begin{aligned}
&+\sum_{s=1}^{n} q_s^0 \left\{ \sum_{j\in P} H_s([\mathbf{0}^j, x_j^1 - x_j^0]\right\} + \sum_{s=1}^{n} q_s^1 \left\{ \sum_{j\in P} H_s([\mathbf{0}^j, x_j^0 - x_j^1]\right\} \\
&= \sum_{j\in P} \{ r_j \langle D_j F_j(\mathbf{x}^0) - D_j F_j(\mathbf{x}^1), x_j^1 - x_j^0 \rangle_j \} \\
&\quad + \sum_{l=1}^{k} p_l^0 \langle DG_l(\mathbf{x}^0), \mathbf{x}^1 - \mathbf{x}^0 \rangle + \sum_{l=1}^{k} p_l^1 \langle DG_l(\mathbf{x}^1), \mathbf{x}^0 - \mathbf{x}^1 \rangle \\
&\quad + \sum_{s=1}^{n} q_s^0 \left(H_s(\mathbf{x}^1) - H_s(\mathbf{x}^0) \right) + \sum_{s=1}^{n} q_s^1 \left(H_s(\mathbf{x}^0) - H_s(\mathbf{x}^1) \right) \\
&= \alpha + \beta + \gamma,
\end{aligned}
$$

in which $DG_l(\mathbf{x}^i)$ denotes the Gateaux derivative of G_l at $\mathbf{x}^i$. As a consequence of our strict diagonal convexity hypothesis, we see that $\alpha < 0$. Further, the linearity of $H_s(\cdot)$ and the fact that both $\mathbf{x}^0$ and $\mathbf{x}^1$ are feasible show that $\gamma = 0$. Finally, the convexity of the functions G_l, the complementary slackness condition, $p_l^i G(\mathbf{x}^i) = 0$, and the feasibility of $\mathbf{x}^0$ and $\mathbf{x}^1$ give us $\beta \leq 0$. Clearly, we arrive at the desired contradiction since this implies that the right-hand side above is strictly negative and not zero. □

In the above section we successfully extended the concept of a normalized equilibrium found in Rosen's seminal paper [7] to an infinite-dimensional setting. In addition to extending his existence and uniqueness theorems, we allowed for the possibility of linear equality constraints. As we shall see below, this fact will enable us to apply the above theory to dynamic games with dynamics described by control systems.

3 Examples

3.1 The Variational Case

In this example we consider the case where each player's dynamics are described by an absolutely continuous function $x_j : [0, T] \to \mathbb{R}^{n_j}$, and the cost functional for each player is described by an integral of the form

$$
F_j(\mathbf{x}) = \int_0^T L_j(t, \mathbf{x}(t), \dot{\mathbf{x}}(t))\, dt, \qquad \text{for} \quad j \in P. \tag{12}
$$

Here we assume that $L_j : [0, T] \times \mathbb{R}^n \times \mathbb{R}^n \to \mathbb{R}$, where $n = \sum_{j\in P} n_j$, is continuous and such that the map $(x_j, z_j) \to L_j(t, [\mathbf{x}^j, x_j], [\mathbf{z}^j, z_j])$ is convex. For simplicity, we assume that the only equality constraints are described by the end conditions

$$
\mathbf{x}(0) = \mathbf{v}^0 \qquad \text{and} \qquad \mathbf{x}(T) = \mathbf{v}^1, \tag{13}
$$

in which $\mathbf{v}^0$ and $\mathbf{v}^1$ are given fixed points in $\mathbb{R}^n$. The inequality constraints are described as classical isoperimetric constraints of the form

$$G_l(\mathbf{x}) = \int_0^T g_l(t, \mathbf{x}(t), \dot{\mathbf{x}}(t))\, dt \leq 0, \qquad \text{for} \quad l = 1, 2, \ldots, k. \tag{14}$$

Here we assume that $g_l : [0, T] \times \mathbb{R}^n \times \mathbb{R}^n \to \mathbb{R}$ is a continuous function such that $g_l(t, \cdot, \cdot)$ is a convex function. To cast this problem in the appropriate setting, we introduce the Hilbert spaces $E_j \doteq W^{12}([0, T]; \mathbb{R}^{n_j})$, the Sobolev space of n_j-vector-valued functions that are absolutely continuous with square-integrable derivatives. The inner product defined on $E_j \times E_j$ is given by

$$\langle x, y \rangle_j = \int_0^T x(t) \cdot y(t)\, dt + \int_0^T \dot{x}(t) \cdot \dot{y}(t)\, dt.$$

To ensure the existence of a Nash equilibrium using the theory presented in the previous section, we must provide conditions under which the cost functionals of each player are weakly continuous. It is here that we encounter our first difficulties since it is well known (see, e.g., [5], Theorem 10.8.v) that to ensure the desired continuity we would need the functions $L_j(\cdot, \cdot, \cdot)$ to have the form

$$L_j(t, \mathbf{x}, \mathbf{z}) = A_{j0}(t, \mathbf{x}) + \sum_{i=1}^{n} A_{ji}(t, \mathbf{x}) z_i,$$

in which the functions $A_{ji} : [0, T] \times \mathbb{R}^n \to \mathbb{R}$ are Carathéodory functions for $j \in P$ and $i = 1, 2, \ldots, n$ which are each bounded above by a square-integrable function. Under these conditions the weak continuity is assured. Therefore, we can conclude that there exists a Nash equilibrium whenever the set of feasible trajectories is closed, bounded, and convex (or more generally when it is a weakly compact set). These conditions can be assured by assuming that one of the many appropriate growth condition holds. For a discussion of this, we refer the reader to Cesari [5].

To conclude our discussion of this example, we address the question of uniqueness. As we have seen in the previous section, this requires the strict diagonal convexity hypotheses to hold. For brevity we assume that the functions $L_j(t, \cdot, \cdot)$ have continuous first partial derivatives. To express the desired condition (9), we first recall that the jth partial Gateaux derivative of F_j at $\mathbf{x}$ in the direction $v_j \in E_j$ is given by the familiar formula

$$\langle D_j F_j(\mathbf{x}), v_j \rangle_j = \int_0^T \left[\nabla_{x_j} L_j(t, \mathbf{x}(t), \dot{\mathbf{x}}(t)) \cdot v_j(t) + \nabla_{\dot{x}_j} L_j(t, \mathbf{x}(t), \dot{\mathbf{x}}(t)) \cdot \dot{v}_j(t) \right] dt.$$

From this we see that the above problem meets the requirements of uniqueness if, for each $\mathbf{x}^0(\cdot)$ and $\mathbf{x}^1(\cdot)$, we have

$$0 > \sum_{j \in P} r_j \langle D_j F_j(\mathbf{x}^0) - D_j F_j(\mathbf{x}^1), x_j^0 - x_j^1 \rangle_j$$

$$
\begin{aligned}
&= \sum_{j\in P}\int_0^T r_j\left[\nabla_{x_j} L_j(t,\mathbf{x}^0(t),\dot{\mathbf{x}}^0(t))\cdot(x_j^0(t)-x_j^1(t))\right.\\
&\qquad \left.+\nabla_{\dot{x}_j} L_j(t,\mathbf{x}^0(t),\dot{\mathbf{x}}^0(t))\cdot(\dot{x}_j^0(t)-\dot{x}_j^1(t))\right]dt\\
&\qquad -\sum_{j\in P}\int_0^T r_j\left[\nabla_{x_j} L_j(t,\mathbf{x}^1(t),\dot{\mathbf{x}}^1(t))\cdot(x_j^0(t)-x_j^1(t))\right.\\
&\qquad \left.+\nabla_{\dot{x}_j} L_j(t,\mathbf{x}^1(t),\dot{\mathbf{x}}^1(t))\cdot(\dot{x}_j^0(t)-\dot{x}_j^1(t))\right]dt\\
&= \int_0^T \sum_{j\in P} r_j\left[\nabla_{(x_j,\dot{x}_j)} L_j(t,\mathbf{x}^0(t),\dot{\mathbf{x}}^0(t))\cdot\left((x_j^0(t)-x_j^1(t)),(\dot{x}_j^0(t)-\dot{x}_j^1(t))\right)\right.\\
&\qquad \left.-\nabla_{(x_j,\dot{x}_j)} L_j(t,\mathbf{x}^1(t),\mathbf{x}^1(t))\cdot\left((x_j^0(t)-x_j^1(t)),(\dot{x}_j^0(t)-\dot{x}_j^1(t))\right)\right]dt.
\end{aligned}
$$

From this we immediately see that uniqueness can be assured if the combined Lagrangian

$$
\mathcal{L}(t,\mathbf{x},\mathbf{y},\mathbf{u},\mathbf{v}) \doteq \sum_{j\in P} r_j L_j(t,[\mathbf{x}^j,y_j],[\mathbf{u}^j,v_j])
$$

is strictly diagonally convex. Under additional smoothness hypotheses, sufficient conditions for this to hold are well known (see [7]).

3.2 An Open-Loop Differential Game-Control Problem

In this example we consider an open-loop differential game in which the dynamics have separated linear controls. That is, the dynamics of each player are described by a control system of the form

$$
\dot{x}_j(t) = f_j(\mathbf{x}(t)) + B_j u_j(t) \quad \text{a.e. } 0\le t\le T, \tag{15}
$$

in which $f_j : \mathbb{R}^n \to \mathbb{R}^{n_j}$ is continuously differentiable with respect to x_j where $n=\sum_{j=0}^{p} n_j$ and B_j is an $n_j\times m_j$ matrix having full rank $m_j\le n_j$. This implies that B_j has a left inverse. That is, there exists an $m_j\times n_j$ matrix B_j^* such that $B_j^*B_j = I_{m_j}$, where I_k denotes the $k\times k$ identity matrix. The costs accumulated by player j are given by the integral functional

$$
F_j(\mathbf{x}(\cdot),u_j(\cdot)) = \int_0^T f_j^0(\mathbf{x}(t),u_j(t))\,dt, \tag{16}
$$

where $f^0 : \mathbb{R}^n\times\mathbb{R}^{n_j}\to\mathbb{R}$ is a given function that is continuously differentiable with respect to x_j and u_j. For this game problem we immediately obtain the "deparametrized" Lagrangians $L_j : \mathbb{R}^n\times\mathbb{R}^{n_j}\to\mathbb{R}$ by the formulas

$$
L_j(\mathbf{x},z_j) \doteq f_j^0\left(\mathbf{x}, B_j^* z_j - B_j^* f(\mathbf{x})\right), \tag{17}
$$

for $j\in P$. Consequently, using the framework from the previous example, it is easy to write down the conditions for strict diagonal convexity in this case. Specifically, we must have that for each each pair of points $(\mathbf{x}^0, \mathbf{z}^0)$ and $(\mathbf{x}^1, \mathbf{z}^1)$,

$$\begin{aligned}
0 &< \sum_{j=1}^{p} r_j \left[\nabla_{(x_j,z_j)} L_j(\mathbf{x}^0, z_j^0) \cdot \left((x_j^0 - x_j^1), (z_j^0 - z_j^1)\right)\right. \\
&\quad \left. - \nabla_{(x_j,z_j)} L_j(\mathbf{x}^1, z_j^1) \cdot \left((x_j^0 - x_j^1), (z_j^0 - z_j^1)\right)\right] \\
&= \sum_{j=1}^{p} r_j \left[\left(\nabla_{x_j} f_j^0(\mathbf{x}^0, u_j^0) - \nabla_{u_j} f_j^0(\mathbf{x}^0, u_j^0) B_j^* D_{x_j} f_j(\mathbf{x}^0)\right) \cdot (x_j^0 - x_j^1)\right. \\
&\quad + \nabla_{u_j} f_j^0(\mathbf{x}^0, u_j^0) \cdot B_j^*(z_j^0 - z_j^1) \\
&\quad - \left(\nabla_{x_j} f_j^0(\mathbf{x}^1, u_j^1) - \nabla_{u_j} f_j^0(\mathbf{x}^1, u_j^1) B_j^* D_{x_j} f_j(\mathbf{x}^1)\right) \cdot (x_j^0 - x_j^1) \\
&\quad \left. - \nabla_{u_j} f_j^0(\mathbf{x}^1, u_j^1) \cdot B_j^*(z_j^0 - z_j^1)\right] \\
&= \sum_{j=1}^{p} r_j \left[\nabla_{(x_j,u_j)} f_j^0(\mathbf{x}^0, u_j^0) \cdot \left((x_j^0 - x_j^1), (u_j^0 - u_j^1)\right)\right. \\
&\quad \left. - \nabla_{(x_j,u_j)} f_j^0(\mathbf{x}^1, u_j^1) \cdot \left((x_j^0 - x_j^1), (u_j^0 - u_j^1)\right)\right] \\
&\quad + \sum_{j=1}^{p} r_j \nabla_{u_j} f_j^0(\mathbf{x}^0, u_j^0) B_j^* \left[f_j(\mathbf{x}^0) - f_j(\mathbf{x}^1) + D_{x_j} f_j(\mathbf{x}^0)(x_j^1 - x_j^0)\right] \\
&\quad + \sum_{j=1}^{p} r_j \nabla_{u_j} f_j^0(\mathbf{x}^1, u_j^1) B_j^* \left[f_j(\mathbf{x}^1) - f_j(\mathbf{x}^0) + D_{x_j} f_j(\mathbf{x}^1)(x_j^0 - x_j^1)\right],
\end{aligned} \tag{18}$$

where $u_j^i = B_j^*\big(z_j^i - f_j(\mathbf{x}^i)\big)$ for $i = 0, 1$ and $j = 1, 2, \ldots, p$. From the above computation it is relatively easy to determine if the desired strict diagonal convexity condition is satisfied. In particular, we have the following result when the dynamics for each player are completely separated from the explicit influence of the other players. That is, the interactions between the players occur only through each player's payoff.

Theorem 3.1. *For $j = 1, 2, \ldots, p$, let $f_j^0 : \mathbb{R}^n \times \mathbb{R}^{m_j} \to \mathbb{R}$ be a set of continuously differentiable functions such that the combined objective function $\sigma_r : \mathbb{R}^n \times \mathbb{R}^n \times \mathbb{R}^m :\to \mathbb{R}$, defined by the formula*

$$\sigma_r(\mathbf{x}, \mathbf{y}, \mathbf{u}) = \sum_{j=1}^{p} f_j^0([\mathbf{x}^j, y_j], u_j),$$

is strictly diagonally convex for some set of positive weights $r_j > 0$, $j \in P$. Here the notation $[\mathbf{x}^j, y_j]$ means

$$[\mathbf{x}^j, y_j] = (x_1, x_2 \ldots, x_{j-1}, y_j, x_{j+1}, \ldots, x_p).$$

For $j = 1, 2, \ldots, p$, let $f_j : \mathbb{R}^{n_j} \to \mathbb{R}^{n_j}$ be continuously differentiable and such that each of its components is a concave function. For $j = 1, 2, \ldots, p$, let B_j be an $n_j \times m_j$ matrix with full rank $m_j \leq n_j$ having left inverse B_j^*, and assume that the components of $(\nabla_{u_j} f_j^0(\mathbf{x}, u_j))B_j^*$ are nonnegative. Then for the dynamic game *with costs described by* (16) *and dynamics by* (15), *the combined Lagrangian*

$$\mathcal{L}(\mathbf{x}, \mathbf{y}, \mathbf{z}) \doteq \sum_{j=1}^{p} L([\mathbf{x}^j, y_j], z_j) = \sum_{j=1}^{p} f_j^0([\mathbf{x}^j, y_j], B_j^*(z_j - f_j([\mathbf{x}^j, y_j])))$$

is strictly diagonally convex.

Proof. The proof of this result is an immediate consequence of the computation given above. □

Remark 3.1. The most difficult condition to verify above is the strict diagonal convexity of the combined costs. In [7] explicit conditions for the strict diagonal convexity of the combined payoffs are given. Specifically, Rosen shows that strict diagonal convexity holds if the Jacobian of the vector-valued function

$$g(\mathbf{x}, \mathbf{u}) = \begin{bmatrix} \nabla_{(x_1,u_1)} f_1^0(\mathbf{x}, u_1) \\ \nabla_{(x_2,u_2)} f_2^0(\mathbf{x}, u_2) \\ \vdots \\ \nabla_{(x_p,u_p)} f_p^0(\mathbf{x}, u_p) \end{bmatrix}$$

is positive definite. It might also be difficult to check that the components of $(\nabla_{u_j} f_j^0(\mathbf{x}, u_j))B_j^*$ are nonnegative, particularly since B_j^* need not be unique. If one wishes to be more explicit, it is known that one candidate for B_j^* is given by

$$B_j^* = (B_j' B_j)^{-1} B_j'.$$

One could then verify these conditions explicitly. These are several special cases in which these conditions are particularly easy to verify. The first is when the number of states equals the number of controls (i.e., $n_j = m_j$) and when B_j is either the $n_j \times n_j$ identity matrix or its negative. In this case it is enough to have $u_j \to f_j^0(\mathbf{x}, \cdot)$ be an increasing function in u_j when B_j is the identity matrix or to have it be a decreasing function of u_j when B_j is the negative of the identity The latter case is precisely the conditions imposed in a neoclassical growth model, an important example from macroeconomics. Consequently, under sufficient smoothness, we see that the conditions of the above theorem are readily checked. Another special case in which the conditions are readily verified is when the dynamics are linear. That is when

$$f_j(x_j) = A_j x_j, \qquad \text{for } j = 1, 2, \ldots, p,$$

in which A_j is an $n_j \times n_j$ constant matrix. In this case $D_{x_j} f_j(\mathbf{x}) = A_j$, so that the terms involving the matrices B_j^* in (19) become zero.

3.3 A Distributed-Parameter Example

In this example we consider a distributed parameter a control system whose dynamics can be described by an abstract evolution equation in a Hilbert space.

Specifically, we let X be a real separable Hilbert space that denotes the state space of the dynamics, and for each player $j \in P$ we let E_j be a real separable space. With these spaces we let A be a densely defined linear operator on X that generates a strongly continuous semigroup of operators $\{S(t) : t \geq 0\}$, and for $j \in P$, we let B_j, be a bounded linear operator from E_j into X. With these notations the dynamics of the game are described by the differential equation

$$\dot{x}(t) = Ax(t) + \sum_{j \in P} B_j u_j(t) \quad \text{a.e.} \quad 0 \leq t \leq T,$$

with initial condition

$$x(0) = x_0.$$

Here we assume that $x_0 \in X$ is fixed, that $x : [0, T] \to X$ is continuous, and that for $j \in P$ the controls $u_j(\cdot) \in L^2([0, T]; E_j)$. As is well known, for each set of control functions $\mathbf{u} = (u_1, u_2, \ldots u_p)$ there is a mild solution to the above initial-value problem given by the variation-of-constants formula

$$x(t) = S(t)x_0 + \int_0^t S(t-s) \sum_{j \in P} B_j u_j(s)\, ds, \qquad 0 \leq t \leq T.$$

In particular, we notice that each set of control functions gives rise to a unique state trajectory.

The cost objective for each player is given by an integral functional of the form

$$F_j(\mathbf{u}) \doteq \int_0^T f_j(x(t), \mathbf{u}(t))\, dt, \qquad j \in P,$$

in which $f_j : X \times E \to \mathbb{R}$, $j \in P$, are assumed to be Gateaux-differentiable and convex. Once again our goal is the discussion of the uniqueness of Nash equilibria. Consequently, we need not go further in our description of the dynamic game. We require, of course, that the strict diagonal convexity condition holds. To do this we first compute the jth partial Gateaux directional derivative of $F_j(\mathbf{u})$ in the direction $v_j \in L^2([0, T]; E_j)$. This is given by

$$\begin{aligned}
D_j F_j(\mathbf{u}; v_j) &= \left.\frac{d}{d\epsilon}\right|_{\epsilon=0} F_j([\mathbf{u}^j, u_j + \epsilon v_j]) \\
&= \int_0^T \left.\frac{d}{d\epsilon}\right|_{\epsilon=0} f_j(x(t;\epsilon), [\mathbf{u}^j, u_j + \epsilon v_j])\, dt \\
&= \int_0^T \Bigg[\langle D_x f_j(x(t), \mathbf{u}(t)), \left.\frac{d}{d\epsilon}\right|_{\epsilon=0} x(t;\epsilon) \rangle \\
&\qquad + \langle D_{u_j} f_j(x(t), \mathbf{u}(t)), v_j(t) \rangle_j \Bigg] dt
\end{aligned}$$

$$= \int_0^T \left[\langle D_x f_j(x(t), \mathbf{u}(t)), \int_0^t S(t-s) B_j v_j(s)\, ds \rangle \right] dt$$
$$+ \int_0^T \left[\langle D_{u_j} f_j(x(t), \mathbf{u}(t)), v_j(t) \rangle_j \right] dt,$$

in which $x(\cdot;\epsilon)$ is the admissible trajectory that results from the control $[\mathbf{u}^j, u_j + \epsilon v_j]$. With this result we now have that the strict diagonal convexity condition becomes

$$\begin{aligned}
0 &> \sum_{j \in P} r_j \langle D_j F_j(\mathbf{u}^0) - D_j F_j(\mathbf{u}^1), u_j^0 - u_j^1 \rangle_j \\
&= \sum_{j \in P} r_j \Bigg[\int_0^T \Big[\langle D_x f_j(x^0(t), \mathbf{u}^0(t)) - D_x f_j(x^1(t), \mathbf{u}^1(t)), \\
&\qquad \int_0^t S(t-s) B_j (u_j^0(s) - u_j^1(s))\, ds \rangle \Big] dt \\
&\quad + \sum_{j \in P} r_j \Bigg[\int_0^T \Big[\langle D_{u_j} f_j(x^0(t), \mathbf{u}^0(t)) \\
&\qquad - D_{u_j} f_j(x^1(t), \mathbf{u}^1(t)), u_j^0(t) - u_j^1(t) \rangle_j \Big] dt \Bigg] \\
&= \sum_{j \in P} r_j \Bigg[\int_0^T \Big[\langle D_x f_j(x^0(t), \mathbf{u}^0(t)) - D_x f_j(x^1(t), \mathbf{u}^1(t)), x_j^0(t) - x_j^1(t) \rangle \Big] dt \\
&\quad + \sum_{j \in P} r_j \Bigg[\int_0^T \Big[\langle D_{u_j} f_j(x^0(t), \mathbf{u}^0(t)) \\
&\qquad - D_{u_j} f_j(x^1(t), \mathbf{u}^1(t)), u_j^0(t) - u_j^1(t) \rangle_j \Big] dt \Bigg],
\end{aligned}$$

in which $x_j^i(\cdot)$ is the unique mild solution of the evolution equation

$$\dot{x}(t) = Ax(t) + B_j u_j^i(t) \quad \text{a.e. } 0 \le t \le T,$$

with initial condition,

$$x(0) = 0.$$

To examine the above condition more closely, it is convenient to define, for $j \in P$, the functions $L_j : [0, T] \times \Pi_{j \in P} X \times \Pi_{j \in P} E_j \to \mathbb{R}$ by the formula

$$L_j(t, x_1, x_2, \ldots, x_p, u_1, u_2, \ldots, u_p) \doteq f_j \left(S(t)x_0 + \sum_{j \in P} x_j, \mathbf{u} \right).$$

With this notation, the above strict diagonal convexity condition may be written as

$$0 > \int_0^T \Bigg[\sum_{j \in P} r_j \langle D_{x_j} L_j(t, \mathbf{x}^0(t), \mathbf{u}^0(t)) - D_{x_j} L_j(t, \mathbf{x}^1(t), \mathbf{u}^1(t)), x_j^0(t) - x_j^1(t) \rangle \Bigg] dt$$
$$+ \int_0^T \Bigg[\sum_{j \in P} r_j \langle D_{u_j} L_j(t, \mathbf{x}^0(t), \mathbf{u}^0(t)) - D_{u_j} L_j(t, \mathbf{x}^1(t), \mathbf{u}^1(t)), u_j^0(t) - u_j^1(t) \rangle_j \Bigg] dt.$$

Consequently, we see that a sufficient condition for uniqueness is that there exists a vector $\mathbf{r} = (r_1, r_2, \ldots, r_p)$, $r_j > 0$, such that the combined Lagrangian

$$\mathcal{L}(t, \mathbf{x}, \mathbf{y}, \mathbf{u}, \mathbf{v}) \doteq \sum_{j \in P} r_j L_j(t, [\mathbf{x}^j, y_j], [\mathbf{u}^j, v_j])$$

is strictly diagonally convex.

Remark 3.2. The best known open-loop differential game is perhaps the linear quadratic regulator problem. The above example includes this example as a special case. In this case, of course, the costs for each player are described by the integral cost function

$$F_j(\mathbf{x}(\cdot), \mathbf{u}(\cdot)) = \int_0^T \left[\mathbf{x}(t)' Q^j \mathbf{x}(t) + \sum_{l \in P} \sum_{k \in P} u_k(t)' R^j_{k,l} u_l(t) \right] dt.$$

For the special case in which the players' dynamics are separated and for which each player's payoffs are coupled only through the states and not the controls, explicit conditions for strict diagonal convexity have been given by Haurie et al. [6] in the context of feedback Nash equilibria instead of the open-loop case considered here. Specifically, for two players this game is described briefly by

$$F_j(\mathbf{x}(\cdot), u_j(\cdot)) = \frac{1}{2} \int_0^T \{ \sum_{k=1,2} \sum_{\ell=1,2} x_k(t)' Q^j_{k\ell} x_\ell(t) + u_j(t)' R_j u_j(t) \} \, dt,$$

with the dynamics described by the ordinary differential equations

$$\dot{x}_j(t) = A_j x_j(t) + B_j u_j(t) \quad \text{a.e.} \quad 0 \le t \le T.$$

Here we assume that A_j is $n_i \times n_j$, that B_j is $n_j \times m_j$, and that Q^j_{jj} is $n_j \times n_j$ and R_j is $m_j \times m_j$ are supposed to be positive definite with $Q^{j\,\prime}_{k\ell} = Q^j_{\ell k}$. In this special case, the desired strict diagonal convexity conditions, as given in [6], are that the matrix

$$\tilde{Q} = \begin{pmatrix} Q_{11}^1 & Q_{12}^2 \\ Q_{21}^1 & Q_{22}^2 \end{pmatrix} + \begin{pmatrix} Q_{11}^1 & Q_{21}^{1\,\prime} \\ Q_{12}^{2\,\prime} & Q_{22}^2 \end{pmatrix}$$

is positive definite.

4 Conclusions

In the preceding sections we extended Rosen's results on the existence and uniqueness of Nash equilibria for convex games in $\mathbb{R}^n$ to convex games in Hilbert space. It is clear that these results could be further extended to the case of a general separable Banach space in a similar manner. One shortcoming in this theory concerns that of existence. In particular, the requirement of weak continuity is in general too restrictive and should be replaced by some form of weak lower semicontinuity. It is clear from the proof of Theorem 2.2 that for a general abstract problem of the type formulated in Section 2, this will generally not be possible without additional hypotheses (at least the proof of the result given here cannot be obviously modified to weaken this requirement). Consequently, one hopes that by looking at a specific class of models (e.g., the optimal control framework) it will be possible to weaken these continuity requirements. The search for these conditions is a continuing process for this author as well as others.

Acknowledgments

This research was supported by the National Science Foundation (INT–9500782).

REFERENCES

[1] Aubin, J-P. Applied functional analysis, *Pure and Applied Mathematics*, John Wiley and Sons, New York, 1979.

[2] Bohnenblust, H. F. and Karlin, S. On a theorem of ville, in *Contributions to the Theory of Games*, Vol. 1, H. W. Kuhn and A. W. Tucker, eds., Princeton University Press, Princeton, pp. 155–160, 1950.

[3] Carlson, D. A. and Haurie, A. A turnpike theory for infinite horizon open-loop differential games with decoupled controls, in *New Trends in Dynamic Games and Applications*, G. J. Olsder, ed., Annals of the International Society of Dynamic Games, Birkhäuser, Boston, pp. 353–376, 1995.

[4] Carlson, D. A. and Haurie, A. A turnpike theory for infinite horizon open-loop competitive processes, *SIAM J. Cont. Opt.* **34** (4):1405–1419, 1996.

[5] Cesari, L. Optimization-theory and applications: Problems with ordinary differential equations, *Applications of Applied Mathematics*, Vol. 17, Springer-Verlag, New York, 1983.

[6] Haurie, A., Moresino, F., and Pourtalier, O. Oligopolies as Dynamic Games: A computational economics perspective, in P. Krall and H.-J. Luethi, eds., *Operations Research—Ongoing Progress Pro. OR 98, Int. Conf. Oper. Res.*, ETH Zurich, Springer-Verlag, 31 August–3 September 1998.

[7] Rosen, J.B. Existence and uniqueness of equilibrium points for concave n-person games, *Econometrica* **33** (3):520–534, 1965.

PART II
Stochastic Games

The Existence of Equilibrium Payoffs in Two-Player Stochastic Games

Nicolas Vieille
HEC
Departement Finance et Economie
Jouy en Josas, France

Abstract

Every two-player stochastic game has an equilibrium payoff. This paper contains a discussion of the main steps of the proof.

1 Introduction

Stochastic games are games played in stages, over a set K of states. In any stage, the players are fully informed of the past play, including the current state k, and choose actions from given sets I and J. The state of the game changes from one stage to the next one; the law of the next state is a function of the current state and of the actions selected by the players. In any stage n, the players receive a payoff, which also depends on the current state and the actions selected. The game never ends. We assume that K, I, and J are finite.

This model was introduced by Shapley [22], who proved that, when payoffs are zero-sum and the infinite stream of payoffs $(g_n)_{n\geq 1}$ is evaluated according to a geometric average $\lambda \sum_{n=1}^{\infty}(1-\lambda)^{n-1} g_n$, the game has a value v_λ, and, for both players, stationary optimal strategies do exist. This was extended to the nonzero-sum case by Fink [9]. Many results followed, relaxing the finiteness assumptions on K, I, and J; see [18] or [1] for general results. Unlike the one-player case ([6]), the value and the optimal strategies vary with λ even when λ is arbitrarily close to 0. This dependency has been investigated by Bewley and Kohlberg, [3], [4], and [5]. Using the algebraic structure of the graph of the equilibrium correspondence, they proved that v_λ has a Puiseux expansion in a neighborhood of 0 (and similar properties for optimal strategies).

Shortly after Shapley, the search for different properties was initiated by Gillette [13], in the zero-sum case.The question asked by Gillette is roughly as follows: do there exist strategies that are eventually good against any strategy of the opponent? This requirement has been strengthened, first by Aumann and Maschler [2] (in the framework of games with incomplete information). Assume the game is stopped after the nth stage and each player wishes to maximize the arithmetic average of the payoffs he received up to that stage. This defines a finite game Γ_n, which has a value v_n. Do there exist strategies that are optimal (or ϵ-optimal) in *every* Γ_n, for n large enough (and does $\lim v_n$ exist)? In the same volume as Gillette's, Milnor and Shapley [19] and Everett [8] studied particular models of stochastic games. In Milnor and Shapley's games of survival, two players, with some initial wealth, repeatedly play a given zero-sum game until one of them is ruined (which may never happen). In Everett's recursive games, the payoff received is zero until an absorbing state is reached (this is a state the play cannot leave). Everett proves the existence of the value and of ε-optimal stationary strategies in Gillette's sense.

A most important step in this line of research has been the analysis by Blackwell and Ferguson [7] of a game, the "Big Match," exhibited by Gillette. Blackwell and Ferguson's existence result was extended by Kohlberg [15] to the so-called zero-sum games with absorbing states. These are games in which all states are absorbing but one (thus, the state of the game may change only once). The study of the zero-sum case culminated with the proof, by Mertens and Neyman ([16] and [17]), that the value exists in every stochastic game with finite K, I, and J; their result is actually more general and uses both the idea of Blackwell and Ferguson and the algebraic properties obtained by Bewley and Kohlberg.

This left the nonzero-sum case open. Using a variation on the Big Match, Sorin [26] proved that the set of equilibrium payoffs could be disconnected from the limit set of discounted equilibrium payoffs; hence, that the techniques of Mertens and Neyman could not be extended to the nonzero-sum case. First results on the existence of equilibrium payoffs were obtained by Thuijsman and Vrieze: in [28], they prove that there always exists a state, starting from which the game has an equilibrium payoff; [35] contains the existence result for the class of games with absorbing states. Various existence results have also been obtained under specific assumptions on the transition and payoffs structure (see, for instance, [27], [21], and [30]). We prove that every two-player stochastic game has an equilibrium payoff. Finally, let us mention that our result does not subsume Mertens and Neyman's result. Rather, our proof makes strong use of it. Here we try to convey the general idea as well as the main ideas of the proof.

2 The Model and Main Result

2.1 Data and Rules of the Game

A stochastic game Γ is defined by a finite state space K, finite sets of actions I and J for players 1 and 2, a transition function $P : K \times I \times J \to \Delta(K)$ that describes how the state of the game evolves as a function of the actions selected by the players [$\Delta(S)$ is the set of probability distributions over S, for every finite set S], and a payoff function $g : K \times I \times J \to \mathbb{R}^2$ that describes the payoff received by the players at any stage.

The game proceeds as follows, for infinitely many stages. An initial state k_1 is given. At any stage n, the players are informed of the current state k_n and of the sequence $(k_1, i_1, j_1, \ldots, k_{n-1}, i_{n-1}, j_{n-1})$ of states previously visited, and actions previously selected; they then choose independently actions i_n and j_n; the next state k_{n+1} is then chosen according to the probability distribution $P(k_n, i_n, j_n)$, and the game proceeds to the next stage.

We let $H_n = (K \times I \times J)^{n-1} \times K$ be the set of histories up to stage n, and $H_\infty = (K \times I \times J)^{\mathbb{N}}$ denote the set of plays. A strategy of player 1 is a map σ defined over $\cup_{n \geq 1} H_n \to \Delta(I)$, with the usual understanding: $\sigma(h_n)$ is the distribution player 1 uses to choose his action in stage n, when the past history of play is $h_n \in H_n$. Σ_i is the set of strategies of player $i = 1, 2$, and $\Sigma = \Sigma_1 \times \Sigma_2$.

We denote by $\mathcal{H}_n$ the σ-algebra over H_∞, induced by H_n, and $\mathcal{H}_\infty = \sigma(\mathcal{H}_n, n \geq 1)$. Any profile (σ, τ) of strategies, together with an initial state k, induces a probability distribution over $(H_\infty, \mathcal{H}_\infty)$, which we denote $\pi_{k,\sigma,\tau}$. $E_{k,\sigma,\tau}$ stands for the corresponding expectation operator.

Stationary (or Markov) strategies of player 1 are strategies that depend only on the current stage. Thus, a stationary strategy of player 1 can be identified with an element $x = (x_k)_{k \in K} \in \prod_k \Delta(I)$, with the understanding that x_k is the lottery player 1 uses to select its action in state k. We denote by $\mathcal{S}_i$ the set of stationary strategies of player i, $i = 1, 2$, and by $\mathcal{O}_i$ the subset of completely mixed stationary strategies: these are the stationary strategies such that each action is used with strictly positive probability in each state. For $x \in \mathcal{S}_1$ and $k \in K$, we set $S(x_k) = \{i \in I, x_k(i) > 0\}$, and the support of x is $S(x) = \prod_k S(x_k)$.

The mixed extensions of g and P to $K \times \Delta(I) \times \Delta(J)$ are still denoted by g and P.

For $(x, y) \in \mathcal{S}_1 \times \mathcal{S}_2$, we will frequently write $P(k, x, y)$, or $P_{k,x,y}$, instead of $P(k, x_k, y_k)$. If u is an $\mathbb{R}$ or $\mathbb{R}^2$-valued function defined on K, and $P \in \Delta(K)$, we denote by Pu the expectation of u under P, and we use the notation $P_{x,y}u$ for $(P_{k,x,y}u)_{k \in K}$.

2.2 Payoffs and Equilibria

For $n \geq 1$, denote by $g_n = g(k_n, i_n, j_n) \in \mathbf{R}^2$ the vector of the payoffs received in stage n and by

$$\gamma_n(k, \sigma, \tau) = \mathrm{E}_{k,\sigma,\tau}\left[\frac{1}{n}\sum_{k=1}^{n} g_k\right]$$

the expected average payoff up to stage n induced by the profile (σ, τ), given that the initial state is k.

Let G be a zero-sum game, that is, $g^1 + g^2 = 0$. We set $\gamma_n(k, \sigma, \tau) = \gamma_n^1(k, \sigma, \tau)$.

Definition 2.1. Let k be the initial state. The number $v(k)$ is the value of G if, for every $\epsilon > 0$,

- there exist σ_ε and $N \in \mathbb{N}$ such that

$$\forall n \geq N, \forall \tau, \gamma_n(k, \sigma_\epsilon, \tau) \geq v(k) - \epsilon,$$

- there exist τ_ε and $N \in \mathbb{N}$ such that

$$\forall n \geq N, \forall \sigma, \gamma_n(k, \sigma, \tau_\varepsilon) \leq v(k) + \epsilon.$$

We refer to such strategies as ε-*min max strategies*. The strategy σ_ε ensures that the average payoff to player 1 will never fall below $v(k) - \varepsilon$, from a certain stage on. The strategy τ_ε has the symmetric property.

We recall the existence result.

Theorem 2.1 (Mertens-Neyman, [16]; see also [17]). *Every zero-sum stochastic game has a value.*

The corresponding notion for nonzero-sum games is that of equilibrium payoffs.

Definition 2.2. Let k be the initial state. A vector $\gamma(k) \in R^2$ is an equilibrium payoff of G if, for every $\varepsilon > 0$, there exist a pair $(\sigma_\varepsilon, \tau_\varepsilon)$ and $N \in N$ such that, for every $n \geq N$

$$\forall \tau, \quad \gamma_n^2(k, \sigma_\varepsilon, \tau) \leq \gamma^2(k) + \varepsilon,$$
$$\forall \sigma, \quad \gamma_n^1(k, \sigma, \tau_\varepsilon) \leq \gamma^1(k) + \varepsilon,$$

and

$$||\gamma_n(k, \sigma_\varepsilon, \tau_\varepsilon) - \gamma(k)|| \leq \varepsilon.$$

The last condition asserts that the average payoffs induced by the pair $(\sigma_\varepsilon, \tau_\varepsilon)$ depend little on the length of the averaging period. Together with this condition, the first two imply that $(\sigma_\varepsilon, \tau_\varepsilon)$ is a 2ε-equilibrium in the n-stage game, provided $n \geq N$.

The result we report on is the following.

Theorem 2.2. *Every two-player stochastic game has an equilibrium payoff.*

The complete proof is in [31] and [32].

3 Example and Discussion

3.1 An Example

In the view of the achievements for zero-sum games, it was tempting to try to construct equilibrium payoffs in the limit set of discounted equilibrium payoffs. Sorin [26] has shown that this path was bound to fail, by analyzing the following variation of the Big Match:

$1, 0^*$	$0, 2^*$
0,1	1,0

Player 1 (resp., player 2) chooses a row (resp., a column). The two starred entries have to be interpreted as follows: as soon as the corresponding action combination is played, the game moves to an absorbing state with the corresponding payoff. Notice first that if player 2 is only concerned by minimizing player 1's payoffs, one is led to analyze the zero-sum game Γ_1

1*	0*
0	1

obtained by forgetting the payoff function of player 2. In that game, player 1 can secure himself $v^1 = 1/2$.

Likewise, player 2 can secure himself $v^2 = 2/3$, the value of the zero-sum game Γ_2 obtained by forgetting the payoff function of player 1. Therefore, any equilibrium payoff d is such that $d \geq (1/2, 2/3)$. One can prove that, for each λ, the set E_λ of λ-discounted equilibrium payoffs is reduced to the vector $(1/2, 2/3)$. On the other hand, the set of equilibrium payoffs is the segment $E = \{(y, 2(1-y)), y \in [1/2, 2/3]\}$.

We shall only explain why any vector $(\alpha, 2(1-\alpha))$, $\alpha \in [1/2, 2/3]$ is an equilibrium payoff.[1] This is the opportunity to introduce, in a simple case, some of the ideas used in the general case. Imagine that the players follow, respectively, the stationary strategies $x = (\eta, 1-\eta)$ and $y = (\alpha, 1-\alpha)$. Then $\gamma(x, y) = (\alpha, 2(1-\alpha)) \geq (1/2, 2/3)$. (x, y) is not an ε-equilibrium, since $\gamma^2(x, (0, 1)) = 2 > 2(1-\alpha)$. However, player 1 can create an incentive for player 2 to play according to $(\alpha, 1-\alpha)$ by threatening him of a *punishment* if he happens to deviate from this strategy. More precisely, we add the following clause: if, in some day $n \geq N$, the observed frequency of player 2's two actions differs from $(\alpha, 1-\alpha)$ by more than ε, player 1 *punishes* player 2 by starting to use an ε-optimal strategy in Γ_2 in order to lower player 2's future payoffs below $2/3(+\varepsilon)$.

What is the effect of that clause? Two classical types of errors may occur. On the one hand, this clause may lead player 1 to punish an innocent player 2: one has to choose $N \gg 1/\varepsilon$ in order that the probability of this event is quite small. On the other hand, it matters, of course, that the probability that the game stops before the statistical control becomes effective be small (otherwise, player 2 could increase his payoff by a nonnegligible amount by starting to play D many times; the game would then stop in the state 0, 2* before player 1 would "notice" this deviation). Thus, one has to choose $\eta \ll 1/N$.

If these conditions hold, it is easy to check that the strategies defined by (x, y) and the control clause define an $O(\varepsilon)$-equilibrium. Let τ be a strategy of player 2, and, for simplicity, let us discuss only two extreme cases. Assume first that player 1 never notices any deviation by player 2. This means that the empirical frequencies of player 2's actions remain, from stage N on, close to α and $1-\alpha$. Since, each day, the choice of player 1 is independent of the choice of player 2, this implies that

[1] We don't follow the proof in [26].

the probabilities that the game will end in the states $1, 0^*$ and $0, 2^*$ are respectively close to α and $1-\alpha$: such a deviation is not profitable. Assume now that, in some day $n \geq N$, player 2 is about to choose an action that will activate the punishment clause. The probability that the game stops on that day is only η; with probability $1-\eta$, the game will go on and the expected average payoffs of player 2 will fall below $2/3$.

Player 1 cannot increase his payoff. Indeed, if he eventually stops the game, the probability that the game stops in state $1, 0^*$ is exactly α. If he never stops the game, his average payoff will converge to $1-\alpha \leq \alpha$.

The discontinuity that appears in this example casts doubts on the possibility of finding a general existence result based only on vanishing discounting methods. However, Thuijsman and Vrieze [35] manage to prove the existence of equilibrium payoffs in the so-called games with absorbing states by such methods.

3.2 Architecture of the Proof

The previous example highlighted a classical principle to prove the existence, or to obtain a characterization, of equilibrium payoffs: build, as much as possible, upon the results for zero-sum games. We shall push this idea quite far.

The idea is to write down a simple *scenario* (stationary in the previous case), which the players are to follow. Every deviation from this scenario is punished.

If this clause is to enjoy incentive properties, it is of course necessary that the payoffs (d^1, d^2) the scenario induces are at least the amounts v^1, v^2 that the players are able to secure themselves (values of the two zero-sum games, defined as in the above example). This condition is usually called the *individual rationality requirement*.

To check whether a given scenario sustained with punishment threats has incentive properties, one has to consider not only the values of the states visited in the scenario, but also the values of the states that each player may reach by deviating. For instance, in the game

k_1:

$2, 2$	$\rightarrow$	$0, 0^*$
$0, 0^*$	$0, 0^*$	$0, 0^*$

k_2:

$\leftarrow$	$3, 3^*$
$1, 5^*$	$2, 4^*$

one has $v^1(k_1) = v^2(k_1) = 0$, $v^1(k_2) = 1$, and $v^2(k_2) = 3$. Playing the first line and column in state k_1 gives to each player i a payoff of $2 \geq v^i(k_1)$. But $(2, 2)$ is not an equilibrium payoff since player 2 would play the second column to reach k_2, where he can secure 3.

The fundamental difficulty in the search for an equilibrium payoff lies in the fact that the limit of the expected average payoffs does not depend continuously on the strategies. A small perturbation in a transition can lead to a drastic change in the asymptotic behavior of the play. In general, $\gamma(x, y) := \lim_n \gamma_n(x, y)$ is discontinuous at each point (x, y) at which the ergodic decomposition of K is not locally constant.

The basic intuition that underlies this work is that, in a proof of the *existence* of equilibrium payoffs, it is possible to separate the two main issues: control of the transitions, and control of the payoffs. More precisely, it is possible to create a distinction between the two types of states. On the one hand, states, which we call *solvable*, in which the payoffs the players receive are *high*, compared with the level of their values, and no opportunity exists of reaching milder states: these games are bound to play the role of absorbing states for the game. On the other hand are states in which the strategic interaction is concerned only with the transitions: at stake is to control the unfolding of the play until an absorbing state is reached. In this perspective, the payoffs received prior to the absorption can be neglected.

In order to highlight this, we define a class of games in which only transitions do matter. Let us call a *positive recursive game* any recursive game in which the payoffs received by player 2 in the various absorbing states are all strictly positive and in which player 1 cannot prevent the absorption: as soon as player 2 uses a stationary strategy with full support, an absorbing state is reached *a.s.* in finite time, regardless of the starting state.

The proof is organized as follows.

- Solvable states: definition and existence; this enables us to simplify the game by turning solvable states into absorbing states.

- Reduction procedure: once this simplification is done, there exist subsets of K whose exit can be controlled in a robust way; this enables us to further simplify the game by replacing each of these subsets by states with fixed transitions. It is possible to choose these subsets in such a way that the new game is strategically "equivalent" to a recursive game.

- Analysis of positive recursive games.

The first step is simple ([28] and [30]). Paper [31] deals with the second. The third is the most delicate. In [33], we define a compactification of the set of stationary strategies with full support, motivated by the theory of Freidlin and Wentzell [12], and we develop a few technical tools. Paper [32] contains the result of the existence of equilibrium payoffs for positive recursive games.

4 Solvable States

Let $(x, y) \in \mathcal{S}$ and R be a recurrent set for the Markov chain over K, induced by (x, y). Then $\gamma_k(x, y)$ does not depend on $k \in R$ and is equal to $\gamma_R(x, y) = \sum_{k \in R} \mu_k(x, y) g(k, x, y)$, where $\mu(x, y)$ is the distribution over R, invariant for (x, y).

Here we try to draw simple conditions under which $\gamma_R(x, y)$ is an equilibrium payoff (for every initial state in R), and the strategy "play (x, y) as long as the punishment clause is not activated" is a corresponding ε-equilibrium profile.

On the equilibrium path, each state in R is visited infinitely many times. Thus, a necessary condition is $\gamma_R^1(x, y) \geq \max_{k\in R} v^1(k)$. Roughly speaking, there are two types of deviation for player 1. He may, any day, choose, in some state k, an action i that has zero probability under x: $x_k[i] = 0$. In that case, the distribution of the next state is $P_{k,i,y}$, and this deviation is immediately noticed by player 2: the expected level of the punishment is thus given by $P_{k,i,y}v^1$. Such deviations fail to be profitable as soon as $P_{k,i,y}v^1 \leq \gamma_R^1(x, y)$, for all k, i. On the other hand, player 1 can modify the probabilities of the various actions in the support of x. Such a deviation may be detected only statistically. The simplest test is to check that, after some day N suitably chosen, the *empirical* average payoff of player 1 does not exceed $\gamma_R^1(x, y) + \varepsilon$. The effectiveness of this preventive clause follows from $\gamma_R^1(x, y) \geq \max_R v^1$. It is easy to check that these two conditions amount to the one below

$$\forall k \in R, \forall i \in I, P_{k,i,y}v^1 \leq \gamma_R^1(x, y). \tag{1}$$

If this condition is met, together with the symmetric condition for player 2, $\gamma_R(x, y)$ is an equilibrium payoff for Γ, for every initial state in R.

It turns out in the analysis that the main difficulty is to ensure that no player is unilaterally able to reach a state with a higher value. Thus, good candidates for being solvable are the states with high value.

Formally, set $F^1 = \{k \in K, v^1(k) = \max_K v^1\}$, $F = \{k \in F^1, v^2(k) = \max_{F^1} v^2\}$, and let G be the set obtained when exchanging the roles of the two players. It is not difficult to show that there exist (x, y), and a recurrent set for (x, y) contained in F, for which property (1) holds (see [28] and [30]).

We need a more general concept to bring the distinction to the fore. We illustrate it on the following game, where $v^1(k_1) \leq 1$, $v^1(k_2) = 0$, $v^2(k_1) = 1$, $v^2(k_2) = 2/3$.

	G_1	M_1	D_1
H_1	1, 0	0, 2	$\rightarrow$
B_1	2, 2*	1, 0*	5, 0*

k_1

	G_2	D_2
H_2	3, 2	0, 1*
M_2	4, 0	0, 1*
B_2	$\leftarrow$	0, 0*

k_2

Let x and y be the stationary strategies defined by $x = (H_1, (\frac{1}{2}H_2, \frac{1}{2}M_2))$ and $y = ((\frac{1}{2}G_1, \frac{1}{2}M_1), G_2)$. $\{k_1\}$ and $\{k_2\}$ are two recurrent sets for the Markov chain induced by (x, y), and $\gamma_{k_1}(x, y) = (1/2, 1)$, $\gamma_{k_2}(x, y) = (7/2, 1)$. Then $(2, 1) = 1/2\gamma_{k_1}(x, y) + 1/2\gamma_{k_2}(x, y)$ is an equilibrium payoff, for the initial states k_1 and k_2.

A corresponding ε-equilibrium is of the following type: play (x_{k_1}, y_{k_1}) for N days; go to k_2 by playing according to $(x_{k_1}, (1-\eta)y_{k_1} + \eta D_1)$, where $\eta > 0$ is small; then play (x_{k_2}, y_{k_2}) for N days; then go back to k_1 by playing $((1-\eta)x_{k_2} + \eta B_2, y_{k_2})$, and so forth. In a more general framework, one has to choose N large enough so that the transfer phases, from k_1 to k_2 or vice versa, are short compared to phases in which the players play according to (x, y). The controls are those from the above discussion. It is crucial that the transfer from k_1 to k_2 is carried out using a small perturbation of y_{k_1}: if, for instance, player 1 did know in advance the day in which player 2 was to move the play from k_1 to k_2, he would simply play B_1 on that day, and the game would stop in the state 5, 0^*.

Let $E \subset K$, $(x, y) \in \mathcal{S}$, μ a probability distribution over the set of those recurrent sets for (x, y), which are contained in E. We say that E is *solvable* for (x, y) and μ if

- E *communicates* under (x, y) : for any two states k_1 and k_2 of E, there exists a perturbation of (x, y) such that E is stable and, starting from k_1, k_2 is reached in finite time,
- $\forall k \in E, i \in I$, $P_{k,i,y}v^1 \le \sum_R \mu_R \gamma_R^1(x, y)$, together with the symmetric property for player 2.

Then, if the initial state belongs to a solvable state, the game has at least an equilibrium payoff. If one replaces each of these states by an absorbing state with payoff $\sum_R \mu_R \gamma_R(x, y)$, one defines a simplified game. One can then prove that the solvable sets of that game coincide with the absorbing states.

5 Reduction Algorithm

We now deal only with such reduced games and look for ε-equilibrium profiles under which the set A of these absorbing states is reached with probability close to 1. The most straightforward idea is to restrict ourselves to the *absorbing* pairs $(x, y) \in \mathcal{S}$ (these, are the pairs such that, regardless of the initial state, A is reached in finite time) and to look for an ε-equilibrium in the set of these pairs, namely, (x^*, y^*) such that, if (x, y^*) is absorbing, $\gamma^1(x, y^*) \le \gamma^1(x^*, y^*) + \varepsilon$ (plus the symmetric condition for player 2). This approach meets a fundamental difficulty. Consider once more the example in Sorin [26] and permute the players:

0, 1	2, 0^*
1, 0	0, 1^*

These absorbing equilibria here are of the form $(x^*, y^*) = ((1, 0), (1 - \varepsilon, \varepsilon))$, for $\varepsilon > 0$. However, $\gamma(x^*, y^*) = (2, 0)$ is not an equilibrium payoff of the game. The problem stems from the fact that the constraint "choose a strategy that is absorbing against x^*" leads player 2 to absorbing payoffs *below* v^2.

Call $\mathcal{P}$ the following property: for each x, there exist y, such that the pair (x, y) is absorbing,[2] and $\gamma^2(x, y) \ge v^2$. This entails the fact that player 1 cannot prevent the absorption. If the game Γ satisfies this property, the game $\tilde{\Gamma}$ obtained by setting the payoffs in the nonabsorbing states to 0 is a positive recursive game. Let $(\tilde{\sigma}_\varepsilon, \tilde{\tau}_\varepsilon)$ be a ε-equilibrium in $\tilde{\Gamma}$. Given the properties of $\tilde{\Gamma}$, absorption occurs

[2] *I.e.*, the Markov chain defined by (x, y) is absorbing, regardless of the initial state.

with probability close to 1. Choose a day N prior to which absorption occurs with probability close to 1. It is not difficult to check that the pair $(\sigma_\varepsilon, \tau_\varepsilon)$ is an ε-equilibrium of Γ, where $(\sigma_\varepsilon, \tau_\varepsilon)$ is obtained by adding to $(\tilde{\sigma}_\varepsilon, \tilde{\tau}_\varepsilon)$ the punishment clause: punish if A has not been reached by day N.

Thus, if Γ satisfies $\mathcal{P}$, the existence of equilibrium payoffs in Γ follows from the existence in $\tilde{\Gamma}$.

An important part of the proof is devoted to a detailed analysis of the structure of the game when $\mathcal{P}$ is not satisfied. The fundamental notion is the following. Let $B \subset K$ and $x \in \mathcal{S}_1$. We say that player 2 is *blocked* on B by x if

$$\forall k \in B,\ j \in J,\ P_{k,x,j}(B) < 1 \Rightarrow P_{k,x,j}v^2 < \max_B v^2 :$$

any unilateral attempt by player 2 to leave B lowers the value of the current state. The analysis is based on the properties of $\overline{x}, \overline{y}$, limits of optimal stationary strategies x_λ, y_λ in the two zero-sum games $(\Gamma_1)_\lambda, (\Gamma_2)_\lambda$: the function $v^1 : K \to \mathbb{R}$ is excessive for the kernel defined by $(\overline{x}, y)$, whatever y is:

$$\forall k, \forall y,\ P_{k,\overline{x},y}v^1 \geq v^1(k);$$

in other words, the value $v^1(k_n)$ of the current state is a submartingale as soon as player 1 uses the strategy $\overline{x}$. The function v^2 has of course analog properties.

Let x be a strategy such that there is no absorbing y, with $\gamma^2(x, y) \geq v^2$. In particular, the pair $(x, \overline{y})$ is not absorbing. More precisely, it is easy to prove the existence of a subset B of K, which communicates under $(x, \overline{y})$, on which v^2 is constant, and player 2 is blocked by x.

One then has to discuss according to the exit opportunities offered to player 1.

If player 1 is not blocked on B by $\overline{y}$, there exist $k \in B$, $i \in I$, with $P_{k,i,\overline{y}}(B) < 1$ and $P_{k,i,\overline{y}}v^1 \geq \max_B v^1$. Among these pairs, choose a pair (k_0, i_0) that maximizes $P_{k,i,\overline{y}}v^1$. Imagine for a moment that the game starts from some state $k \in B$ and that the players follow this scenario: in each visit to k_0, actions are chosen according to $(1-\eta)x_{k_0} + \eta i_0$ and $\overline{y}_{k_0}$; between two visits to k_0, the players slightly perturb x and $\overline{y}$ in order to reach k_0 in finite time, without leaving B. According to this scenario, the play leaves B in finite time, and the distribution of the exit state is $P_{k_0,i_0,\overline{y}_{k_0}}$. One adds the following control clauses: check every day that the actions chosen are compatible with the scenario; check that the empirical distribution of the actions selected by player 2 in the visits to k_0 is close to $\overline{y}$; check that exit from B occurs before some given N. With these clauses, the control of the exit from B is quite robust.

We define a simplified game Γ_B by replacing the subset B by a *mirror state* b, in which transitions are given by $P_{k_0,i_0,\overline{y}}$, independently of the actions selected: it is the game obtained by assuming that, each time the play enters B, it bounces and leaves B according to $P_{k_0,i_0,\overline{y}}$. If Γ_B has the property $\mathcal{P}$,[3] one may associate to it a positive recursive game $\tilde{\Gamma}_B$ as above and deduce, from any ε-equilibrium $(\tilde{\sigma}_\varepsilon, \tilde{\tau}_\varepsilon)$

[3]Still written with v^1 and v^2, values of Γ_1 and Γ_2, and setting $v^1(b) = \max_B v^1$.

of $\tilde{\Gamma}_B$, a ε-equilibrium $(\hat{\sigma}_\varepsilon, \hat{\tau}_\varepsilon)$ of Γ_B, by simply adding a punishment clause. We define a pair $(\sigma_\varepsilon, \tau_\varepsilon)$ of strategies by the following: when the game is not in B, play according to $(\hat{\sigma}_\varepsilon, \hat{\tau}_\varepsilon)$ (and ignore what happened during the possible visits to B). Whenever the game enters B, start following the scenario described above.

Then, provided the parameters η and N have been suitably chosen, $(\sigma_\varepsilon, \tau_\varepsilon)$ is an $O(\varepsilon)$-equilibrium of Γ.

However, it may happen that player 1 is blocked on B by $\overline{y}$. By repeating the previous process, we can find a subset D of B that communicates under $(\overline{x}, \overline{y})$, on which v is constant, and player 1 is blocked by $\overline{y}$. As above, if player 2 is not blocked on D by $\overline{x}$, one can simplify the game by turning D into a mirror state d.

It remains to discuss the case where player 2 is also blocked on D by $\overline{x}$. Up to now, all the steps are quite elementary and follow quite easily from the submartingale properties of $v^1(k_n)$ and $v^2(k_n)$. In order to highlight the dichotomy, it seems unavoidable to use deeper arguments. Recall that the ε-optimal strategies built by Mertens and Neyman consist of playing $x_{\lambda(s_n^1)}, y_{\lambda(s_n^2)} \simeq \overline{x}, \overline{y}$. Denote by $\tilde{x}_\lambda, \tilde{y}_\lambda$ the strategies obtained, after renormalization, by eliminating the actions i and j such that $P_{k,i,\overline{y}}(D) < 1$, $P_{k,\overline{x},j}(D) < 1$.

One easily checks, by repeating the computation of Mertens and Neyman, that the strategies (σ, τ), which call for playing $\tilde{x}_{\lambda(s_n^1)}, \tilde{y}_{\lambda(s_n^2)}$, satisfy $E_{k,\sigma,\tau}[\overline{g}_n] \geq v(k) - \varepsilon$, for n large enough and $k \in D$. Essentially, two cases then need to be discussed. If, under (σ, τ), exit from D occurs with probability close to 0, one can show that D is solvable, which contradicts our assumptions on Γ. If exit from D occurs with probability close to 1, the exit distribution Q_D can be written as a convex combination $Q_D = \sum_{l \in L} \alpha_l P_l$, where P_l is an *elementary* exit distribution $P_{k,i,j}$, where $k \in D$, and i, j are such that $P_{k,i,\overline{y}}(D) = P_{k,\overline{x},j}(D) = 1$. In that case, it is still possible to write an exit scenario D, which leads to the exit distribution Q_D and can be controlled.

Thus, if Γ does not have property $\mathcal{P}$, it can be simplified by turning a subset of states into a mirror state. It may well happen that this first step does not suffice. This procedure needs to be iterated. One then has to deal with some technical issues.

We illustrate this algorithm on the game

0, 1	2, 0*
1, 0	0, 1*

Here, $v^1 = 2/3$, $v^2 = 1/2$, $x_\lambda = (1/3, 2/3)$, and $y_\lambda = (1/1 + \lambda, \lambda/1 + \lambda)$. Therefore, $\overline{x} = (1/3, 2/3)$ and $\overline{y} = (1, 0)$.

Let x be the stationary strategy $(\alpha, 1 - \alpha)$, where $\alpha > 1/2$. Of course, the only way for player 2 to leave the nonabsorbing state is to play D, which results in a payoff $\gamma^2(x, D) = 1 - \alpha < v^2$; hence, player 2 is blocked in the nonabsorbing state. Therefore, the game does not satisfy property $\mathcal{P}$.

Given that player 2 plays $\overline{y}$, the game remains in the nonabsorbing state no matter what strategy player 1 follows. Hence, player 1 is blocked in that state by $\overline{y}$.

Finally, player 2 is not blocked by $\overline{x}$ in the nonabsorbing state, since $P_{\overline{x},D}v^2 = 2/3$. Thus, this algorithm here selects the equilibrium payoff $\gamma(\overline{x}, D) = (2/3, 2/3)$.

6 Positive Recursive Games

At that point, the situation seems quite clear. Recursive games are, in many respects, much easier to handle. The zero-sum case is encouraging, since players then have stationary ε-optimal strategies. The additional properties at our disposal should further simplify the analysis. Despite all this, all attempts to find a simple proof of existence for these games have failed. In the case of two nonabsorbing states, a proof slightly simpler than ours has been obtained by Solan [23]. A recent example by Flesch et al. [10] shows that, in general, recursive games have no stationary ε-equilibrium. This is still true with our additionnal properties, as is shown by the next example:

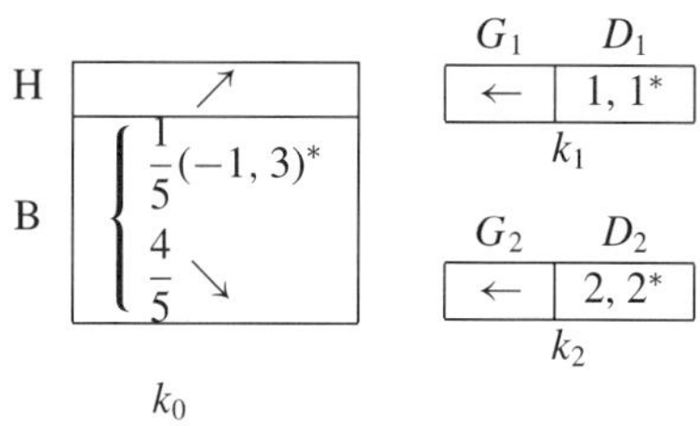

In that game, the players play in turn. Player 1 is the row player. The description of state k_0 is as follows: If player 1 chooses to play T, the game moves to state k_1. If he chooses to play B, the transition is nondeterministic: with probability $1/5$, the game moves to an absorbing state with payoff $(-1, 3)$; it otherwise moves to state k_2.

This game has no stationary ε-equilibrium.

We now describe the general characteristics of our equilibrium scenarios. A pair $(x, y) \in \mathcal{S}$ is given, which serves as a basis for the behavior of the players, together with a partition $(T, E_1, \dots, E_N)$ of $K \setminus A$, an exit distribution Q_n from E_n, for each n, and the payoffs γ induced by the scenario. States in T are transient for (x, y), and each of the sets $E_1, \dots, E_N$ communicates under (x, y). Whenever the play is in T, the players play according to (x, y). As soon as the play enters E_n, the players start following a complex auxiliary scenario, based upon perturbations of (x, y), which enables them to quit E_n according to the distribution Q_n. Because of the complexity of this auxiliary scenario, its incentive properties depend on the *continuation payoff*, payoff "promised" to the players if exit occurs without any deviation being detected.

Given the stationarity of the main scenario, these continuation payoffs depend only, at any time, on the current state of the game.

The minimal coherency between these data is the following. For each $k \in E_n$, the continuation payoff γ_k has to be equal to the expected continuation payoff once exit has occurred: $\gamma_k = Q_n\gamma$. Likewise, for $k \in T$, γ_k has to be equal to the expected continuation payoff of the next day: $\gamma_k = P_{k,x,y}\gamma$.

Moreover, for this to completely describe the unfolding of the play until the absorption, it must be the case that the Markov chain over $\{E_1, \ldots, E_N\} \cup T$, defined by the exit distributions and the transitions $P_{k,x,y}$, $k \in T$, is absorbing.

In that case, it is easy to prove that γ is an equilibrium payoff, as soon as

- the auxiliairy (exit) scenarios have incentive properties,
- for each $k \in T$, (x_k, y_k) is an equilibrium of the one-shot game with payoff function $P_{k,\ldots,\cdot}\gamma$.

It is clear that γ, the partition of K, and (x, y) have to be defined simultaneously and endogenously. To achieve this, we need a fixed-point argument. The first step is to define a convenient auxiliary strategy space

6.1 A New Strategy Space: Consistent Conditional Systems

The definition of the right strategy space is motivated by the expression of $\gamma(x, y)$, for $(x, y) \in \mathcal{O}$. We illustrate this with the following game:

k_1	y_1	y_2
x_1	$\rightarrow$	.
x_2	.	$1, 1^*$

k_2	$\tilde{y}_1$	$\tilde{y}_2$
$\tilde{x}_1$	$0, 3^*$	.
$\tilde{x}_2$	$3, 0^*$	$\leftarrow$

The payoff $\gamma(x, y)$ satisfies a system of linear equations. The theory of Freidlin and Wentzell gives a simple expression of this payoff. Let us call A-graph any oriented graph over K such that, starting from any state k in $K \setminus A$, there is exactly one path leading from k to a state in A. In the present case, there are 5 of these graphs:[4]

$$\begin{aligned} g_1 &= \{k_1 \rightarrow k_2; k_2 \rightarrow 0, 3^*\}, \quad g_2 = \{k_1 \rightarrow k_2; k_2 \rightarrow 3, 0^*\}, \\ g_3 &= \{k_1 \rightarrow 1, 1^*; k_2 \rightarrow 0, 3^*\}, \quad g_4 = \{k_1 \rightarrow 1, 1, k_2 \rightarrow 3, 0^*\}, \\ g_5 &= \{k_1 \rightarrow 1, 1^*; k_2 \rightarrow k_1\}. \end{aligned}$$

Let G be the set of these graphs and, for $k \in \{k_1, k_2\}$, $k' \in A$, let $G(k \rightarrow k')$ be the set of A-graphs such that the path starting from k ends up in k'. The probability of a graph g under (x, y) is defined as the product of the probabilities of the different transitions. For instance, $\pi_{x,y}(g_1) = P_{x,y}(k_1, k_2)P_{x,y}(k_2, (0, 3^*))$. The theory of Freidlin and Wentzell [12] tells us that the probability that the play will end up in k', starting from k, is given by

$$\frac{\sum_{G(k \rightarrow k')} \pi_{x,y}(g)}{\sum_G \pi_{x,y}(g)}. \tag{2}$$

[4]Other A-graphs exist, but only graphs g_1 through g_5 are compatible with the transitions of the game.

In our case this imples that

$$\gamma^1_{k_1}(x,y) = \frac{x_2y_2\tilde{x}_2\tilde{y}_2 + x_2y_2\tilde{x}_1\tilde{y}_1 + x_2y_2\tilde{x}_2\tilde{y}_1 + 3x_1y_1\tilde{x}_2\tilde{y}_1}{x_1y_1\tilde{x}_1\tilde{y}_1 + x_1y_1\tilde{x}_2\tilde{y}_1 + x_2y_2\tilde{x}_2\tilde{y}_2 + x_2y_2\tilde{x}_1\tilde{y}_1 + x_2y_2\tilde{x}_2\tilde{y}_1}.$$

More generally, for $l = (i_k, j_k)_{k\in K}$, and $(x, y) \in \mathcal{O}$, we set $x \otimes y(l) = \prod_k x_k(i_k)y_k(j_k)$. We define the one-to-one continuous map $\phi : (x, y) \in \mathcal{O} \mapsto (x \otimes y(l)/x \otimes y(l'))_{l,l'\in(I\times J)^K} \in]0, +\infty[^{(|I||J|)^{2K}}$ and the auxiliary strategy space Θ as the closure of $\phi(\mathcal{O})$.

This space Θ is well known and appears in the study of Nash equilibrium refinements (see [14] and [20]). This space is not needed per se in the proof but is a convenient way to think of limits of sequences of stationary profiles.

6.2 Θ and Perturbations of Markov Chains

Let $s \in \Theta$. It is a limit of a sequence $(x^\varepsilon, y^\varepsilon)$ in $\mathcal{O}$, in the sense that $\phi(x^\varepsilon, y^\varepsilon) \to_{\varepsilon\to 0} s$. This implies the convergence of $\gamma(x^\varepsilon, y^\varepsilon)$, to a limit denoted by $\gamma(s)$, and that of $(x^\varepsilon, y^\varepsilon)$, to a pair $(x(s), y(s)) \in \mathcal{S}$. But we do have much more. If a subset B of K is given, and some initial state $k \in B$, the exit from B occurs in finite time under $(x^\varepsilon, y^\varepsilon)$. Moreover, the exit distribution is given by a formula analog to (2). Therefore, this exit distribution has a limit when ε goes to zero. We then obtain, from a different angle, a decomposition of K into cycles, which is well known to the probabilists (see [29], for instance). The pair $(x(s), y(s))$ defines a Markov chain over K, and thus a partition into transient states T and ergodic sets $E_1, \ldots, E_N$. The exit distribution from E_n induced by s is well defined and does not depend on the initial state $k \in E_n$. Using these exit distributions, we may define a transition function over $\{E_1, \ldots, E_N\} \cup T$. Thus, we define a quite natural Markov chain at the second level. If this Markov chain has ergodic sets distinct from absorbing states, we may define a third-level Markov chain without the slightest difficulty, and so on. There appears an hierarchy of sets, which correspond to different levels of ergodicity. These sets may be characterized by the following property: a set E belongs to the hierarchy if, and only if, given any two states k_1, k_2 in E, the probability that, starting from k_1, k_2 is reached prior to the exit from E goes to 1 as ε goes to 0.

Therefore, the space Θ is perfectly suited to the type of analysis we are looking for: each $s \in \Theta$ induces a pair $(x(s), y(s)) \in \mathcal{S}$, a partition of $K \setminus A$ into maximal cycles (communicating sets) and transient states, together with exit distributions from the cycles, which are obtained as perturbations of $(x(s), y(s))$.

It is highly convenient to show that the sequence $(x^\varepsilon, y^\varepsilon)$ can be chosen in such a way that the probabilities $x^\varepsilon_k[i]$, $y^\varepsilon_k[j]$ have exponential equivalents $x^\varepsilon_k[i] \sim x_k[i]\varepsilon^{d_k(i)}$, $y^\varepsilon_k[j] \sim y_k[j]\varepsilon^{d_k(j)}$ in the neighborhood of 0. The proof of this fact uses tools from real algebraic geometry.[5] This enables us to associate to every transition

[5] In general, there is a large latitude in the choice of these valuations: the set of vectors $d = (d_k(i), d_k(j))_{k,i,j}$ that are suitable is a convex set.

$k \to k'$ a valuation $d(k \to k') = \min\{d_k(i) + d_k(j),\ \text{where } P_{k,i,j}(k') > 0\}$ (so that $P_{k,x^\varepsilon,y^\varepsilon}(k')$ is of the order $\varepsilon^{d(k \to k')}$). We may then characterize the cycles in a purely combinatoric way, using these valuations. The valuation of a graph is defined as the sum of the valuations of its arrows. We may then define the valuation of a subset B as $d_B = \min\{d(g),\ g \in G_B\}$: d_B gives an estimate of the exit speed from B under $(x_\varepsilon, y_\varepsilon)$. If B belongs to the hierarchy, the support of the exit distribution from B (induced by s) coincides with the states $k \in K \setminus B$ such that there exists a graph g in G_B, with valuation d_B, such that all paths of g end up in k (it is easy to convince oneself that if two paths of a graph in G_B end up in different states, the valuation of this graph exceeds d_B).

6.3 Design a Correspondence Over Θ

There remains to prove the existence of $s \in \Theta$ for which the incentive properties are satisfied. Traditionnally, the existence of an equilibrium in a game where the strategy sets are Σ_1 and Σ_2 is proven by analyzing the properties of a set-valued map $B : \Sigma_1 \times \Sigma_2 \to \Sigma_1 \times \Sigma_2$, $B(\sigma_1, \sigma_2) = B_1(\sigma_2) \times B_2(\sigma_1)$, where $B_1(\sigma_2)$ (resp., $B_2(\sigma_1)$) is interpreted as the set of best replies of player 1 to σ_2. Any fixed-point of B is then an equilibrium. It is clear that such an approach has no meaning here since Θ is not a product space. Moreover, any symmetric definition of a "best-reply" correspondence seems bound to fail: one could then obtain a game of the type "choose the smallest perturbation," with no equilibrium.

Our approach consists first in a preliminary separate analysis of the two players. For $\delta > 0$, denote by $(x_\delta(s), y_\delta(s)) \in \mathcal{O}$ a point δ-close to s (which depends continuously on s). One may show that the set of x that maximize the vector $\gamma^1(x, y_\delta(s))$ is a face of $\mathcal{S}_1$, whose set of vertices is of the form $\prod_k \tilde{I}_k(s)$, where $\tilde{I}_k(s) \subset I$ depends upper hemicontinuously on $y_\delta(s)$, and therefore on s.

As for player 2, we compute, for each k, j, the continuation payoff induced by j, and we compare it to the maximum in that state; namely, we define the cost of an action as $c(j) = \max_J P_{k,x(s),\cdot}\gamma^2(s) - P_{k,x(s),j}\gamma^2(s)$. This indicator allows us to compare actions in different states of K: we denote by $M_1, \ldots, M_N$ the partition of $\cup_k J_k$ into level sets for c, ordered by increasing level of cost.

We then associate to s the set $\Psi_\delta(s)$ of those $s' \in \Theta$ for which any valuation function d has the following properties:

1. for every k_1, k_2, i, j, if $i \notin \tilde{I}_{k_1}(s)$, then $d_{k_1}(i) > |K| d_{k_2}(j)$;
2. for $k_1, k_2, j_1 \in J_{k_1}, j_2 \in J_{k_2}$, if $c_s(j_1) > c_s(j_2)$, then $d_{k_1}(j_1) > |K| d_{k_2}(j_2)$.

We analyze the asymptotics of a family of fixed points $(s_\delta)_\delta$, as δ goes to 0. This description is heuristic in nature. It can easily be made precise (see [32], Sections 4.2 through 4.4).

7 Extensions

This section briefly reviews some known results on games with more than two players. The method presented above is quite specific to the case of two players. None of its steps extends easily to more general setups, except the existence of solvable states, which holds for any number of players (see [34]).

Whether or not every n-player stochastic game has an equilibrium payoff is not known yet. Positive results are only scattered. The initial impetus for these came through a three-player example devised by Flesch et al. [11]. It motivated the study of three-player games with absorbing states by Solan [24] (remember these are games where the state may change at most once during any play). The ε-equilibrium profiles designed by Solan have the same features as the ε-equilibrium profiles obtained for two-player games: stationary profiles (plus threats) or (nonstationary) perturbations of a given stationary profile. This suggests that appropriate modifications of the proofs for two-player games might yield strong positive results. However, a simple four-player quitting game, for which no ε-equilibrium profile of one of the above types exists, was found by Solan and Vieille [25]. Quitting games are sequential games in which, at any stage, each player has the choice between *continuing* and *quitting*. The game ends as soon as *at least* one player chooses to quit; player i then receives a payoff, which depends on the set of players that did choose to quit. If the game never ends, the payoff to each player is 0.

Quitting games constitute one of the simplest classes of stochastic games with nontrivial transitions. However, the existence of equilibrium payoffs for these games is still partly an open problem.

REFERENCES

[1] Altman, E., Hordijk, A., and Spieksma, F. M. Contraction conditions for average and α-discount optimality in countable state Markov games with unbounded rewards, *Math. Oper. Res.*, **3**, pp. 588–618, 1997.

[2] Aumann, R. J., and Maschler, M. with the collaboration of R. E. Stearns. *Repeated Games with Incomplete Information*, MIT Press, Cambridge, MA, 1995.

[3] Bewley, T. and Kohlberg, E. The asymptotic solution of a recursion equation occurring in stochastic games, *Math. Oper. Res.*, **1**, pp. 321–336, 1976.

[4] Bewley, T. and Kolhberg, E. The asymptotic theory of stochastic games, *Math. Oper. Res.*, **1**, pp. 197–208, 1976.

[5] Bewley, T. and Kohlberg, E. On stochastic games with stationary optimal strategies, *Math. Oper. Res.*, **3**, pp. 104–125, 1978.

[6] Blackwell, D. Discrete dynamic programming, *Ann. Math. Stat.*, **331**, pp. 719–726, 1962.

[7] Blackwell, D. and Ferguson, T. S. The big match, *Ann. Math. Stat.*, **39**, pp. 159–163, 1968.

[8] Everett, H. Recursive games, in *Contributions to the Theory of Games*, **3**, pp. 47–78, Annals of Mathematical Studies, **39**, Princeton Univ. Press, Princeton, 1957.

[9] Fink, A. M. Equilibrium in a stochastic n-person game, *J. Sci. Hiroshima Univ. Ser. A*, **28**, pp. 89–93, 1964.

[10] Flesch, J., Thuijsman, F., and Vrieze, O. J., Recursive repeated games with absorbing states, *Math. Oper. Res.*, **21**, pp. 1016–1022, 1996.

[11] Flesch, J., Thuijsman, F., and Vrieze, O.J. Cyclic Markov equilibria in stochastic games, *Int. J. Game Th.*, **26**, pp. 303–314, 1997.

[12] Freidlin, M. and Wentzell, A. *Random Perturbations of Dynamical Systems*, Springer-Verlag, Berlin, 1984.

[13] Gillette, D. Stochastic games with zero-stop probabilities, in *Contributions to the Theory of Games*, **3**, pp. 179–187, Annals of Mathematical Studies, **39**, Princeton Univ. Press, Princeton, 1957.

[14] Hammond, P. Extended probabilities for decision theory and games, *Mimeo, Stanford University*, Stanford, CA, 1987.

[15] Kohlberg, E. Repeated games with absorbing states, *Ann. Stat.*, **2**, pp. 724–738, 1974.

[16] Mertens, J. F. and Neyman, A. Stochastic games, *Int. J. Game Th.*, **10** pp. 53–66, 1981.

[17] Mertens, J. F. and Neyman, A. Stochastic games have a value, *Proc. NAS*, **79**, pp. 2145–2146, 1982.

[18] Mertens, J. F. and Parthasarathy, T. Non-zero sum stochastic games, in *Stochastic Games and Related Topics, Shapley Honor Volume*, Kluwer Publishing Co., Dordrecht, 1989.

[19] Milnor, J. and Shapley, L. S. On games of survival, in *Contributions to the Theory of Games, 3*, pp. 15–45, Annals of Mathematical Studies, **39**, Princeton Univ. Press, Princeton, 1957.

[20] Myerson, R. B. Multistage games with communication, *Econometrica*, **54**, pp. 323–358, 1986.

[21] Raghavan, T. E. S., Tijs, S. H., and Vrieze, O. J. On Stochastic Games with additive reward and transition structure, *J. Opt. Th. Appl.*, **47** pp. 451–464, 1985.

[22] Shapley, L. S. Stochastic games, *Proc. Nat. Acad. Sci. USA*, **39**, pp. 1095–1100, 1953.

[23] Solan, E. Stochastic Games with Two Non-Absorbing States, *Israel J. Math.*, **119**, pp. 29–54, 2000.

[24] Solan, E. Three-person absorbing games, *Math. Oper. Res.*, **24**(3):669–698, 1999.

[25] Solan, E. and Vieille, N. Quitting games, D.P. #1227, Center for Mathematical Studies in Economics and Management Science, Northwestern Univ., *Math. Oper. Res.*, to appear.

[26] Sorin, S. Asymptotic properties of a non-zero sum stochastic game, *Int. J. Game Theory*, **11**, pp. 147–160, 1986.

[27] Thuijsman, F. and Raghavan, T. E. S. Perfect information stochastic games and related classes, *Int. J. Game Theory*, **26**, pp. 403–408, 1997.

[28] Thuijsman, F. and Vrieze, O. J. Easy initial states in stochastic games, in *Stochastic Games and Related Topics, in Honor of L. S. Shapley*, pp. 85–100, Kluwer Academic Publishers, Dordrecht, 1991.

[29] Trouvé, A. Cycle decompositions and simulated annealing, *SIAM J. Cont. Opt.*, **34**, pp. 966–986, 1996.

[30] Vieille, N. Solvable states in stochastic games, *Int. J. Game Theory*, **21**, pp. 395–404, 1993.

[31] Vieille, N. Two-player stochastic games I: A reduction, *Israel J. Math.*, **119**, pp. 55–91, 2000.

[32] Vieille, N. Two-player stochastic games II: The case of recursive games, *Israel J. Math.*, **119**, pp. 93–126, 2000.

[33] Vieille, N. Small perturbations and stochastic games, *Israel J. Math.*, **119**, pp. 127–142, 2000.

[34] Vieille, N. Solvable states in n-player stochastic games, *SIAM J. Cont. Opt.*, **38**, *6*, pp. 1794–1804, 2000.

[35] Vrieze, O. J. and Thuijsman, F. On equilibria in stochastic games with absorbing states, *Int. J. Game Theory*, **18**, pp. 293–310, 1989.

Persistently Good Strategies for Nonleavable Stochastic Games with Finite State Space

Piercesare Secchi
Politecnico di Milano
Dipartimento di Matematica
Milano, Italy

William D. Sudderth
University of Minnesota
School of Statistics
Minneapolis, Minnesota, USA

Abstract

The notion of a persistently optimal strategy in gambling theory is analogous to that of subgame-perfect equilibria in game theory. We prove the existence of persistently ϵ-optimal strategies for the players engaged in a nonleavable stochastic game with finite state space.

1 Introduction and Preliminaries

The purpose of this paper is to prove the existence of persistently ϵ-optimal strategies for the players engaged in a nonleavable stochastic game. This is a two-person, multistage, zero-sum competition that Maitra and Sudderth [4] defined by means of five elements: a state space S; two finite sets of actions A and B; for players I and II, respectively; a law of motion q that maps $S \times A \times B$ onto the set of probability distributions defined on S; and a bounded utility function $u : S \to \Re$. The dynamics of a nonleavable stochastic game $\mathcal{N}(u)(x)$ with initial state $x \in S$ are the following: player I chooses an action $a_1 \in A$, while player II simultaneously chooses an action $b_1 \in B$, and the game moves to state X_1 according to the probability distribution $q(\cdot|x, a_1, b_1)$. This ends the first stage of the game; in the second stage the players choose actions $a_2 \in A$ and $b_2 \in B$, respectively, and the game moves to state X_2 according to the probability distribution $q(\cdot|X_1, a_2, b_2)$. Iterating this procedure for an infinite number of stages produces an infinite sequence $\{X_n\}$ of states; the payoff of the game from player II to player I is defined to be the expected value of

$$u^* = \limsup u(X_n).$$

Maitra and Sudderth [4] proved the existence of ϵ-optimal strategies for the players engaged in a nonleavable game $\mathcal{N}(u)$ by showing that the game always has a value V in the sense that, for every initial state $x \in S$,

$$\inf_{\beta} \sup_{\alpha} E_{x,\alpha,\beta} u^* = \sup_{\alpha} \inf_{\beta} E_{x,\alpha,\beta} u^* = V(x),$$

where α and β range over the set of strategies available to players I and II, respectively. For a complete introduction to nonleavable stochastic games, we refer to [5]; in what follows we use the same notation and terminology as in that book and in [8] without further comment.

Given $\epsilon > 0$ and the initial state $x \in S$ of the game $\mathcal{N}(u)$, a strategy α is said to be persistently ϵ-optimal for I at x if it is ϵ-optimal and, for every $n \geq 1$ and every partial history $p = ((x_1, a_1, b_1), \ldots, (x_n, a_n, b_n))$ of length n, the conditional strategy $\alpha[p]$ is ϵ-optimal for I for the game $\mathcal{N}(u)(x_n)$. A family of strategies $\bar{\alpha}$ is persistently ϵ-optimal if $\bar{\alpha}(x)$ is persistently ϵ-optimal for all $x \in S$. Analogous definitions hold for II. We note that, for all $x \in S$, $n \geq 1$, and all partial histories $p = ((x_1, a_1, b_1), \ldots, (x_n, a_n, b_n))$, the game $\mathcal{N}(u)(x_n)$ is a proper subgame of $\mathcal{N}(u)(x)$; therefore, when a player uses a persistently ϵ-optimal strategy α, he cannot gain more than ϵ by deviating from α in any subgame of $\mathcal{N}(u)(x)$. In this sense the definition of "persistently ϵ-optimal strategy" coincides with that of "subgame perfect ϵ-equilibria" of Selten [9]. Our terminology stems from gambling theory and from the analogous definitions due in that setting to Dubins and Sudderth [2].

The notion of persistently optimal strategy is related to that of stationarity. Recall that a stationary family of strategies $\bar{\alpha}$ assigns a strategy $\bar{\alpha}(x)$, for player I say, to each state x in such a way that, for every $x \in S$, $n \geq 1$, and partial history $p = ((x_1, a_1, b_1), \ldots, (x_n, a_n, b_n)), \bar{\alpha}(x)_n[p] = \mu(x_n)$, where μ is a mapping from S to the set of probability distributions defined on A. Thus, the action chosen at any stage depends only on the current state. Suppose that $\bar{\alpha}(x)$ is ϵ-optimal for every $x \in S$. Then $\bar{\alpha}(x)$ is also persistently ϵ-optimal because $\bar{\alpha}(x)[p] = \bar{\alpha}(x_n)$. In [8] it is shown that an ϵ-optimal stationary family is always available to player II when the state space S of the nonleavable game $\mathcal{N}(u)$ is finite; hence, player II always has persistently ϵ-optimal strategies. However, Example 7.13.4 of [5] a variation of an example of [3], shows that player I need not have an ϵ-optimal stationary strategy at a fixed initial state of the game, even if S is finite. Therefore, the following theorem takes on interest; together with the results in [8] it affirmatively settles the question regarding the existence of persistently ϵ-optimal strategies for the players engaged in a nonleavable game with finite state space.

Theorem 1.1. *Let $\epsilon > 0$. For any nonleavable game $\mathcal{N}(u)$ defined on a finite state space S, a persistently ϵ-optimal family of strategies $\bar{\mu}$ is available to player I.*

When the state space of the game S is infinite and the utility u is unbounded, player I need not have a persistently ϵ-optimal strategy. This follows from the next example, which appears in [10] and is a modification of an example given by Blackwell [1]. Incidentally, the example shows that subgame perfect ϵ-equilibria need not exist for all zero-sum games with a value.

Example 1.1. Let $S = \{\ldots, -2, -1, 0, 1, 2, \ldots\}$, $A = \{0, 1\}$, and $B = \{0\}$; player II is therefore a dummy and the game is really a one-player game. If $x < 0$, set $u(x) = 2^{-x}$ and $q(x|x, 0, 0) = q(x|x, 1, 0) = 1$; if $x > 0$, set $u(x) = 0$, $q(x+1|x, 1, 0) = 1 - q(0|x, 1, 0) = b(x)/(2b(x+1))$, with

$$b(x) = 1 + \frac{1}{x},$$

and $q(-x|x, 0, 0) = 1$. Finally, define $u(0) = 0$ and $q(0|0, 0, 0) = q(0|0, 1, 0) = 1$. Therefore, all nonpositive states are absorbing and the utility u is nonnull only on the strictly negative states.

For all strategies σ available to player I, let $\pi_0(\sigma)$ be the probability that action 1 is chosen at the first stage of the game. Given an initial state $x \in S$, we also define the partial histories $p_n = ((x+1, 1, 0), (x+2, 1, 0), \ldots, (x+n, 1, 0))$, $n \geq 1$. Let $\underline{0}$ be the only strategy available to player II. Then, for all $x > 0$ and all strategies σ of player I, we may compute

$$\begin{aligned} E_{x,\sigma,\underline{0}}u^* &= (1 - \pi_0(\sigma))2^x + \pi_0(\sigma)(1 - \pi_0(\sigma[p_1]))\frac{b(x)}{2b(x+1)}2^{x+1} + \ldots \quad (1)\\ &\quad \ldots + \pi_0(\sigma)[\Pi_{i=1}^{n-1}\pi_0(\sigma[p_i])](1 - \pi_0(\sigma[p_n]))\frac{b(x)}{2b(x+1)} \\ &\quad \ldots \frac{b(x+n-1)}{2b(x+n)}2^{x+n} + \ldots . \end{aligned}$$

Hence,

$$\sup_\sigma E_{x,\sigma,\underline{0}}u^* \leq \sup\left\{\frac{b(x)}{b(x+n)}2^x : n \geq 0\right\} = b(x)2^x.$$

Let σ^* be the strategy selecting action 1 with probability 1 for the first $n+1$ stages of the game, $n \geq 1$, and then action 0 forever; then

$$E_{x,\sigma^*,\underline{0}}u^* = \frac{b(x)}{b(x+n)}2^x.$$

This shows that, for all $x > 0$,

$$\sup_\sigma E_{x,\sigma,\underline{0}} \geq \sup\{\frac{b(x)}{b(x+n)}2^x : n \geq 0\} = b(x)2^x.$$

Therefore, the game has value $V(x) = b(x)2^x$. Obviously, $V(0) = 0$ and $V(x) = 2^{-x}$ for all $x < 0$.

Before proving that no persistently ϵ-optimal strategy is available to player I if the initial state of the game is big enough, we find a necessary condition for a strategy σ to be ϵ-optimal for I when the initial state of the game is strictly positive. Let $0 < \epsilon \leq 1$ and $x > 0$. If σ is ϵ-optimal for I at x, then

$$\begin{aligned} b(x)2^x - \epsilon &= V(x) - \epsilon \\ &\leq (1 - \pi_0(\sigma))2^x + \pi_0(\sigma)\frac{b(x)}{2b(x+1)}E_{x,\sigma[p_1],\underline{0}}u^* \end{aligned}$$

$$\leq (1 - \pi_0(\sigma))2^x + \pi_0(\sigma)\frac{b(x)}{2b(x+1)}V(x+1)$$
$$= (1 - \pi_0(\sigma))2^x + \pi_0(\sigma)b(x)2^x.$$

Hence,

$$1 - \pi_0(\sigma) \leq \frac{2^{-x}}{b(x)-1}\epsilon \leq x2^{-x}.$$

Now let $\epsilon = 1$, $x \geq 5$, and assume there is a strategy μ persistently 1-optimal for I at x. Then for all $n \geq 1$ and all partial histories $p = ((x_1, a_1, 0), \ldots, (x_n, a_n, 0))$ with $x_i > 0$, and $a_i \in A$, $i = 1, \ldots, n$,

$$1 - \pi_0(\mu[p]) \leq x_n 2^{-x_n}.$$

This inequality and Eq. (1) imply that

$$E_{x,\mu,\underline{0}}u^* \leq (1 - \pi_0(\mu))2^x + \frac{b(x)}{2b(x+1)}(1 - \pi_0(\mu[p_1])2^{x+1} + \ldots$$
$$\ldots + \frac{b(x)}{2b(x+1)} \cdots \frac{b(x+n-1)}{2b(x+n)}(1 - \pi_0(\mu[p_n])2^{x+n} + \ldots$$
$$\leq x + \frac{2}{3}(x+1) + \ldots + (\frac{2}{3})^n(x+n) + \ldots$$
$$= 3x + 6,$$

where we used the fact that $b(x+n)/(2b(x+n+1)) \leq 2/3$ for all $n \geq 0$ if $x \geq 1$. But $3x + 6 < V(x) - 1$ if $x \geq 5$, and thus μ is not persistently 1-optimal at x since it is not 1-optimal.

When S is countably infinite and u is bounded, it is still an open question whether a player engaged in a nonleavable game $\mathcal{N}(u)$ has a persistently ϵ-optimal family of strategies. A referee has pointed out that under the assumptions of continuity and compactness of [6], persistently optimal strategies exist for at least one of the players.

2 Proof of Theorem 1.1

Before constructing a family of strategies persistently good for player I in the nonleavable game $\mathcal{N}(u)$, we begin by considering a leavable game $\mathcal{L}(u)$ defined on the finite state space S by means of the same law of motion q, utility u, and sets of actions A and B for I and II, respectively, as $\mathcal{N}(u)$, but with the additional feature that player I is now allowed to stop the game according to a stop rule t of her choice and the payoff from II to I is the expected value of the utility $u(X_t)$ of the state when the game is stopped. Leavable games have a value, and an optimal stationary family of strategies is always available to player II [4]; for all $\epsilon > 0$, the existence of an ϵ-optimal stationary family for player I was proved in [8].

Set $U_0 = u$ and, for all $n \geq 1$, define $U_n = u \vee GU_{n-1}$, where G is the one-day operator defined for any bounded, real-valued function ϕ on S and for every $x \in S$, by

$$(G\phi)(x) = \inf_{\nu} \sup_{\mu} E_{x,\nu,\mu}\phi, \tag{2}$$

with μ and ν ranging over $\mathcal{P}(A)$ and $\mathcal{P}(B)$, the sets of probability distributions defined on A and B, respectively. For all $n \geq 1$, $U_n \geq U_{n-1}$ and the sequence $\{U_n\}$ converges to the value U of $\mathcal{L}(u)$ ([4]); the convergence is uniform since S is finite.

Fix $\epsilon > 0$ sufficiently small and such that, for all $x \in S$, $|U(x) - u(x)| \leq \epsilon$ if and only if $U(x) = u(x)$. Let m be an integer such that, $\sup_{x \in S} |U_m(x) - U(x)| < \epsilon/2$, and let

$$0 = \epsilon_0 < \epsilon_1 < \ldots < \epsilon_m = \frac{\epsilon}{2}.$$

Define, for all $x \in S$,

$$\phi(x) = \max\{U_k(x) - \epsilon_k : 0 \leq k \leq m\}.$$

Note that $u \leq \phi \leq U$; when $\phi(x) = u(x)$,

$$U(x) - \epsilon \leq U_m(x) - \frac{\epsilon}{2} \leq \phi(x) = u(x).$$

Therefore, $\phi(x) = u(x)$ if and only if $u(x) = U(x)$.

We now introduce a function $\mu_\epsilon \colon S \to \mathcal{P}(A)$ generating a stationary family μ_ϵ^∞ that is 2ϵ-optimal for I in $\mathcal{L}(u)$. If $\phi(x) > u(x)$, let $k = k(x)$ be such that $\phi(x) = U_k(x) - \epsilon_k$ and set $\mu_\epsilon(x) \in \mathcal{P}(A)$ to be optimal for I in the one-day game $\mathcal{A}(U_{k-1})(x)$. If $\phi(x) = u(x)$, let $\mu_\epsilon(x)$ be optimal for I in the one-day-game $\mathcal{A}(U)(x)$ (cf. section 7.6 of [4]).

For all $k \geq 0$, define

$$\tau(k) = \inf\{n \geq k : \phi(X_n) = u(X_n)\} = \inf\{n \geq k : U(X_n) = u(X_n)\}. \tag{3}$$

The next two inequalities (4) and (5) regarding the stopping time $\tau(0)$ were demonstrated in the course of the proof of Theorem 3.2 of [7] and will also be very useful in the present context. The first implies that, no matter what player II does, the game will reach the set of states where $U = u$ with probability 1 if player I uses the stationary family μ_ϵ^∞. In fact, set

$$\zeta = \frac{1}{2} \min\{\epsilon_k - \epsilon_{k-1} : 0 \leq k \leq m\}.$$

Then, for all $x \in S$ and $n \geq 1$,

$$P_{x,\mu_\epsilon^\infty(x),\beta}[\tau(0) \geq n] \leq \frac{2\|u\|}{n\zeta}, \tag{4}$$

for all strategies β of II where $\|u\| = \max\{|u(x)| : x \in S\}$; therefore, $\tau(0)$ is finite with probability 1. Now choose n big enough so that $4\|u\|^2(n\zeta)^{-1} \leq \epsilon$. Then, for all $x \in S$,

$$E_{x,\mu_\epsilon^\infty(x),\beta} u(X_{\tau(0)\wedge n}) \geq U(x) - 2\epsilon \tag{5}$$

against any strategy β of player II; this immediately implies that μ_ϵ^∞ is 2ϵ-optimal for I in $\mathcal{L}(u)$.

When player I is not allowed to stop the game $\mathcal{L}(u)$ immediately, but must play at least for one day, the following result is of use.

Lemma 2.1. *For all $x \in S$ and every strategy β of II,*

$$P_{x,\mu_\epsilon^\infty(x),\beta}[\tau(1) < \infty] = 1 \tag{6}$$

and

$$E_{x,\mu_\epsilon^\infty(x),\beta} u(X_{\tau(1)}) \geq GU(x) - 2\epsilon. \tag{7}$$

Proof. Equation (6) follows from (4) and the fact that the conditional stopping time $\tau(1)[X_1]$ is equal to $\tau(0)$. In fact, for all $x \in S$,

$$P_{x,\mu_\epsilon^\infty(x),\beta}[\tau(1) < \infty] = \int P_{X_1,\mu_\epsilon^\infty(X_1),\beta[X_1]}[\tau(0) < \infty] dP_{x,\mu_\epsilon^\infty(x),\beta} = 1,$$

for all β of II.

In order to prove (7), assume first that $U(x) > u(x)$. Then $\tau(1) = \tau(0)$ and, if $4\|u\|^2(n\zeta)^{-1} \leq \epsilon$,

$$E_{x,\mu_\epsilon^\infty(x),\beta} u(X_{\tau(1)\wedge n}) = E_{x,\mu_\epsilon^\infty(x),\beta} u(X_{\tau(0)\wedge n}) \geq U(x) - 2\epsilon = GU(x) - 2\epsilon$$

against any β of II, where the last equality follows from the fact that $U = u \vee GU$ ([4]). Since u is bounded and $\tau(1)$ is finite with probability 1, we get (7) in this case by letting $n \to \infty$. If $U(x) = u(x)$,

$$\begin{aligned} &E_{x,\mu_\epsilon^\infty(x),\beta} u(X_{\tau(1)}) \\ &\quad = \int E_{X_1,\mu_\epsilon^\infty(X_1),\beta[X_1]} u(X_{\tau(0)}) dP_{x,\mu_\epsilon^\infty(x),\beta} \\ &\quad \geq \int (U(X_1) - 2\epsilon) dP_{x,\mu_\epsilon^\infty(x),\beta} \\ &\quad \geq GU(x) - 2\epsilon, \end{aligned}$$

for any β of II; the next-to-last inequality follows from (5) by letting $n \to \infty$, whereas the last inequality is a consequence of the definition of μ_ϵ. □

We are now ready to return to the nonleavable game $\mathcal{N}(u)$ defined on the finite state space S. Lemma 4.1 of [7] shows that the value V of $\mathcal{N}(u)$ is also the value of the leavable game $\mathcal{L}(u \wedge V)$ that is defined on S and has the same law of motion q and action spaces A and B for I and II, respectively, as the nonleavable game $\mathcal{N}(u)$, but utility $u \wedge V$. Applying the previous arguments to the game $\mathcal{L}(u \wedge V)$, set $\epsilon > 0$ small enough and such that, for all $x \in S$, $V(x) - (u \wedge V)(x) < \epsilon$ if and only if $V(x) = (u \wedge V)(x)$. Then, for all $k \geq 0$, the stopping time $\tau(k)$ defined by (3) becomes

$$\tau(k) = \inf\{n \geq k : V(X_n) \leq u(X_n)\}.$$

Lemma 2.1 implies that there is a stationary family μ_ϵ^∞ for player I such that, for every $x \in S$ and every strategy β of II,

$$P_{x,\mu_\epsilon^\infty(x),\beta}[\tau(1) < \infty] = 1 \tag{8}$$

and

$$E_{x,\mu_\epsilon^\infty(x),\beta}(u \wedge V)(X_{\tau(1)}) \geq GV(x) - 2\epsilon = V(x) - 2\epsilon, \tag{9}$$

where the last equality follows from the fact that $GV = V$ ([5]).

In order to introduce the family of strategies $\bar{\mu}_\epsilon$ that will be proved to be persistently 4ϵ-optimal for I in $\mathcal{N}(u)$, define a sequence of stopping times τ_n by setting $\tau_0 = \underline{0}$ and, for all $n \geq 0$, $\tau_{n+1} = \tau(\tau_n + 1)$. The family $\bar{\mu}_\epsilon$ is constructed by sequential composition of the strategies $\{\mu^\infty_{\epsilon/2^{n-1}}\}$ along the sequence $\{\tau_n\}$, $n = 1, 2, \ldots$. That is, for all $x \in S$, $\bar{\mu}_\epsilon(x)$ follows $\mu_\epsilon^\infty(x)$ up to time τ_1, then uses $\mu^\infty_{\epsilon/2}(X_{\tau_1})$ up to time τ_2, and so on. We will begin by proving, with the next two lemmas, that $\bar{\mu}_\epsilon$ is 4ϵ-optimal for player I.

Lemma 2.2. *For every $x \in S$ and β of II,*

$$P_{x,\bar{\mu}_\epsilon(x),\beta}[\bigcap_{n=1}^{\infty}[\tau_n < \infty]] = 1.$$

Proof. Note that, for every $n \geq 0$, $x \in S$, and β of player II,

$$P_{x,\bar{\mu}_\epsilon(x),\beta}[\tau_{n+1} < \infty] = \int P_{X_{\tau_n},\mu^\infty_{\frac{\epsilon}{2^n}}(X_{\tau_n}),\beta[p_{\tau_n}]}[\tau(1) < \infty]\, dP_{x,\bar{\mu}_\epsilon(x),\beta}$$

if $P_{x,\bar{\mu}_\epsilon(x),\beta}[\tau_n < \infty] = 1$. Now use (8) and induction on n. □

Lemma 2.3. *$\bar{\mu}_\epsilon$ is 4ϵ-optimal for I in $\mathcal{N}(u)$.*

Proof. Fix $x \in S$ and a strategy β for II. For ease of notation, in the following we use P for the probability $P_{x,\bar{\mu}_\epsilon(x),\beta}$ and E for the expected values computed with respect to P.

Define $Y_0 = V(x)$ and, for every $n \geq 1$, set

$$Y_n = V(X_{\tau_n}) + 2\epsilon \sum_{j=0}^{n-1} 2^{-j}.$$

The sequence $\{Y_n\}$ is a bounded submartingale with respect to P. In fact, for every $n \geq 0$, $|Y_n| < \|u\| + 4\epsilon$ and

$$\begin{aligned} &E[Y_{n+1}|X_1, \ldots, X_{\tau_n}] \\ &\quad = E[V(X_{\tau_{n+1}})|X_1, \ldots, X_{\tau_n}] + 2\epsilon \sum_{j=0}^{n} 2^{-j} \end{aligned}$$

$$
\begin{aligned}
&= E_{X_{\tau_n},\mu^{\infty}_{\frac{\epsilon}{2^n}}(X_{\tau_n}),\beta[p_{\tau_n}]} V(X_{\tau_{n+1}[p_{\tau_n}]}) + 2\epsilon\sum_{j=0}^{n} 2^{-j} \\
&\geq E_{X_{\tau_n},\mu^{\infty}_{\frac{\epsilon}{2^n}}(X_{\tau_n}),\beta[p_{\tau_n}]} u \wedge V(X_{\tau(1)}) + 2\epsilon\sum_{j=0}^{n} 2^{-j} \\
&\geq V(X_{\tau_n}) - 2\frac{\epsilon}{2^n} + 2\epsilon\sum_{j=0}^{n} 2^{-j} \\
&= Y_n,
\end{aligned}
$$

where the first inequality holds because $\tau_{n+1}[p_{\tau_n}] = \tau(1)$ and $V \geq u \wedge V$, whereas the second inequality follows from (9).

Being a bounded submartingale, $\{Y_n\}$ converges with P probability 1 and

$$E[\lim_{n\to\infty} Y_n] \geq Y_0 = V(x).$$

Therefore,

$$E[\lim_{n\to\infty} V(X_{\tau_n})] \geq V(x) - 4\epsilon.$$

This proves the lemma since

$$Eu^* = E[\limsup_{n\to\infty} u(X_n)] \geq E[\limsup_{n\to\infty} u(X_{\tau_n})] \geq E[\lim_{n\to\infty} V(X_{\tau_n})],$$

where the first inequality follows from Lemma 2.2 and the second from the definition of τ_n. □

Final steps of the proof of Theorem 1.1. In order to prove that $\bar{\mu}_\epsilon$ is persistently 4ϵ-optimal, it is enough to show that, for every $x \in S$, $j \geq 1$, and partial history $p = ((x_1, a_1, b_1), \ldots, (x_j, a_j, b_j))$ of length j,

$$E_{x_j,\bar{\mu}_\epsilon(x)[p],\beta} u^* \geq V(x_j) - 4\epsilon$$

against any strategy β of II.

Let $k \geq 0$ be such that, for every history $h' = ((x_1', a_1', b_1'), (x_2', a_2', b_2'), \ldots)$,

$$\tau_k(ph') \leq j < \tau_{k+1}(ph'),$$

with $ph' - ((x_1, a_1, b_1), \ldots, (x_j, a_j, b_j), (x_1', a_1', b_1'), (x_2', a_2', b_2'), \ldots)$. Now $\bar{\mu}_\epsilon(x)[p]$ is the sequential composition of the strategies $\{\mu^{\infty}_{\epsilon/2^{k+n-1}}\}$ along the sequence $\{\tau_{k+n}\}$, $n = 1, 2, \ldots$, and, hence, equals $\bar{\mu}_{\epsilon/2^k}(x_j)$. The desired result now follows from Lemma 2.3. □

Acknowledgments

This paper was completed while the first author was visiting the International Institute for Applied System Analysis in Vienna: his research was partly supported by CNR. The research of Sudderth was supported by National Science Foundation

Grant DMS-9703285. We thank Paolo Bertoletti for some stimulating discussions on subgame perfect equilibria.

REFERENCES

[1] Blackwell, D. Positive dynamic programming, in *Proc. 5th Berkeley Symposium Math. Stat. Prob.*, **1**, pp. 415–418, 1967.

[2] Dubins, L. E. and Sudderth, W. D. Persistently ϵ-optimal strategies, *Math. Oper. Res.*, **2**, pp. 125–134, 1977.

[3] Kumar, P. R. and Shiau, T. H. Existence of value and randomized strategies in zero-sum discrete time stochastic dynamic games, *SIAM J. Cont. Opt.*, **19**, 617–634, 1981.

[4] Maitra, A. and Sudderth, W. D. An operator solution of stochastic games, *Israel J. Math.*, **78**, pp. 33–49, 1992.

[5] Maitra, A. and Sudderth, W. D. *Discrete Gambling and Stochastic Games*, Springer-Verlag, New York, 1996.

[6] Nowak, A. S. Semicontinuous nonstationary stochastic games, *J. Math. Anal. Appl.*, **117**, pp. 84–99, 1986.

[7] Secchi, P. Stationary strategies for recursive games, *Math. Oper. Res.*, **22**, 494–512, 1997.

[8] Secchi, P. On the existence of good stationary strategies for nonleavable stochastic games, *Int. J. Game Th.*, **27**, 61–81, 1998.

[9] Selten, R. Spieltheoretische Behandlung eines Oligopolmodells mit Nachfrage-tragheit, *Zeitscrift fur die gesante Staatswissenschaft*, **12**, pp. 301–324, 1965.

[10] van der Wal, J. *Stochastic Dynamic Programming*. The Mathematical Centre, Amsterdam, 1981.

On Stochastic Hybrid Zero-Sum Games with Nonlinear Slow Dynamics

Minh-Tuan Nguyen
DSTO
Salisbury, Australia

Eitan Altman
INRIA
Sophia Antipolis, France

Vladimir Gaitsgory
School of Mathematics
University of South Australia
The Levels 5095, Australia

Abstract

This paper considers a continuous-time stochastic hybrid system, controlled by two players with opposite objectives (zero-sum game). The parameters of the system may jump at discrete moments of time according to a Markov decision process, namely, a Markov chain that is directly controlled by both players and has finite state and action spaces. Assuming that the length of the intervals between the jumps is defined by a small parameter ϵ, the value of this game is shown to have a limit as the small parameter tends to 0. This limit is established to coincide with the viscosity solution of some Hamilton–Jacobi-type equations.

1 Introduction and Statement of the Problem

Consider the following hybrid stochastic controlled system. The state $Z_s \in \mathbb{R}^n$ evolves according to the following dynamics:

$$\frac{d}{ds} Z_s = f(Z_s, Y_s), \qquad s \in [t, T], \qquad Z_t = z, \tag{1}$$

where $Y_s \in \mathbb{R}^k$ is the "control" and $f : \mathbb{R}^n \times \mathbb{R}^k \to \mathbb{R}^n$ is a vector function. Y_s is not chosen directly by the controllers, but is obtained as a result of controlling the following underlying stochastic discrete event system.

- Let ϵ be the basic time unit. Time is discretized, that is, transitions occur at times $s = l\epsilon, l = 0, 1, 2, \ldots, \lfloor (T-t)\epsilon^{-1} \rfloor$, where $\lfloor x \rfloor$ stands for the greatest integer that is smaller than or equal to x.

- There is a finite state space $\mathbf{X}$ and two players having finite action spaces $\mathbf{A}^1$ and $\mathbf{A}^2$, respectively. Let $\mathbf{A} = \mathbf{A}^1 \times \mathbf{A}^2$.
- If the state is v and actions $a = (a^1, a^2)$ are chosen by the players, then the next state is w with probability P_{vaw}. Denote $\mathcal{P} = \{P_{vaw}\}$.
- A policy $u^i = \{u_0^i, u_1^i, \ldots\}$ in the set of policies U^i for player i, $i = 1, 2$, is a sequence of probability measures on $\mathbf{A}^i$ conditioned on the history of all previous states (the $\mathbf{X}$-component only) and actions of both players, as well as the current state. More precisely, define the set of histories
$$\mathbf{H} := \bigcup_l \mathbf{H}_l, \qquad \text{where } \mathbf{H}_l := \left\{(x_0, a_0^1, a_0^2, x_1, a_1^1, a_1^2, \ldots, x_l)\right\}$$
are the sets of all sequences of $3l+1$ elements describing the possible samples of previous states and actions prior to l as well as the currents state at stage l (i.e., at time $l\epsilon$). (The range of l will be either $l = 0, 1, \ldots, \lfloor (T-t)\epsilon^{-1} \rfloor$ or, in other contexts, all nonnegative integers, depending on whether we consider finite or infinite horizon problems.) The policy at stage l for player i, u_l^i, is a map from $\mathbf{H}_l$ to the set of probability measures over the action space $\mathbf{A}^i$. (Hence, at each time $t = l\epsilon$, player i, observing the history h_l, chooses action a_l^i with probability $p(a_l^i | h_l)$.)
- Let $\mathcal{F}_l$ be the discrete σ-algebra corresponding to $\mathbf{H}_l$. Each initial distribution ξ and policy pair u for the players uniquely defines a probability measure P_ξ^u over the space of samples $\mathbf{H}$ (equipped with the discrete σ-algebra); see, e.g., [10]. Denote by E_ξ^u the corresponding expectation operator. On the above probability space the random processes X_l and $A_l = (A_l^1, A_l^2)$, denoting the state and actions processes are now defined. When the initial distribution is concentrated on a single state x, we shall denote the corresponding probability measure and expectation by P_x^u and E_x^u, respectively.

Remark 1.1. The reason we do not include the Z-part of the state in the definition of the policies is that the trajectory of this component is fully determined by the trajectories of the $\mathbf{X}$-component together with the actions, for a fixed initial state z. The latter is assumed to be fixed and common knowledge for the players.

Let $g : \mathbf{X} \times \mathbf{A} \to \mathbb{R}^k$ be some given vector-valued bounded function and X_n and $A_n = (A_n^1, A_n^2)$ denote the state and actions processes, respectively. Then Y_s in (1) is given by
$$Y_s = g(X_{\lfloor s/\epsilon \rfloor}, A_{\lfloor s/\epsilon \rfloor}). \tag{2}$$
Y_s and thus Z_s are well-defined stochastic processes and are both $\mathcal{F}_{\lfloor (T-t)\epsilon^{-1} \rfloor}$-measurable.

We are especially interested in the following classes of policies:

- the stationary policies, denoted by $\mathcal{S}^1$, for player 1, and $\mathcal{S}^2$, for player 2. A policy u is called *stationary* if u_l depends only on the current state (the $\mathbf{X}$-component) and not on previous states and actions nor on the time. Let $\mathcal{S} := \mathcal{S}^1 \times \mathcal{S}^2$.

- the Markov policies $\mathcal{M}^1, \mathcal{M}^2$: these are policies where u_l^i depends only on the current $\mathbf{X}$-component of the state (at time $t = l\epsilon$) and on stage l and not on previous states and actions. Denote $\mathcal{M} = \mathcal{M}^1 \times \mathcal{M}^2$.

Let us define the payoff of the game by the equation

$$J_\epsilon(t, z, x; u^1, u^2) = E_x^{(u^1,u^2)}\left\{\int_t^T F(Z_s, Y_s)\,ds + G(Z_T)\right\},$$

when policies u^1, u^2 are used by the players; the initial state of the system is z; the initial state of the controlled Markov chain is x; Z_s is obtained through (1), and $F\colon \mathbb{R}^n \times \mathbb{R}^k \to \mathbb{R}^1$, $G\colon \mathbb{R}^n \to \mathbb{R}^1$ are the running cost and terminal cost function, respectively.

In our dynamic game, player 1 wishes to maximize $J_\epsilon(t, z, x; u^1, u^2)$ and player 2 wants to minimize it. We define the upper- and lower-value functions of the hybrid game as

$$B_\epsilon^{\rm up}(t, z, x) = \inf_{u^2 \in U^2} \sup_{u^1 \in U^1} J_\epsilon(t, z, x; u^1, u^2),$$

$$B_\epsilon^{\rm lo}(t, z, x) = \sup_{u^1 \in U^1} \inf_{u^2 \in U^2} J_\epsilon(t, z, x; u^1, u^2).$$

It can be shown (see the Appendix) that the stochastic hybrid game has value $B_\epsilon(t, z, x)$. That is, for all $(t, z, x) \in [0, T] \times \mathbb{R}^n \times \mathbf{X}$,

$$B_\epsilon(t, z, x) \stackrel{\rm def}{=} B_\epsilon^{\rm up}(t, z, x) = B_\epsilon^{lo}(t, z, x).$$

Our model is characterized by the fact that ϵ is supposed to be a small parameter and our objective is to show that the value of the game has a *limit* as $\epsilon \to 0$, this limit is a viscosity solution of some Hamilton–Jacobi-type equations.

Notice that this result can be viewed as an extension of viscosity solutions for deterministic, singularly perturbed zero-sum differential games (see [9]) to the stochastic case under consideration.

This paper is a continuation and generalization of previous works [16] which solves a hybrid problem restricted to a single controller and [1] which considers a linear hybrid game with linear cost. As in [16] and [1], the fact that ϵ is small means that the variables Y_s can be considered fast with respect to Z_s since, by (2), they may have a finite (not tending with ϵ to zero) change at each interval of the length ϵ. This, along with the dynamic equation of system (1), allows us to decompose the game into stochastic subgames on a sequence of intervals that are short with respect to the variables Z_s (in the sense that Z_s remain almost unchanged on these intervals) and long enough with respect to Y_s (so that the corresponding stochastic subgames show on their intervals their limit properties).

The type of model we introduce is natural in the control of inventories or of production (see, for example, [18]), where we deal with material whose quantity may change continuously. Breakdowns, repairs, and other control decisions yield the underlying controlled Markov chain. In particular, repair or preventive maintenance decisions are typical actions of a player who minimizes costs. If there is some unknown parameter (disturbance) of the dynamics of the system (e.g., the probability of breakdowns) that may change depending on the current and past states in a way that is unknown and unpredictable by the minimizer, we may formulate this situation as a zero-sum game, where the minimizer wishes to guarantee the best performance (lowest expected cost) under the worst-case behavior of nature. Nature may then be modeled as the maximizing player.

Our model may also be used in the control of highly loaded queueing networks for which the fluid approximation holds (see [14], p. 56). The quantities Z_t may then represent the number of customers in the different queues, whereas the underlying controlled Markov chain may correspond to routing, or flow control, of, say, some on–off traffic, with nature again controlling some disturbances in quantities such as service rates.

The remainder of this paper is organized as follows. Section 2 gives all imposed assumptions, and then the associated subgame and the existence of the value of such game are introduced in Section 3. The limit Hamilton–Jacobi–Isaacs equations for the stochastic hybrid game are defined in Section 4. The main results are presented in Section 5. Finally, the most tedious proofs are gathered in the last two sections.

2 Basic Assumptions

In our consideration, we use the following assumptions.

Assumption 2.1. The stochastic process $\{X_n, A_n\}$, which is known as the *competitive Markov decision process* or *stochastic game*, has a unichain structure: under any pair of stationary policies for the two players, the state process constitutes a single ergodic class of states.

Assumption 2.2. There exists a compact subset $D_1 \in \mathbb{R}^n$ that contains all solutions Z_s of system (1) obtained with different admissible controls Y_s defined as a piecewise constant function of time taking their values in a finite subset of $\mathbb{R}^k$. Denote this subset by D_2.

Assumption 2.3. All functions used in the definitions of the stochastic hybrid and associated games are continuous on $D_1 \times D_2$, also, they satisfy the local Lipschitz conditions in (z, y) with Lipschitz constant $L \geq 0$. That is, for any $(z_i, y_i) \in (D_1 \times D_2)$, $i = 1, 2$,

$$\begin{aligned} \big\| f(z_1, y_1) - f(z_2, y_2)\big\| &\leq L\big\{\|z_1 - z_2\| - \|y_1 - y_2\|\big\}, \\ \big\| F(z_1, y_1) - F(z_2, y_2)\big\| &\leq L\big\{\|z_1 - z_2\| - \|y_1 - y_2\|\big\}, \\ \big\| G(z_1) - G(z_2)\big\| &\leq L\|z_1 - z_2\|. \end{aligned}$$

Notice that from Assumptions 2.2 and 2.3, it follows that all the functions used f, and F and G are bounded. That is, there exists a constant $M \geq 0$ such that for all $(z, y) \in D_1 \times D_2$,

$$\|f(z, y)\| \leq M, \qquad \|F(z, y)\| \leq M, \qquad \text{and} \qquad \|G(z)\| \leq M.$$

3 (z, λ)-Associated Games

Consider a family of infinite-horizon stochastic games, all with the same state and action spaces $\mathbf{X}$ and $\mathbf{A}$ as above and the same transition probabilities $\mathcal{P}$, parameterized by a pair $(z, \lambda) \in \mathbb{R}^n \times \mathbb{R}^n$. Let $r : \mathbb{R}^n \times \mathbb{R}^n \times \mathbf{X} \times \mathbf{A} \to \mathbb{R}$ be the immediate cost defined as follows:

$$r(z, \lambda, x, a) = F\big(z, g(x, a)\big) + \lambda^T f\big(z, g(x, a)\big). \tag{3}$$

With the same definition of the set of policies $\mathcal{U} = (U^1, U^2)$ as above, let

$$\overline{\sigma}(z, \lambda, x, u) := \varliminf_{m\to\infty} \frac{1}{m} E_x^u \sum_{i=0}^{m-1} r(z, \lambda, X_i, A_i). \tag{4}$$

A policy pair $u_{z,\lambda} = (u^1_{z,\lambda}, u^2_{z,\lambda}) \in \mathcal{U}$ is said to be a saddle point or an equilibrium policy pair for (z, λ)-associated game with infinite-horizon expected average cost criterion if, for all $u^1 \in U^1$, $u^2 \in U^2$,

$$\overline{\sigma}(z, \lambda, x, u^1, u^2_{z,\lambda}) \le \overline{\sigma}(z, \lambda, x, u^1_{z,\lambda}, u^2_{z,\lambda}) \le \overline{\sigma}(z, \lambda, x, u^1_{z,\lambda}, u^2).$$

Let $\hat{f}_{z,\lambda} = (\hat{f}^1_{z,\lambda}, \hat{f}^2_{z,\lambda})$, where $\hat{f}^1_{z,\lambda}, \hat{f}^2_{z,\lambda}$ are some stationary equilibrium policy pair for the expected average problem. The existence of such a stationary equilibrium policy pair is well known under our unichain assumption 2.1; see [15] and [17] (this extends to the countable case under simultaneous Doeblin condition introduced in [11, Section 11.1], with a communicating condition, or under contraction conditions, see, e.g., [7] and [3], respectively). The function

$$\overline{\sigma}(z, \lambda) := \overline{\sigma}(z, \lambda, x, \hat{f}^1_{z,\lambda}, \hat{f}^2_{z,\lambda}) \tag{5}$$

is then defined to be the value of the (z, λ)-associated game and is known to be independent on x (which we shall thus omit from the notation). It can be computed using value iteration (see, e.g., [19, ch. 13]).

Remark 3.1. In the proof of main results (Section 6), we will need a result of another class of games called "stochastic games with asymptotic average criteria" (see [5]) that consider first a finite m-stage game with the total expected average cost evaluated by

$$\sigma^m(z, \lambda, x, u) := \frac{1}{m} E_x^u \sum_{i=0}^{m-1} r(z, \lambda, X_i, A_i)$$

and then study the asymptotic behavior of its value as m goes to infinity. As shown in [6], the limit of the value of this m-stage game as m tends to infinity converges to $\overline{\sigma}(z, \lambda)$ defined in (5).

4 Limit Hamilton–Jacobi–Isaacs Equations for the Stochastic Hybrid Game

Let us consider Hamilton–Jacobi equations

$$-\frac{\partial B(t,z)}{\partial t} + H\left(z, \frac{\partial B(t,z)}{\partial z}\right) = 0, \qquad (t,z) \in [0,T) \times \mathbb{R}^n, \tag{6}$$

with Hamiltonian $H(z,\lambda)$ equal to $-\overline{\sigma}(z,\lambda)$ defined in (5). These equations will be referred to as limit Hamilton–Jacobi–Isaacs (LHJI) equations for the stochastic hybrid game. Let us denote by $B(t,z)$ the viscosity solutions (see Definition 6.1 ahead) of this equation which satisfy the boundary condition

$$B(T,z) = G(z), \qquad \forall z \in \mathbb{R}^n. \tag{7}$$

In the following sections, we establish that the value of our hybrid game converges to $B(t,z)$ as ϵ tends to 0.

As in stochastic hybrid, optimal control problems (e.g., [2], [16]), the above results can be considered a justification of a decomposition of the stochastic hybrid game into the associated fast game, allowing us to describe an asymptotically optimal behavior of the players if the slow parameters are fixed and the LHJI equations are responsible for a "near-optimality" of the slow dynamics.

5 Main Result

Our main result is now formulated as Theorem 5.1.

Theorem 5.1. *Let Assumption* 2.1–2.3 *be true. Let equation* (6) *with* $H(z,\lambda) = -\overline{\sigma}(z,\lambda)$ *have the unique continuous viscosity solution* $B(t,z)$ *satisfying the boundary condition* (7). *Then the stochastic hybrid game have a value in the limit. That is,*

$$\lim_{\epsilon \to 0} B_\epsilon(t,z,x) \stackrel{\text{def}}{=} B(t,z), \tag{8}$$

with the convergence being uniform on compact set $[0,T] \times D_1 \times \mathbf{X}$.

Remark 5.1. Conditions for the uniqueness of the viscosity solutions of the LHJI equation (6) satisfying boundary condition (7) can be found in [8, 12, 4].

The following equi-continuous type property of the value function is crucial in our consideration.

Lemma 5.1. *Corresponding to any compact set* $[0, T] \times D_1 \times \mathbf{X}$ *there exists continuous functions* $\omega(\alpha)$ *and* $\mu(\alpha)$ *tending to zero as* α *tends to zero such that for any* $(t^i, z^i, x^i) \in [0, T] \times D_1 \times \mathbf{X}$, $i = 1, 2$

$$\left|B_\epsilon(t^1, z^1, x^1) - B_\epsilon(t^2, z^2, x^2)\right| \leq \omega(|t^1 - t^2| + |z^1 - z^2|) + \mu(\epsilon), \tag{9}$$

with

$$B_\epsilon(T, z, x) = G(z), \quad \forall (z, x) \in D_1 \times \mathbf{X}. \tag{10}$$

This lemma is established in Appendix. Let us introduce the notation

$$V_\epsilon(t, z) \stackrel{\text{def}}{=} B_\epsilon(t, z, x^*)$$

where x^* is some fixed (but arbitrary) state. By Lemma 5.1, if (t, z, x) belongs to a compact set $[0, T] \times D_1 \times \mathbf{X}$, then

$$\left|B_\epsilon(t, z, x) - V_\epsilon(t, z)\right| \leq \mu(\epsilon). \tag{11}$$

Hence, to prove (8) it is sufficient to show that

$$\lim_{\epsilon \to 0} V_\epsilon(t, z) = B(t, z), \tag{12}$$

where the convergence is uniform with respect to (t, z) from any compact subset of $[0, T] \times \mathbb{R}^n$. For the sake of brevity we shall refer to this sort of convergence as U-convergence and the corresponding limits will be called U-limits.

From Lemma 5.1, it follows that for $(t^i, z^i) \in [0, T] \times D_1$, $i = 1, 2$

$$\left|V_\epsilon(t^1, z^1) - V_\epsilon(t^2, z^2)\right| \leq \omega\big(|t^1 - t^2| + |z^1 - z^2|\big) + \mu(\epsilon). \tag{13}$$

Lemma 5.2. *Given any sequence* ϵ_i *tending to zero, one can find a subsequence* $\epsilon_{i_l} = \epsilon_l$ *of this sequence such that there exists the* U*-limit*

$$\lim_{\epsilon_l \to 0} V_{\epsilon_l}(t, z) \stackrel{\text{def}}{=} V(t, z). \tag{14}$$

The proof of the lemma is also given in the Appendix.

Let us show that any function obtained as U-limit in (14) coincides with $B(t, z)$. Notice that, by (13), any such function $V(t, z)$ is continuous on $[0, T] \times \mathbb{R}^n$ and, by (10) and (11), it satisfies the condition

$$V(T, z) = G(z), \quad \forall z \in \mathbb{R}^n.$$

Thus, to show that it coincides with $B(t, z)$ it is enough to show that it is a viscosity solution of (6) with $H(z, \lambda) = -\overline{\sigma}(z, \lambda)$.

6 Proof of the Main Result

To begin this section, let us recall the definition of viscosity solutions.

Definition 6.1.

1. A function $V(t, z)$ is called a *viscosity subsolution* of (6) if
$$-\frac{\partial v(\bar{t}, \bar{z})}{\partial t} + H\left(\bar{z}, \frac{\partial v(\bar{t}, \bar{z})}{\partial z}\right) \leq 0,$$
for any $(\bar{t}, \bar{z}) \in [0, T) \times \mathbb{R}^n$ and for each function $v(t, z)$ that has continuous partial derivatives on $[0, T) \times \mathbb{R}^n$ and satisfies the conditions $v(\bar{t}, \bar{z}) = V(\bar{t}, \bar{z})$ and $v(t, z) \geq V(t, z)$ in some neighborhood of $(\bar{t}, \bar{z})$.
2. A function $V(t, z)$ is called a *viscosity supersolution* of (6) if
$$-\frac{\partial v(\bar{t}, \bar{z})}{\partial t} + H\left(\bar{z}, \frac{\partial v(\bar{t}, \bar{z})}{\partial z}\right) \geq 0,$$
for any $(\bar{t}, \bar{z}) \in [0, T) \times \mathbb{R}^n$ and for each function $v(t, z)$ that has continuous partial derivatives on $[0, T) \times \mathbb{R}^n$ and that satisfies the conditions: $v(\bar{t}, \bar{z}) = V(\bar{t}, \bar{z})$ and $v(t, z) \leq V(t, z)$ in some neighborhood of $(\bar{t}, \bar{z})$.
3. A function $V(t, z)$ that is both a viscosity sub- and supersolution is called a *viscosity solution* of eq. (6).

Proof of Theorem 5.1. We first note that the hybrid game has the value (see Lemma 7.1 in the last section). It allows us to consider the value function $B_\epsilon(t, z, x)$ instead of its upper and lower values when dealing with the dynamic programming approach.

Let $\Delta \stackrel{\text{def}}{=} \epsilon K(\epsilon)$ and $K(\epsilon)$ be such that
$$\lim_{\epsilon \to 0} \Delta = 0, \qquad \lim_{\epsilon \to 0} K(\epsilon) = \infty,$$
where $K(\epsilon)$ has taking integer values.

Take $\bar{t} \in [0, T]$ with initial state of the system $Z_{\bar{t}} = \bar{z}$ and initial state of the controlled Markov chain $X_{\lfloor \bar{t}/\epsilon \rfloor} = \bar{x}$. Then, the optimality principle implies that
$$\begin{aligned} B_\epsilon^{up}(\bar{t}, \bar{z}, \bar{x}) &= \inf_{u^2 \in U^2} \sup_{u^1 \in U^1} E_x^{(u^1, u^2)} \left\{ \int_{\bar{t}}^{T} F(Z_s, Y_s)\, ds + G(Z_T) \right\} \\ &= \inf_{u^2 \in U^2} \sup_{u^1 \in U^1} E_x^{(u^1, u^2)} \left\{ \int_{\bar{t}}^{\bar{t}+\Delta} F(Z_s, Y_s)\, ds \right. \\ &\qquad \left. + B_\epsilon\big(\bar{t} + \Delta, z(\bar{t} + \Delta), x(\bar{t} + \Delta)\big) \right\}, \end{aligned} \tag{15}$$
where $z(\bar{t} + \Delta) = Z_{\bar{t}+\Delta}$ and $x(\bar{t} + \Delta) = X_{\lfloor (\bar{t}+\Delta)/\epsilon \rfloor}$. Note that for brevity, we haven't distinguished the policy sets. It should be understood that U^1, U^2 in the second line of the above equation are restricted policy sets defined on interval $[\bar{t}, \bar{t} + \Delta]$.

Let $(\bar{t}, \bar{z}, \bar{x}) \in [0, T) \times D_1 \times \mathbf{X}$. Then, by Assumption 2.2, $\big(z(\bar{t} + \Delta), x(\bar{t} + \Delta)\big) \in D_1 \times \mathbf{X}$, where D_1 are compact sets in $\mathbb{R}^n$. Since the convergence in (14) is uniform with respect to (t, z) from any compact subset of $[0, T] \times \mathbb{R}^n$, there exists a function $\tilde{\nu}(\epsilon_l)$,

$$\lim_{\epsilon_l \to 0} \tilde{\nu}(\epsilon_l) = 0, \tag{16}$$

such that

$$|V_{\epsilon_l}(t, z) - V(t, z)| \leq \tilde{\nu}(\epsilon_l), \qquad \forall (t, z) \in [0, T] \times D_1.$$

Using this and (11), one obtains from (15)

$$V(\bar{t}, \bar{z}) = \inf_{u^2 \in U^2} \sup_{u^1 \in U^1} E_x^{(u^1, u^2)} \left\{ \int_{\bar{t}}^{\bar{t}+\Delta} F(Z_s, Y_s)\, ds + V\big(\bar{t} + \Delta, z(\bar{t} + \Delta)\big) \right\} + O(\tilde{\mu}(\epsilon_l)), \tag{17}$$

where

$$\tilde{\mu}(\epsilon_l) = \max\{\mu(\epsilon_l),\ \tilde{\nu}(\epsilon_l), \epsilon_l\}. \tag{18}$$

Now let $v(t, z)$ have continuous partial derivatives and satisfy the condition: $v(\bar{t}, \bar{z}) = V(\bar{t}, \bar{z})$ and $v(t, z) \geq V(t, z)$ for (t, z) in some neighborhood of $(\bar{t}, \bar{z})$. From (17), it then follows

$$v(\bar{t}, \bar{z}) \leq \inf_{u^2 \in U^2} \sup_{u^1 \in U^1} E_x^{(u^1, u^2)} \left\{ \int_{\bar{t}}^{\bar{t}+\Delta} F(Z_s, Y_s)\, ds + v\big(\bar{t} + \Delta, z(\bar{t} + \Delta)\big) \right\} + O(\tilde{\mu}(\epsilon_l)). \tag{19}$$

By definition,

$$z(\bar{t} + \Delta) = \bar{z} + \int_{\bar{t}}^{\bar{t}+\Delta} f(Z_s, Y_s)\, ds. \tag{20}$$

By Assumptions 2.2 and 2.3, the function f is continuous and its arguments belong to compact sets; the second term in the right-hand side of (20) is of the order $O\big(\Delta(\epsilon_l)\big)$. Thus, substituting (20) into (19) and taking into account that $v(t, z)$ has continuous partial derivatives, one obtains

$$\frac{\partial v(\bar{t}, \bar{z})}{\partial t} + \inf_{u^2 \in U^2} \sup_{u^1 \in U^1} E_x^{(u^1, u^2)} \times \left\{ \frac{1}{\Delta(\epsilon_l)} \int_{\bar{t}}^{\bar{t}+\Delta} \left[F(Z_s, Y_s) + \Big(\frac{\partial v(\bar{t}, \bar{z})}{\partial z} \Big)^T f(Z_s, Y_s) \right] ds \right\} + \frac{O\big(\tilde{\mu}(\epsilon_l)\big)}{\Delta(\epsilon_l)} + \frac{o\big(\Delta(\epsilon_l)\big)}{\Delta(\epsilon_l)} \geq 0. \tag{21}$$

Notice that for any $s \in [\bar{t}, \bar{t} + \Delta(\epsilon_l)]$,

$$Z_s = \bar{z} + \int_{\bar{t}}^{\bar{t}+\Delta(\epsilon_l)} f(Z_s, Y_s)\, ds.$$

Hence,

$$\|Z_s - \bar{z}\| \leq M\Delta(\epsilon_l).$$

This and Assumption 2.3 imply that

$$\frac{1}{\Delta(\epsilon_l)}\int_{\bar{t}}^{\bar{t}+\Delta}\left[F(Z_s, Y_s) + \left(\frac{\partial v(\bar{t},\bar{z})}{\partial z}\right)^T f(Z_s, Y_s)\right]ds$$
$$= \frac{1}{\Delta(\epsilon_l)}\int_{\bar{t}}^{\bar{t}+\Delta}\left[F(\bar{z}, Y_s) + \left(\frac{\partial v(\bar{t},\bar{z})}{\partial z}\right)^T f(\bar{z}, Y_s)\right]ds + O\big(\Delta(\epsilon_l)\big). \quad (22)$$

Let

$$m(\epsilon_l) = \left\lfloor \frac{\bar{t} + \Delta(\epsilon_l)}{\epsilon_l}\right\rfloor - \left\lfloor \frac{\bar{t}}{\epsilon_l}\right\rfloor,$$

and note that

$$\left| m(\epsilon_l) - \frac{\Delta(\epsilon_l)}{\epsilon_l}\right| \leq 1;$$

then

$$\left|\frac{1}{m(\epsilon_l)} - \frac{\epsilon_l}{\Delta(\epsilon_l)}\right| \leq \frac{\epsilon_l^2}{\Delta^2(\epsilon_l)}\left(\frac{1}{1 - \epsilon_l/\Delta(\epsilon_l)}\right).$$

From this and (3), it follows that there exist positive constants L_1 and L_2 such that

$$\left\| \frac{1}{\Delta(\epsilon_l)}\int_{\bar{t}}^{\bar{t}+\Delta(\epsilon_l)}\left[F(\bar{z}, Y_s) + \lambda^T f(\bar{z}, Y_s)\right]ds - \frac{\epsilon_l}{\Delta(\epsilon_l)}\sum_{i=\lfloor \bar{t}/\epsilon_l\rfloor+1}^{\lfloor \bar{t}/\epsilon_l\rfloor + m(\epsilon_l)} r(\bar{z}, \lambda, X_i, A_i)\right\| \leq L_1\frac{\epsilon_l}{\Delta(\epsilon_l)}$$

$$\left\| \frac{\epsilon_l}{\Delta(\epsilon_l)}\sum_{i=\lfloor \bar{t}/\epsilon_l\rfloor+1}^{\lfloor \bar{t}/\epsilon_l\rfloor + m(\epsilon_l)} r(\bar{z}, \lambda, X_i, A_i) - \frac{1}{m(\epsilon_l)}\sum_{i=\lfloor \bar{t}/\epsilon_l\rfloor+1}^{\lfloor \bar{t}/\epsilon_l\rfloor + m(\epsilon_l)} r(\bar{z}, \lambda, X_i, A_i)\right\| \leq L_2\frac{\epsilon_l}{\Delta(\epsilon_l)}. \quad (23)$$

Using (22) and (23), one may obtain from (21),

$$\frac{\partial v(\bar{t},\bar{z})}{\partial t} + \inf_{u^2\in U^2}\sup_{u^1\in U^1} E_x^{(u^1,u^2)}\frac{1}{m(\epsilon_l)}\sum_{i=\lfloor \bar{t}/\epsilon_l\rfloor+1}^{\lfloor \bar{t}/\epsilon_l\rfloor + m(\epsilon_l)} r\left(\bar{z}, \frac{\partial x(\bar{t},\bar{z})}{\partial z}, X_i, A_i\right)$$
$$+O\left(\frac{\epsilon_l}{\Delta(\epsilon_l)}\right) + O\big(\Delta(\epsilon_l)\big) + \frac{O\big(\tilde{\mu}(\epsilon_l)\big)}{\Delta(\epsilon_l)} + \frac{o\big(\Delta(\epsilon_l)\big)}{\Delta(\epsilon_l)} \geq 0. \quad (24)$$

Now, define now $\Delta(\epsilon_l)$ as follows:

$$\Delta(\epsilon_l) = \sqrt{\tilde{\mu}(\epsilon_l)} \Rightarrow \frac{O\big(\tilde{\mu}(\epsilon_l)\big)}{\Delta(\epsilon_l)} = O\Big(\sqrt{\tilde{\mu}(\epsilon_l)}\Big).$$

Hence, passing to the limit in (24) as ϵ_l tends to 0, and taking into account the associated game with the existence of the stationary equilibrium policy pair as stated in (5) and Remark 3.1, one obtains

$$\frac{\partial v(\bar{t},\bar{z})}{dt} + \overline{\sigma}\left(\bar{z}, \frac{\partial v(\bar{t},\bar{z})}{\partial z}\right) \geq 0 \Rightarrow -\frac{\partial v(\bar{t},\bar{z})}{dt} + H\left(\bar{z}, \frac{\partial v(\bar{t},\bar{z})}{\partial z}\right) \leq 0.$$

This establishes that $V(t,z)$ is a viscosity subsolution of (6) on $[0,T)\times\mathbb{R}^n$.

Similarly, letting $v(t,z)$ have continuous partial derivatives and satisfy the conditions $v(\bar{t},\bar{z}) = V(\bar{t},\bar{z})$ and $v(t,z) \leq V(t,z)$ in some neighborhood of $(\bar{t},\bar{z}) \in [0,T)\times\mathbb{R}^n$, one can obtain

$$-\frac{\partial v(\bar{t},\bar{z})}{dt} + H\left(\bar{z}, \frac{\partial v(\bar{t},\bar{z})}{\partial z}\right) \geq 0,$$

which means that $V(t,z)$ is a viscosity super-solution of (6) on $[0,T)\times\mathbb{R}^n$. Thus, $V(t,z)$ is a viscosity solution (6) on $[0,T)\times\mathbb{R}^n$ and, consequently, it coincides with $B(t,z)$.

This proves that $B_\epsilon(t,z,x)$ U-converges (as ϵ tends to 0) to $B(t,z)$ since, otherwise, by Lemma 5.2, one would be able to choose a subsequence ϵ_l tending to 0 such that the U-limit (14) does not coincide with $B(t,z)$. □

7 Appendix

In this appendix we first present some general properties of the original game with fixed ϵ, which allows us to obtain some properties of the limit game. We show that the original game is equivalent to a *stochastic game with finite state and action spaces*. This will allow us to use standard results to obtain the representation of the value and optimal policies.

7.1 An Equivalent Stochastic Game

Lemma 7.1. *For a fixed ϵ, the original hybrid game is equivalent to a finite-stage stochastic (Markov) game with finite state and action spaces and has a value $B_\epsilon(t,z,x)$. That is,*

$$B_\epsilon(t,z,x) = B_\epsilon^{\mathrm{up}}(t,z,x) = B_\epsilon^{\mathrm{lo}}(t,z,x).$$

Let's look at the following stochastic game:

- *State space*: consists of the *histories*

$$\mathcal{X} := \bigcup_{l=0}^{\lfloor (T-t)\epsilon^{-1} \rfloor} \mathbf{H}_l, \quad \text{where } \mathbf{H}_l := \left\{(x_0, a_0^1, a_0^2, x_1, a_1^1, a_1^2, \ldots, x_l)\right\}.$$

An element of the state space is denoted by h; $n(h)$ denotes the length of the horizon.

- *Action spaces*: unchanged, i.e., $\mathbf{A} = \mathbf{A}^1 \times \mathbf{A}^2$.
- *Transition probabilities* are obvious; for

  $h^1 = \{(x_0, a_0^1, a_0^2, x_1, a_1^1, a_1^2, \ldots, x_l)\}, h^2 = \{(y_0, b_0^1, b_0^2, y_1, b_1^1, b_1^2, \ldots, y_k)\}$,

  we have

$$\hat{\mathcal{P}}_{h^1,\alpha^1,\alpha^2,h^2} = \mathcal{P}_{x_l,\alpha^1,\alpha^2,y_k},$$

  for

$$\begin{aligned} k &= l+1, & x_0 &= y_0, & a_0^1 &= b_0^1, & a_0^2 &= b_0^2, \\ x_1 &= y_1, \ldots, x_l = y_l, & & & b_k^1 &= \alpha^1, & b_k^2 &= \alpha^2, \end{aligned}$$

  and 0 otherwise.
- *Immediate costs*:

$$c(t, z, h; a^1, a^2) = \int_{\epsilon n(h)}^{\epsilon n(h)+\epsilon} F(Z_s, Y_s)\, ds,$$

  for $n(h) < \lfloor (T-t)\epsilon^{-1} \rfloor$, and

$$c(t, z, h; a^1, a^2) = \int_{\epsilon n(h)}^{T} F(Z_s, Y_s)\, ds + G(Z_T),$$

  for $n(h) = \lfloor (T-t)\epsilon^{-1} \rfloor$.
  Note that the immediate cost is parametrized by the initial z and t. We did not write the immediate cost explicitly; however, the random variables Z_s, Y_s, and Z_T appearing in the immediate cost are fully determined by h and the actions a^1, a^2.

Let's define the payoff of the new game: for any h such that $n(h) \le \lfloor (T - t)\epsilon^{-1} \rfloor$, we set

$$\hat{J}_\epsilon(t, z, h; v^1, v^2) = E_h^{(v^1,v^2)} \left\{ \int_{t+\epsilon n(h)}^{T} F(Z_s, Y_s)\, ds + G(Z_T) \right\}.$$

Note that each policy u^i for player i in the original game has an obvious equivalent Markov policy v^i in this new game that achieves the same costs. It is thus simple to show that one may restrict to Markov policies in the new game (optimal Markov policies will depend, of course, on z and t). The original policies generate the same costs in the original game as their equivalent new policies in the new game:

$$\hat{J}_\epsilon(t, z, h; v^1, v^2) = J_\epsilon(t, z, x; u^1, u^2),$$

where $h = x$.

Since the new game is a standard stochastic game with a finite number of states and actions, it has a *value* (see, e.g., Van Der Wal [19, ch. 10]). We conclude that the lower and upper values in the original game coincide and are equal to this value. Note also that dynamic programming can be used to characterize the value and optimal policies for both players.

7.2 Proof of Lemma 5.1

It follows from arguments as in [9] that there exists some real number ω such that for any policies u^1 and u^2 for the two players and any z^1, z^2, t^1, t^2, and x,

$$|J_\epsilon(t^1, z^1, x : u^1, u^2) - J_\epsilon(t^2, z^2, x : u^1, u^2)| \leq \omega(|t^1 - t^2| + |z^1 - z^2|) + O(\epsilon). \tag{25}$$

This implies that (9) holds for the case where $x^1 = x^2$. To conclude the proof, it thus suffices to show that for any z and t,

$$|J_\epsilon(t, z, x^1; u^1, u^2) - J_\epsilon(t, z, x^2; u^1, u^2)| < \mu(\epsilon),$$

where μ is as in Lemma 5.1. We do this next. Choose some arbitrary x^* and denote $\eta = \inf\{n : X_n = x^*\}$. Then

$$\sup_{u^1, u^2, x} E_x^{u^1, u^2} \eta < \infty. \tag{26}$$

Indeed, there exists a pure stationary pair u^1, u^2 that achieves this sup, since achieving the sup is equivalent to a problem of maximizing some total expected cost in a transient MDP with finite state and action spaces (see, e.g., [13] or [11]). The fact that the sup is finite follows from the unichain assumption 2.1.

Note that

$$\begin{aligned}
&B_\epsilon(t^1, z^1, x) - B_\epsilon(t^2, z^2, x) \\
&\quad = \sup_{u^1 \in U^1} \inf_{u^2 \in U^2} J_\epsilon(t^1, z^1, x; u^1, u^2) - \sup_{u^1 \in U^1} \inf_{u^2 \in U^2} J_\epsilon(t^2, z^2, x; u^1, u^2) \\
&\quad \leq \sup_{u^1 \in U^1} \left(\inf_{u^2 \in U^2} J_\epsilon(t^1, z^1, x; u^1, u^2) - \inf_{u^2 \in U^2} J_\epsilon(t^2, z^2, x; u^1, u^2) \right) \\
&\quad \leq \sup_{u^1, u^2} |J_\epsilon(t^1, z^1, x; u^1, u^2) - J_\epsilon(t^2, z^2, x; u^1, u^2)|.
\end{aligned}$$

Since the same holds for $B_\epsilon(t^2, z^2, x) - B_\epsilon(t^1, z^1, x)$, we conclude from the last equation and from (25) that

$$|B_\epsilon(t^1, z^1, x) - B_\epsilon(t^2, z^2, x)| \leq \omega(|t^1 - t^2| + |z^1 - z^2|) + O(\epsilon). \tag{27}$$

Denote $\tau = \min(T, t + \eta\epsilon)$ and $\sigma = t + \eta\epsilon$. Now, the optimality principle implies that

$$\begin{aligned}
&B_\epsilon(t, z, x) \\
&\quad = \sup_{u^1 \in U^1} \inf_{u^2 \in U^2} E_x^{u^1, u^2} \left(\int_t^T F(Z_s, Y_s)\, ds + G(Z_T) \right) \\
&\quad = \sup_{u^1 \in U^1} \inf_{u^2 \in U^2} E_x^{u^1, u^2} \left(\int_t^\tau F(Z_s, Y_s)\, ds + \int_\tau^T F(Z_s, Y_s)\, ds + G(Z_T) \right)
\end{aligned}$$

$$= \sup_{u^1 \in U^1} \inf_{u^2 \in U^2} E_x^{u^1,u^2} \left(\int_t^\tau F(Z_s, Y_s)\,ds + G(Z_T) 1\{\sigma > T\} \right.$$
$$\left. + B_\epsilon(\tau, Z_\tau, x^*) 1\{\sigma \le T\} \right).$$

Thus,

$$\begin{aligned}
&B_\epsilon(t, z, x^1) - B_\epsilon(t, z, x^2) \\
&\quad = \sup_{u^1 \in U^1} \inf_{u^2 \in U^2} E_{x^1}^{u^1,u^2} \Big(\int_t^\tau F(Z_s, Y_s)\,ds \\
&\qquad\qquad + G(Z_T) 1\{\sigma > T\} + B_\epsilon(\tau, Z_\tau, x^*) 1\{\sigma \le T\} \Big) \\
&\quad\quad - \sup_{u^1 \in U^1} \inf_{u^2 \in U^2} E_{x^2}^{u^1,u^2} \Big(\int_t^\tau F(Z_s, Y_s)\,ds + G(Z_T) 1\{\sigma > T\} \\
&\qquad\qquad + B_\epsilon(\tau, Z_\tau, x^*) 1\{\sigma \le T\} \Big) \\
&\quad \le \sup_{u^1,u^2} E_{x^1}^{u^1,u^2} \left| \int_t^\tau F(Z_s, Y_s)\,ds + G(Z_T) 1\{\sigma > T\} \right| \\
&\quad\quad + \sup_{u^1,u^2} E_{x^2}^{u^1,u^2} \left| \int_t^\tau F(Z_s, Y_s)\,ds + G(Z_T) 1\{\sigma > T\} \right| \\
&\quad\quad + \sup_{u^1,u^2} \left| E_{x^1}^{u^1,u^2} B_\epsilon(\tau, Z_\tau, x^*) 1\{\sigma \le T\} - E_{x^2}^{u^1,u^2} B_\epsilon(\tau, Z_\tau, x^*) 1\{\sigma \le T\} \right|.
\end{aligned}$$

Since the same bound also holds for $B_\epsilon(t, z, x^2) - B_\epsilon(t, z, x^1)$, we conclude that

$$\left| B_\epsilon(t, z, x^1) - B_\epsilon(t, z, x^2) \right| \tag{28}$$
$$\le 2 \sup_{u^1,u^2,x} \left| E_x^{u^1,u^2} \left\{ \int_t^\tau F(Z_s, Y_s)\,ds + G(Z_T) 1\{\sigma > T\} \right\} \right| \tag{29}$$
$$+ \sup_{u^1,u^2} \left| E_{x^1}^{u^1,u^2} B_\epsilon(\tau, Z_\tau, x^*) 1\{\sigma \le T\} - E_{x^2}^{u^1,u^2} B_\epsilon(\tau, Z_\tau, x^*) 1\{\sigma \le T\} \right|. \tag{30}$$

The first term above is $O(\epsilon)$ because F and G are bounded, since $\tau \le t + \eta\epsilon$ and due to (26). Next we bound the second term. We have

$$\begin{aligned}
&\left| E_{x^1}^{u^1,u^2} B_\epsilon(\tau, Z_\tau, x^*) 1\{\sigma \le T\} - E_{x^2}^{u^1,u^2} B_\epsilon(\tau, Z_\tau, x^*) 1\{\sigma \le T\} \right| \\
&\quad = \Big| E_{x^1}^{u^1,u^2} B_\epsilon(\tau, Z_\tau, x^*) - E_{x^2}^{u^1,u^2} B_\epsilon(\tau, Z_\tau, x^*) \\
&\qquad - E_{x^1}^{u^1,u^2} B_\epsilon(\tau, Z_\tau, x^*) 1\{\sigma > T\} + E_{x^2}^{u^1,u^2} B_\epsilon(\tau, Z_\tau, x^*) 1\{\sigma > T\} \Big| \\
&\quad \le \left| E_{x^1}^{u^1,u^2} B_\epsilon(\tau, Z_\tau, x^*) - E_{x^2}^{u^1,u^2} B_\epsilon(\tau, Z_\tau, x^*) \right| \\
&\qquad + \sup_{x,u^1,u^2} 2M(T - \tau + 1) P_x^{u^1,u^2}(\sigma > T).
\end{aligned}$$

Due to (26), it follows that $P_x^{u^1,u^2}(\sigma > T)$ is of the order of $O(\epsilon)$. It remains to estimate the first term in the right-hand side of the above inequality.

Now consider an arbitrary augmented probability space on which the two state and action trajectories are defined simultaneously: those that start from initial states x^1 and x^2, respectively, and for which the marginal distribution of each trajectory separately is given by the corresponding probabilities $P_{x^1}^{u^1,u^2}$ and $P_{x^2}^{u^1,u^2}$, respectively. Let $\overline{P}_{x^1,x^2}^{u^1,u^2}$ be the probability measure governing the augmented probability space, and we denote by $\overline{E}_{x^1,x^2}^{u^1,u^2}$ the corresponding expectation.

Let t^1 and t^2 be the times at which we reach state x^* starting from states x^1 and state x^2, respectively. Let z^1 and z^2 be the value of the z-variable at those instances. Then by (27) we have

$$
\begin{aligned}
&|E_{x^1}^{u^1,u^2} B_\epsilon(\tau, Z_\tau, x^*) - E_{x^2}^{u^1,u^2} B_\epsilon(\tau, Z_\tau, x^*)| \\
&\quad = |\overline{E}_{x^1,x^2}^{u^1,u^2}(B_\epsilon(t^1, z^1, x^*) - B_\epsilon(t^2, z^2, x^*))| \le \omega \overline{E}_{x^1,x^2}^{u^1,u^2}(|t^1 - t^2| + |z^1 - z^2|).
\end{aligned}
$$

We also have

$$
\begin{aligned}
\sup_{x^1,x^2,u^1,u^2} \overline{E}_{x^1,x^2}^{u^1,u^2}|t^1 - t^2| \le &\sup_{x^1,x^2,u^1,u^2} \overline{E}_{x^1,x^2}^{u^1,u^2}|t^1 - t| \\
&+ \sup_{x^1,x^2,u^1,u^2} \overline{E}_{x^1,x^2}^{u^1,u^2}|t^2 - t| \le 2\epsilon \sup_{u^1,u^2,x} E_x^{u^1,u^2}\eta.
\end{aligned}
$$

Due to the bounded function f, one may get

$$
\sup_{x^1,x^2,u^1,u^2} E_x^{u^1,u^2}|z^1 - z^2| \le 2M\epsilon \sup_{u^1,u^2,x} E_x^{u^1,u^2}\eta.
$$

Thus, we see that (30) is bounded by $2\epsilon(M+1)\sup_{u^1,u^2,x} E_x^{u_1,u_2}\eta$. This concludes the proof. □

Note that, in the preceding proof, in order to be able to compute expressions such as $E(t^1 - t^2)$, we couple in some sense two systems that start in different initial states. However, we did not have to make any particular assumption on the joint distribution between those two systems in order to obtain the required bounds.

7.3 Proof of Lemma 5.2

First, one can establish that, given a compact subset $[0, T] \times D$, $D \subset \mathbb{R}^n$, the following space

$$
\mathcal{V} = \left\{ V_\epsilon(t, z),\ \epsilon \to 0 \right\},
$$

where $V_\epsilon(t, z)$ are continuous functions on $[0, T] \times D$ satisfying property (13) under sup-metric, is complete and totally bounded. Thus it is compact. Hence, every sequence in $\mathcal{V}$ in the form $\left\{V_{\epsilon_n},\ n \to \infty\right\}$ contains a subsequence $\left\{V_{\epsilon_l}\right\}$ such that the convergence (14) will be uniform with respect to $(t, z) \in [0, T] \times D$.

Notice that our claim is stronger than that since U-convergence is uniform with respect to (t, z) from *any compact subset* of $[0, T] \times \mathbb{R}^n$. So, we just remove the dependence of compact set D by choosing a sequence of subsets $[0, T] \times D^l$, where $D^l \subset \mathbb{R}^n$ is the closed ball centered at 0 and having the radius $l, l = 1, 2, \ldots$. Then, using a diagolization procedure, one can construct a subsequence providing the required U-convergence. □

Acknowledgments

The work of M.-T. Nguyen and V. Gaitsgory was supported by the Australian Research Council grant (ARC A69703141).

REFERENCES

[1] Altman E. and Gaitsgory, V. A hybrid (differential-stochastic) zero-sum game with fast stochastic part, in *Preprint Volume Sixth Int. Symp. Dynamic Games and Appl.*, pp. 475–480, St-Jovite, Quebec, July 13–15, 1994.

[2] Altman, E. and Gaitsgory, V. Asymptotic optimization of a nonlinear hybrid system governed by a markov decision process, *SIAM J. Cont. Opt.*, **35** (6):2070–2085, 1997.

[3] Altman, E., Hordijk, A., and Spieksma, F. M. Contraction conditions for average and α-discount optimality in countable state markov games with unbounded rewards, *MOR*, **22**, pp. 588–618, 1997.

[4] Bardi, M. and Capuzzo-Dolcetta, I. *Optimal Control and Viscosity Solutions of Hamilton–Jacobi–Bellman Equations*, Birkhäuser, Boston, 1997.

[5] Bewley, T. and Kohlberg, E. The asymptotic theory of stochastic games, *Math. Oper. Res.*, **1** (3):197–208, 1976.

[6] Bewley, T. and Kohlberg, E. On stochastic games with stationary optimal strategies. *Math. Oper. Res.*, **3** (2):104–125, 1978.

[7] Federgruen, A. On N-person stochastic games with denumerable state space, *Adv. Appl. Prob.*, **10**, pp. 452–471, 1978.

[8] Fleming, W. H. and Soner, H. M. *Controlled Markov Processes and Viscosity Solutions*. Springer-Verlag, New York, 1993.

[9] Gaitsgory, V. Limit Hamilton–Jacobi–Isaacs equations for singularly perturbed zero-sum differential games, *Math. Anal. Appl.*, **202** (0352):862–899, 1996.

[10] Hinderer, K. *Foundations of Non-Stationary Dynamic Programming with Discrete Time Parameter*, vol. **33** of *Lecture Notes in Operations Research and Mathematical Systems*, Springer-Verlag, New York, 1970.

[11] Hordijk, A. *Dynamic Programming and Markov Potential Theory*, 2nd ed., Mathematical Centre Tracts **51**, Math. Cntr., Amsterdam, 1977.

[12] Ishii, H. Uniqueness of unbounded viscosity solutions of Hamilton–Jacobi equations, *Indiana Univ. Math. J.*, **26**, pp. 721–748, 1984.

[13] Kallenberg, L. C. M. *Linear Programming and Finite Markovian Control Problems*. Mathematical Centre Tracts **148**, Mathe. Cntr., Amsterdam, 1983.

[14] Kleinrock, L. *Queuing Systems, Vol. II: Computer Applications*, John Wiley, New York, 1976.

[15] Rogers, P. Nonzero-Sum Stochastic Games, Technical Report Report ORC 69-8, Berkeley, Univ. California, Operation Research Center, 1969.

[16] Shi, P., Altman, E., and Gaitsgory, V. On asymptotic optimization for a class of nonlinear stochastic hybrid systems, *Math. Meth. Oper. Res.*, **47** (2):229–315, 1998.

[17] Sobel, M. Noncooperative stochastic games, *Ann. Math. Stat.*, **42** pp. 1930–1935, 1971.

[18] Sethi, S. P. and Zhang, Q. *Hierarchical Decision Making in Stochastic Manufacturing Systems*, Birkhauser, Boston, Basel, Berlin, 1994.

[19] Van Der Wal, J. *Stochastic Dynamic Programming: Successive Approximations and Nearly Optimal Strategies for Markov Decision Processes and Markov Games*, Mathematical Centre Tracts **139**. Math. Cntr., Amsterdam, 1981.

On Multichain Markov Games

Heinz-Uwe Küenle
Brandenburgische Technische Universitäet Cottbus
Institut für Mathematik
Cottbus, Germany

Abstract

Two-person, zero-sum Markov games with arbitrary (Borel) state and action spaces, unbounded stage costs, and the average cost criterion are considered. The assumption on the transition probabilities implies some n-stage contraction property and lets the Markov chain of the states under a given strategy pair have several periodic recurrence classes, but the recurrence structure of all possible resulting Markov chains is identical. Under this assumption, some results are presented concerning the existence of ε-optimal strategies.

1 Introduction

The two-person stochastic game can be described in the following way: the state x_n of a dynamic system is periodically observed at times $n = 1, 2, \ldots$. After an observation at time n, the first player chooses an action a_n from the action set $\boldsymbol{A}(x_n)$, after, the second player chooses an action b_n from the action set $\boldsymbol{B}(x_n)$ dependent on the complete history of the system at this time. The first player must pay cost $k^1(x_n, a_n, b_n)$; the second player must pay $k^2(x_n, a_n, b_n)$; and the system moves to a new state x_{n+1} from the state space $\mathbf{X}$ according to the transition probability $p(\cdot \mid x_n, a_n, b_n)$.

In this paper we consider stochastic games with standard Borel state space, standard Borel action spaces, unbounded stage costs, and an average cost criterion. The assumption on the transition probabilities implies some n-stage contraction property and allows for the Markov chain of the states under a given strategy pair to have several periodic recurrence classes, but the recurrence structure of all possible resulting Markov chains is identical. We prove that the average costs of a stationary strategy pair are the unique solution of a certain fixed-point equation. For a Markov game with perfect information, we show under the assumption of countable action space of one player that the optimality equation has a unique solution and that ε-optimal stationary deterministic strategies exist for every $\varepsilon > 0$.

Several authors consider stochastic games with Borel state space. Related results are given by Maitra and Sudderth [10], [11], [12], Nowak [13], Rieder [15], and Küenle [6] in the case of bounded costs (payoffs). The case of unbounded payoffs is treated by Nowak [14] under a certain ergodicity assumption. Kurano [8], [9] considers Markov decision processes where the assumptions concerning the transition probabilities are weaker than the assumptions in this paper, but bounded stage costs are assumed in his papers . (Nowak [14] gives many additional comments on the existing literature on related topics.)

The paper is organized as follows: Section 2 presents the mathematical model of Markov games with arbitrary state and action spaces. Section 3 contains the assumptions on the transition probabilities and the stage costs as well as some preliminary results on the average costs. A fixed-point equation for the average costs of a stationary strategy pair is also presented in this section. In Section 4 we show the solvability of the optimality equation and the existence of ε-optimal stationary strategies under certain additional assumptions.

2 The Mathematical Model

Stochastic games considered in this paper are defined by 10 objects, as follows.

Definition 2.1. $\mathcal{M} = ((\mathbf{X}, \sigma_{\mathbf{X}}), (\mathbf{A}, \sigma_{\mathbf{A}}), \boldsymbol{A}, (\mathbf{B}, \sigma_{\mathbf{B}}), \boldsymbol{B}, p, k^1, k^2, \mathbf{E}, \mathbf{F})$ is called a *Markov game* if the elements of this tuple have the following meaning:

- $(\mathbf{X}, \sigma_{\mathbf{X}})$ is a standard Borel space, called the *state space*.
- $(\mathbf{A}, \sigma_{\mathbf{A}})$ is a standard Borel space and $\boldsymbol{A} : \mathbf{X} \to \sigma_{\mathbf{A}}$ is a set-valued map that has a $(\sigma_{\mathbf{X}} - \sigma_{\mathbf{A}})$-measurable selector. $\mathbf{A}$ is called the *action space of the first player*, and $\boldsymbol{A}(x)$ is called the *admissible action set of the first player at state* $x \in \mathbf{X}$.
  We assume that $\{(x, a) : x \in \mathbf{X}, a \in \boldsymbol{A}(x)\} \subseteq \sigma_{\mathbf{X}\times\mathbf{A}}$.
- $(\mathbf{B}, \sigma_{\mathbf{B}})$ is a standard Borel space, and $\boldsymbol{B} : \mathbf{X}\times\mathbf{A} \to \sigma_{\mathbf{B}}$ is a set-valued map that has a $(\sigma_{\mathbf{X}} - \sigma_{\mathbf{B}})$-measurable selector. $\mathbf{B}$ is called the *action space of the second player*, and $\boldsymbol{B}(x)$ is called the *admissible action set of the second player at state* $x \in \mathbf{X}$.
  We assume that $\{(x, b) : x \in \mathbf{X}, a \in \boldsymbol{B}(x)\} \subseteq \sigma_{\mathbf{X}\times\mathbf{B}}$.
- p is a transition probability from $\sigma_{\mathbf{X}\times\mathbf{A}\times\mathbf{B}}$ to $\sigma_{\mathbf{X}}$, the *transition law*.
- k^i, $i = 1, 2$, are $\sigma_{\mathbf{X}\times\mathbf{A}\times\mathbf{B}}$-measurable functions, called *stage cost functions*.
- Assume that $(\mathbf{Y}, \sigma_{\mathbf{Y}})$ is a standard Borel space. Then we denote by $\overline{\sigma}_{\mathbf{Y}}$ the σ-algebra of the $\sigma_{\mathbf{Y}}$-universally measurable sets. Let $\mathbf{H}_n := (\mathbf{X} \times \mathbf{A} \times \mathbf{B})^n \times \mathbf{X}$ for $n \geq 1$, $\mathbf{H}_0 := \mathbf{X}$. $h \in \mathbf{H}_n$ is called the *history at time* n.
  A transition probability π_n from $\overline{\sigma}_{\mathbf{H}_n}$ to $\overline{\sigma}_{\mathbf{A}}$ with $\pi_n(\boldsymbol{A}(x_n) \mid x_0, a_0, b_0, \ldots, x_n) = 1$ for all $(x_0, a_0, b_0, \ldots, x_n) \in \mathbf{H}_n$ is called a *decision rule of the first player at time* n.
  A transition probability ρ_n from $\overline{\sigma}_{\mathbf{H}_n\times\mathbf{A}}$ to $\overline{\sigma}_{\mathbf{B}}$ with $\rho_n(\boldsymbol{B}(x_n) \mid x_0, a_0, b_0, \ldots, x_n, a_n) = 1$ for all $(x_0, a_0, b_0, \ldots, x_n, a_n) \in \mathbf{H}_n \times \mathbf{A}$ is called a *decision rule of the second player at time* n.
  A decision rule of the first [second] player is called *Markov* iff a transition probability $\tilde{\pi}_n$ from $\overline{\sigma}_{\mathbf{X}}$ to $\overline{\sigma}_{\mathbf{A}}$ [$\tilde{\rho}_n$ from $\overline{\sigma}_{\mathbf{X}\times\mathbf{A}}$ to $\overline{\sigma}_{\mathbf{B}}$] exists with $\pi_n(\cdot \mid x_0, a_0, b_0, \ldots, x_n) = \tilde{\pi}_n(\cdot \mid x_n)$ [$\rho_n(\cdot \mid x_0, a_0, b_0, \ldots, x_n, a_n) = \tilde{\rho}_n(\cdot \mid$

$x_n, a_n)]$ for all $(x_0, a_0, b_0, \ldots, x_n, a_n) \in \mathbf{H}_n \times \mathbf{A}$. (Note: we identify π_n as $\tilde{\pi}_n$ and ρ_n as $\tilde{\rho}_n$.)
E and **F** denote nonempty sets of Markov decision rules.

A decision rule of the first [second] player is called *deterministic* iff a function e_n: $\mathbf{H}_n \to \mathbf{A}$ [f_n: $\mathbf{H}_n \times \mathbf{A} \to \mathbf{B}$] exists with $\pi_n(e_n(h_n) \mid h_n) = 1$ for all $h_n \in \mathbf{H}_n$ [$\rho_n(f_n(h_n, a_n) \mid h_n, a_n) = 1$ for all $(h_n, a_n) \in \mathbf{H}_n \times \mathbf{A}$]. In this case we put also $\delta_{e_n} := \pi_n$ and $\delta_{f_n} := \rho_n$.

A sequence $\Pi = (\pi_n)$ or $P = (\rho_n)$ of decision rules of the first or second player is called a *strategy* of that player. Strategies are called *deterministic*, or *Markov*, iff all their decision rules have the corresponding property. A Markov strategy $\Pi = (\pi_n)$ or $P = (\rho_n)$ is called *stationary* iff $\pi_0 = \pi_1 = \pi_2 = \ldots$ or $\rho_0 = \rho_1 = \rho_2 = \ldots$. (Note: $\Pi = \pi^\infty$ or $P = \rho^\infty$.) We assume in this paper that the sets of all admissible strategies are $\mathbf{E}^\infty$ and $\mathbf{F}^\infty$. Hence, only Markov strategies are allowed. But by means of dynamic programming methods, it is possible to get corresponding results also for Markov games with larger sets of admissible strategies. If **E** and **F** are the sets of all Markov decision rules (in the preceding sense), then we have a Markov game with perfect (or complete) information. In this case the action set of the second player may also depend on the present action of the first player. If **E** is the set of all Markov decision rules but **F** is the set of all Markov decision rules that do not depend on the present action of the first player, then we have a usual Markov game with independent action choice. Let

$$\Omega = \mathbf{X}\times\mathbf{A}\times\mathbf{B}\times\mathbf{X}\times\mathbf{A}\times\mathbf{B}\times\ldots \qquad \text{and} \qquad \mathcal{A} = \sigma_{\mathbf{X}}\otimes\sigma_{\mathbf{A}}\otimes\sigma_{\mathbf{B}}\otimes\sigma_{\mathbf{X}}\otimes\sigma_{\mathbf{A}}\otimes\sigma_{\mathbf{B}}\otimes\ldots.$$

Then for every $x \in \mathbf{X}$ and every strategy pair (Π, P), $\Pi = (\pi_n)$, $P = (\rho_n)$, there is a unique probability measure $\mathsf{P}_{x,\Pi,P}$ on $\mathcal{A}$ defined by the transition probabilities, π_n, ρ_n, and p (see [4]). Let $K^{i,N}(\omega) = \sum_{j=0}^{N} k^i(x_j, a_j, b_j)$ for $\omega = (x_0, a_0, b_0, x_1, \ldots)$. We set

$$V_{\Pi P}^{i,N}(x) = \int_\Omega K^{i,N}(\omega)\mathsf{P}_{x,\Pi,P}(d\omega) \tag{1}$$

and

$$\Phi_{\Pi P}^{i}(x) = \liminf_{N\to\infty} \frac{1}{N+1} V_{\Pi P}^{i,N}(x) \tag{2}$$

if the corresponding integral exists.

Definition 2.2. Let $\varepsilon \geq 0$. A strategy pair (Π^*, P^*) is called an *ε-equilibrium pair* iff

$$\Phi_{\Pi^* P^*}^{1} \leq \Phi_{\Pi P^*}^{1} + \varepsilon,$$
$$\Phi_{\Pi^* P^*}^{2} \leq \Phi_{\Pi^* P}^{2} + \varepsilon,$$

for all strategy pairs (Π, P). A 0-equilibrium pair is called a *Nash equilibrium pair.*

In this paper we will especially consider zero-sum Markov games, that is, where $k^1 = -k^2$. In this case we call an ε-equilibrium pair also an *ε-optimal strategy pair*. We set $k = k^1$, $V_{\Pi P}^N = V_{\Pi P}^{1,N}$, and $\Phi_{\Pi P}^N = \Phi_{\Pi P}^{1,N}$.

3 Assumptions and Preliminary Results

We first present in this section the assumptions concerning the transition probabilities. Let (Π, P) be a Markovian strategy pair. Then $p_{\Pi P}^{(n)}$ denotes the n-step transition probability of the state under this strategy pair, that means $p_{\Pi P}^{(n)}(C \mid x) = \text{Prob}(\hat{x}_n \in C \mid \hat{x}_0 = x)$ for all $x \in \mathbf{X}$, $C \in \overline{\sigma}_{\mathbf{X}}$, where $\hat{x}_j$ is the random state at time j. We put $p_{\Pi P} := p_{\Pi P}^{(1)}$.

Assumption 3.1. There is a measurable partition $\mathbf{D}_0, \ldots, \mathbf{D}_m$ of $\mathbf{X}$ $(m \in \mathbb{N})$ with the following properties.

(a) There exist a natural number $\kappa \geq 1$ and measures μ_i $(i = 1, \ldots, m)$ on $\mathbf{X}$ with

$$p_{\Pi P}^{(\kappa)}(\cdot \mid x) \geq \mu_i \qquad \text{and} \qquad \mu_i(\mathbf{X}) = \mu_i(\mathbf{D}_i) > 0,$$

for all $x \in \mathbf{D}_i$, $\Pi \in \mathbf{E}^\infty$, $P \in \mathbf{F}^\infty$, and $i = 1, \ldots, m$.

(b)
$$p_{\Pi P}^{(\kappa)}(\overline{\mathbf{D}}_i \mid x) = 0,$$

for all $\Pi \in \mathbf{E}^\infty$, $P \in \mathbf{F}^\infty$, and all $x \in \mathbf{D}_i$, $i = 1, \ldots, m$.

A typical situation in which Assumption 3.1 may be fulfilled is the following: for every strategy pair, the corresponding Markov chain has the same set $\mathbf{D}_0$ of transient states and the same recurrent sets, which consist of several cyclic sets $\mathbf{D}_j$, $j \geq 1$. For $\mathbf{D}_1 = \mathbf{X}$ and $\kappa = 1$, Assumption 3.1 is equivalent to the strong ergodicity assumptions used, for instance, in [3], [6], and [18].

Lemma 3.1. *Let $j \in \{1, \ldots, m\}$, $\Pi \in \mathbf{E}^\infty$, and $P \in \mathbf{F}^\infty$. There is a finite sequence $j_0, j_1, \ldots, j_\kappa$, with $j_0 = j_\kappa = j$ and*

$$p_{\Pi P}(\mathbf{D}_{j_i} \mid x) = 1 \textit{ for all } x \in \mathbf{D}_{j_{i-1}}, \tag{3}$$
$$p_{\Pi P}^{(i)}(\mathbf{D}_{j_i} \mid x) = 1 \textit{ for all } x \in \mathbf{D}_j. \tag{4}$$

Furthermore,

$$p_{\Pi P}(\mathbf{D}_i \mid x) \in \{0, 1\} \quad \textit{for all } x \in \mathbf{D}_j, \quad i, j \in \{0, \ldots, m\}, \quad j \geq 1. \tag{5}$$

Proof. Let $\Pi = (\pi_n) \in \mathbf{E}^\infty$, $P = (\rho_n) \in \mathbf{F}^\infty$ arbitrary strategies, $\Pi^{(i)} := (\pi_{n+i})$, and $P^{(i)} := (\rho_{n+i})$. Obviously, from Assumption 3.1(c) it follows that

$$p_{\Pi P}^{(2\kappa)}(\overline{\mathbf{D}}_i \mid x) = 0, \tag{6}$$

for all $x \in \mathbf{D}_i$, $i = 1, \ldots, m$.

Let $j \geq 1$, $\mathbf{E}_l := \{x \in \mathbf{D}_j : p^{(\kappa-1)}_{\Pi^{(\kappa)} P^{(\kappa)}}(\mathbf{D}_l \mid x) > 0\}$. Since $\bigcup_{l=1}^m \mathbf{E}_l = \mathbf{D}_j$, there is $i \in \{1, \ldots, m\}$ with $\mu_j(\mathbf{E}_i) > 0$. Assume that there are an i' with $i \neq i' \in \{1, \ldots, m\}$ and an $x \in \mathbf{D}_{i'}$ with $p_{\Pi P}(\mathbf{D}_j \mid x) > 0$. Then it follows that

$$\begin{aligned} p^{(2\kappa)}_{\Pi P}(\overline{\mathbf{D}}_{i'} \mid x) &\geq p^{(2\kappa)}_{\Pi P}(\mathbf{D}_i \mid x) \\ &\geq \int_{\mathbf{D}_j} p_{\Pi P}(d\xi \mid x) \int_{\mathbf{D}_j} p^{(\kappa)}_{\Pi^{(1)} P^{(1)}}(d\eta \mid \xi) p^{(\kappa-1)}_{\Pi^{(\kappa)} P^{(\kappa)}}(\mathbf{D}_i \mid \eta) \\ &\geq \int_{\mathbf{D}_j} p_{\Pi P}(d\xi \mid x) \int_{\mathbf{E}_i} \mu_j(d\eta) p^{(\kappa-1)}_{\Pi^{(\kappa)} P^{(\kappa)}}(\mathbf{D}_i \mid \eta) \\ &> 0, \end{aligned}$$

which contradicts (6). Hence, there is one and only one $j_{\kappa-1} \in \{1, \ldots, m\}$, which includes an element x' with $p_{\Pi^{(\kappa-1)} P^{(\kappa-1)}}(\mathbf{D}_j \mid x') > 0$. It follows successively that there is one and only one $j_1 \in \{1, \ldots, m\}$ with $p^{(\kappa-1)}_{\Pi^{(1)} P^{(1)}}(\mathbf{D}_j \mid x'') > 0$ for some $x'' \in \mathbf{D}_{j_1}$.

It holds that

$$1 = p^{(\kappa)}_{\Pi P}(\mathbf{D}_j \mid x) = \int_{\mathbf{D}_{j_1}} p_{\Pi P}(d\xi \mid x)\, p^{(\kappa-1)}_{\Pi^{(1)} P^{(1)}}(\mathbf{D}_j \mid \xi),$$

for all $x \in \mathbf{D}_j$. Therefore, $p_{\Pi P}(\mathbf{D}_{j_1} \mid x) = 1$. Since $j \in \{1, \ldots, m\}$ and $\Pi = (\pi_n) \in \mathbf{E}^\infty$, $P = (\rho_n) \in \mathbf{F}^\infty$ are arbitrary, we get (3) and (5). (4) follows analogously. □

Unbounded costs are allowed in this paper. The following assumptions concerning stage costs are related to the assumptions in [1] and [14].

Assumption 3.2. There are a measurable function $v : \mathbf{X} \to [1, \infty)$ and a constant $\alpha < 1$ with the following properties:

(a) $$\int_{\mathbf{X}} p(dy \mid x, a, b) v(y) \leq \text{const} \cdot v(x),$$

for all $x \in \mathbf{X}$, $a \in \boldsymbol{A}(x)$, and $b \in \boldsymbol{B}(x)$.

(b) $$\left| \frac{k(x, a, b)}{v(x)} \right| \leq \text{const},$$

for all $x \in \mathbf{X}$, $a \in \boldsymbol{A}(x)$, $b \in \boldsymbol{B}(x)$.

(c) $$\sup_{\Pi P} \int_{\mathbf{X}} p^{(\kappa)}_{\Pi P}(dy \mid x) v(y) \leq \alpha v(x) + \int_{\mathbf{X}} \mu_i(dy) v(y),$$

for all $x \in \mathbf{D}_i$, $i = 0, \ldots, m$, where $\mu_0 \equiv 0$.

In the rest of this paper, we assume that Assumptions 3.1 and 3.2 are fulfilled.

Let u be a real function on $\mathbf{X}$. Then we put $\|u\|_v = \sup_{x \in \mathbf{X}}(|u(x)|/v(x))$. Furthermore, we denote by $\mathcal{V}$ the set of all universally measurable functions u on $\mathbf{X}$ with $\|u\|_v < \infty$. Then we can consider $\mathcal{V}$ as a complete metric space with the metric given by $\|\cdot\|_v$. We denote furthermore by $\mathcal{V}'$ the set of all universally measurable real functions u' on $\mathbf{X}\times\mathbf{A}$ with $\sup_{x\in\mathbf{X}}(|u'(x,a)|/v(x)) \le \text{const}$ for each $a\in\mathbf{A}$ and by $\mathcal{V}''$ the set of all universally measurable real functions u'' on $\mathbf{X}\times\mathbf{A}\times\mathbf{B}$ with $\sup_{x\in\mathbf{X}}(|u''(x,a,b)|/v(x)) \le \text{const}$ for all $a\in\mathbf{A}$, $b\in\mathbf{B}$. Let $\pi\in\mathbf{E}$, $\rho\in\mathbf{F}$, and μ be the substochastic kernel from $\mathbf{X}$ to $\mathbf{X}$ with $\mu(\cdot\mid x)=\mu_i(\cdot)$ for all $x\in\mathbf{D}_i$, $i=0,\dots,m$, where $\mu_0\equiv 0$. Then we define the operators p, T, μ, π, ρ, U, and L as follows:

$$\begin{aligned}
pu(x,a,b) &= \int_{\mathbf{X}} p(dy\mid x,a,b)u(y) \text{ for all } x\in\mathbf{X}, a\in\mathbf{A}, b\in\mathbf{B}, u\in\mathcal{V};\\
Tu &= k+pu \text{ for all } u\in\mathcal{V};\\
\mu u(x) &= \int_{\mathbf{X}} \mu(dy\mid x)u(y) \text{ for all } x\in\mathbf{X}, u\in\mathcal{V};\\
\pi u'(x) &= \int_{\mathbf{A}} \pi(da\mid x)u'(x,a) \text{ for all } x\in\mathbf{X}, u'\in\mathcal{V}';\\
\rho u''(x,a) &= \int_{\mathbf{B}} \rho(db\mid x,a)u''(x,a,b) \text{ for all } x\in\mathbf{X}, a\in\mathbf{A}, u''\in\mathcal{V}'';\\
Lu'(x) &= \inf_{\pi\in\mathbf{E}} \pi u'(x) \text{ for all } x\in\mathbf{X}, u'\in\mathcal{V}';\\
Uu''(x,a) &= \sup_{\rho\in\mathbf{F}} \rho u''(x,a) \text{ for all } x\in\mathbf{X}, a\in\mathbf{A}, u''\in\mathcal{V}''.
\end{aligned}$$

We will use these notations if the functions are restricted on a Borel subset $\mathbf{Y}\subseteq\mathbf{X}$ instead of $\mathbf{X}$.

We give some properties of these operators. The statements in the next lemma are well known (see [5], for instance) and can be concluded from Assumption 3.2.

Lemma 3.2. *Let $\Pi=(\pi_n)\in\mathbf{E}^\infty$ and $P=(\rho_n)\in\mathbf{F}^\infty$. Then the following conditions hold.*

(a) *$p\mathcal{V}\subseteq\mathcal{V}''$, $T\mathcal{V}\subseteq\mathcal{V}''$, $\mu\mathcal{V}\subseteq\mathcal{V}$, $\pi\mathcal{V}'\subseteq\mathcal{V}$, and $\rho\mathcal{V}''\subseteq\mathcal{V}'$. Furthermore, we have $Lu'\in\mathcal{V}$ for $u'\in\mathcal{V}'$ if Lu' is universally measurable, and $Uu''\in\mathcal{V}'$ for $u''\in\mathcal{V}''$ if Uu'' is universally measurable.*
(b) *The operators p, T, μ, π_n, ρ_n, U, and L are order-preserving.*
(c) *$T(u+v)=Tu+pv$.*
(d) *$\pi_0\rho_0T\pi_1\rho_1T\cdots\pi_n\rho_nTu = V^n_{\Pi P}+\pi_0\rho_0p\pi_1\rho_1p\cdots\pi_n\rho_npu$.*

Lemma 3.3. *Let $\Pi=(\pi_n)\in\mathbf{E}^\infty$, $P=(\rho_n)\in\mathbf{F}^\infty$, and $T_{\Pi P} := \pi_0\rho_0T\pi_1\rho_1T\cdots\pi_{\kappa-1}\rho_{\kappa-1}T-\mu$. Then the following hold.*

(a) *$T_{\Pi P}\mathcal{V}\subseteq\mathcal{V}$.*
(b) *$T_{\Pi P}$ is order-preserving.*
(c) *$T_{\Pi P}$ contracts on $\mathcal{V}$ with contraction constant α.*

Proof. Part (a) follows from Lemma 3.2 (a). Part (b) follows from Assumption 3.1 (a) and Lemma 3.2 (c). By means of Lemma 3.2 (c) and Assumption 3.2, we get

$$\begin{aligned} T_{\Pi P}u_1 - T_{\Pi P}u_2 &= (\pi_0\rho_0 p \cdots \pi_{\kappa-1}\rho_{\kappa-1}p - \mu)(u_1 - u_2) \\ &\le \|u_1 - u_2\|_v \cdot (\pi_0\rho_0 p \cdots \pi_{\kappa-1}\rho_{\kappa-1}p - \mu)v \\ &\le \|u_1 - u_2\|_v \cdot \alpha v + \|u_1 - u_2\|_v \cdot \mu v \\ &\quad - \|u_1 - u_2\|_v \cdot \mu v \\ &= \|u_1 - u_2\|_v \cdot \alpha v. \end{aligned}$$

If we permute u_1 and u_2, the statement in part (c) follows. □

Lemma 3.4.

(a) *There are constants C_1 and C_2 with $|V_{\Pi P}^N| \le C_1 v + NC_2$ for all $N \in \mathbb{N}$.*
(b) $\Phi_{\Pi P} = \liminf_{N\to\infty}(1/N\kappa)V_{\Pi P}^{N\kappa}$ *for all $\Pi = (\pi_n) \in \mathbf{E}^\infty$, $P = (\rho_n) \in \mathbf{F}^\infty$.*

Proof. From Assumption 3.2 it follows that there are constants $\hat{C}$ and $\tilde{C}$ with $\pi_n\rho_n|k| \le \hat{C}v$ and $\pi_n\rho_n pv \le \tilde{C}v$. This implies

$$\pi_{j\kappa}\rho_{j\kappa}p\pi_{j\kappa+1}\rho_{j\kappa+1}p\cdots\pi_{j\kappa+l}\rho_{j\kappa+l}|k| \le \hat{C}\tilde{C}^l v,$$

for all $j, l \in \mathbb{N}$, $1 \le l < \kappa$. Furthermore, from Assumption 3.2 we conclude

$$\int_{\mathbf{X}} p_{\Pi P}^{(j\kappa)}(dy \mid x)v(y) \le \alpha^j v(x) + \frac{1}{1-\alpha}\mu v(x),$$

for all $x \in \mathbf{X}$. Hence,

$$\int_{\mathbf{X}} p_{\Pi P}^{(j\kappa+l)}(dy \mid x)\pi_{j\kappa+l}\rho_{j\kappa+l}|k|(y) \le \hat{C}\tilde{C}^l(\alpha^j v(x) + \frac{1}{1-\alpha}\mu v(x)).$$

From this inequality, (a) follows and

$$|V_{\Pi P}^{j\kappa+l} - V_{\Pi P}^{j\kappa}| \le \text{const} \cdot v,$$

which implies (b). □

Lemma 3.5. *Let $\mathbf{D}_0 = \emptyset$, $\pi \in \mathbf{E}$, and $\rho \in \mathbf{F}$. Then there are $u_{\pi\rho} \in \mathcal{V}$ and constants g_i, $i = 1, \ldots, m$, with*

$$u_{\pi\rho}(x) = (\pi\rho T)^\kappa u_{\pi\rho}(x) - \kappa g_i, \tag{7}$$

for $x \in \mathbf{D}_i$, $i = 1, \ldots, m$.

Proof. From Lemma 3.3 and Banach's fixed-point theorem, we conclude that $T_{\pi^\infty\rho^\infty}$ has a unique fixed point $u_{\pi\rho}$. It follows that, $u_{\pi\rho} = (\pi\rho T)^\kappa u_{\pi\rho} - \mu u_{\pi\rho}$. $\mu u_{\pi\rho}$ is constant on $\mathbf{D}_i$. We prove the statement if we put $g_i = (1/\kappa)\mu u_{\pi\rho}(x)$ for $x \in \mathbf{D}_i$, $i = 1, \ldots, m$. □

Lemma 3.6. *Let $\mathbf{D}_0 = \emptyset$, $\pi \in \mathbf{E}$, and $\rho \in \mathbf{F}$. Let $u \in \mathcal{V}$, $\phi \in \mathcal{V}$, with*

$$u = (\pi\rho T)^\kappa u - \kappa\phi, \tag{8}$$

$$\phi = (\pi\rho p)^\kappa \phi. \tag{9}$$

Then it holds (with the notation of Lemma 3.5) that

$$\phi = \phi_{\pi\rho} = \lim_{n\to\infty} \frac{1}{n\kappa} (\pi\rho T)^{n\kappa} u_{\pi\rho}, \tag{10}$$

where $\phi_{\pi\rho}(x) = g_i = \text{const}$ *for* $x \in \mathbf{D}_i$, $i = 1, \ldots, m$, *and*

$$u(x) = u_{\pi\rho}(x) + \text{const}\,, \tag{11}$$

for $x \in \mathbf{D}_i$, $i = 1, \cdots, m$.

Proof. From (8) and (9) and Lemma 3.2, we get for all $n \in \mathbb{N}$ by mathematical induction

$$\begin{aligned} u &= (\pi\rho T)^{n\kappa} u - n\kappa\phi \\ &= (\pi\rho T)^{n\kappa} u_{\pi\rho} + (\pi\rho p)^{n\kappa}(u - u_{\pi\rho}) - n\kappa\phi. \end{aligned}$$

For $n \to \infty$, we conclude (10) because from Assumption 3.2 it follows that

$$(\pi\rho p)^{n\kappa} |u - u_{\pi\rho}| \le \|u - u_{\pi\rho}\|_v \left(\alpha^n v + \frac{1}{1-\alpha}\mu v\right). \tag{12}$$

In particular, we have $g_i = \phi(x)$ for $x \in \mathbf{D}_i$, $i = 1, \ldots, m$.
We set $\tilde{u}(x) = u(x) - d(x)$, where

$$d(x) = \frac{1}{\mu_i(\mathbf{X})} \int_{\mathbf{X}} \mu_i(dx) u(x) = \frac{1}{\mu_i(\mathbf{X})} \mu u(x)$$

for $x \in \mathbf{D}_i$, $i = 1, \ldots, m$. Obviously, $\mu\tilde{u} = 0$. It then follows from (8)

$$\begin{aligned} \tilde{u} + \kappa\phi &= (\pi\rho T)^\kappa u - d = (\pi\rho T)^\kappa \tilde{u} + (\pi\rho p)^\kappa d - d = (\pi\rho T)^\kappa \tilde{u} \\ &= ((\pi\rho T)^\kappa - \mu)\tilde{u} + \mu\tilde{u}. \end{aligned}$$

Since

$$((\pi\rho T)^\kappa - \mu) = T_{\pi^\infty\rho^\infty},$$

we have

$$\tilde{u} + \kappa\phi = T_{\pi^\infty\rho^\infty}\tilde{u}.$$

Analogously as in the proof of Lemma 3.5, we conclude that this fixed-point-equation has a unique solution. Hence, for two solutions u_1, u_2 of (8), it holds that

$$u_1(x) - \frac{1}{\mu_i(\mathbf{X})}\mu u_1(x) = u_2(x) - \frac{1}{\mu_i(\mathbf{X})}\mu u_2(x),$$

for all $x \in \mathbf{D}_i$ $i = 1, \ldots, m$. □

Theorem 3.1. *Let* $\mathbf{D}_0 = \emptyset$, $\pi \in \mathbf{E}$, *and* $\rho \in \mathbf{F}$. *Then there is a solution* $\tilde{u}$, $\tilde{\phi}$ *of the equation*

$$\tilde{u} + \tilde{\phi} = \pi\rho T\tilde{u}, \tag{13}$$

with $\pi\rho p\tilde{\phi} = \tilde{\phi}$. For each such solution, it holds that $\tilde{\phi} = \phi_{\pi\rho}$. Furthermore, if $j_0, j_1, \ldots, j_\kappa$ is a sequence as in Lemma 3.1 *then $g_{j_1} = g_{j_2} = \cdots = g_{j_\kappa}$.*

Proof. Without loss of generality, we can assume that $\mathbf{X} = \mathbf{D}_{j_1} \cup \mathbf{D}_{j_2} \cup \ldots \cup \mathbf{D}_{j_\kappa}$. From Lemma 3.6, the existence of a $g = g_{j_\kappa} = \text{const}$ with

$$u_{\pi\rho}(x) = (\pi\rho T)^\kappa u_{\pi\rho}(x) - \kappa g$$

for $x \in \mathbf{D}_{j_\kappa}$ is shown. We set $\hat{u}(x) = u_{\pi\rho}(x)$ for $x \in \mathbf{D}_{j_\kappa}$. Clearly,

$$\hat{u}(x) = (\pi\rho T)^\kappa \hat{u}(x) - \kappa g, \tag{14}$$

for $x \in \mathbf{D}_{j_\kappa}$.

We now assume that (14) is true for a certain $\hat{u}$ and all $x \in \mathbf{D}_{j_i}$. Let $\hat{u}(x) = \pi\rho T\hat{u}(x)$ for $x \in \mathbf{D}_{j_{i-1}}$. Then it holds, for $x \in \mathbf{D}_{j_{i-1}}$, that

$$\begin{aligned}\hat{u}(x) &= \pi\rho T((\pi\rho T)^\kappa u_{\pi\rho} - \kappa g)(x) \\ &= (\pi\rho T)^{\kappa+1} u_{\pi\rho}(x) - \kappa\pi\rho p g \\ &= (\pi\rho T)^\kappa \pi\rho T u_{\pi\rho}(x) - \kappa g \\ &= (\pi\rho T)^\kappa \hat{u}(x) - \kappa g.\end{aligned}$$

We get (14) inductively for all $x \in \mathbf{X}$. From Lemma 3.6 we have $\phi_{\pi\rho} \equiv g$. We set $\tilde{u} = \pi\rho T u_{\pi\rho}$. Then $(\pi\rho T)^\kappa \tilde{u} = (\pi\rho T)^{\kappa+1} u_{\pi\rho} = \pi\rho T(u_{\pi\rho} + \kappa g) = \pi\rho T u_{\pi\rho} + \kappa g = \tilde{u} + \kappa g$. From Lemma 3.6 it follows that there is a constant $\tilde{g}_i$ with $\pi\rho T u_{\pi\rho}(x) = \tilde{u}(x) = u_{\pi\rho}(x) + \tilde{g}_i$ for $x \in \mathbf{D}_{j_i}$. Hence, $(\pi\rho T)^\kappa(x) = u_{\pi\rho}(x)$. Using Lemma 3.6 once more, we get $\tilde{g}_i = g$. Therefore, $\tilde{u} = \pi\rho T(\tilde{u} - g) = \pi\rho T\tilde{u} - g$. □

4 The Existence of ε-Optimal Stationary Strategies

In this section we will prove that the optimality equation has a solution and that ε-optimal strategies exist. For our proof it is essential that a suitable set $\mathcal{W} \subseteq \mathcal{V}$ exist with $LUT\mathcal{W} \subseteq \mathcal{W}$. Furthermore, the selection of "nearly optimal" decision rules must be possible. Therefore, we will assume here that $\mathcal{M}$ is a Markov game with perfect information and that $\mathbf{B}$ is countable. But if we compare the proofs in this section with corresponding proofs in [5], [7], and [15], it is clear that we can get similar results also under certain semicontinuity and compactness conditions and in the case of independent action choice.

We assume in this section that Assumptions 3.1, and 3.2 and the following assumption are satisfied.

Assumption 4.1.

(a) **E** and **F** are the sets of all admissible Markov decision rules.
(b) **B** is countable.

Furthermore, we will assume in the following that $\mathcal{W}$ is the set of all lower semianalytic functions w on $\mathbf{X}$ with $\sup_{x\in\mathbf{X}}(|w(x)|/v(x)) < \infty$, where a function

$u : \mathbf{X} \to \mathbb{R}$ is called *lower semianalytic* iff the set $\{x \in \mathbf{X} : u(x) > c\}$ is analytic for every $c \in \mathbb{R}$.

The following results concerning semianalytic functions can be found in [2] (see also [17]).

Lemma 4.1. *Let (Y, σ_Y) and (Z, σ_Z) be standard Borel spaces.*

(a) *u is lower semianalytic if and only if $\{y \in Y : u(y) \geq c\}$ is analytic for every real c.*
(b) *If u is Borel measurable, then u is lower semianalytic.*
(c) *If u is lower semianalytic, then u is universally measurable.*
(d) *If u_i, $i \in \mathbf{N}$, are lower semianalytic, then γu_1 for $\gamma \geq 0$, $\sum_{i=1}^{\infty} u_i$, $\sup_{i \in \mathbf{N}} u_i$, $\inf_{i \in \mathbf{N}} u_i$, $\limsup_{i \to \infty} u_i$, and $\liminf_{i \to \infty} u_i$ are lower semianalytic.*
(e) *If $u : Y \times Z \to [-\infty, \infty]$ is lower semianalytic and $G : Y \to 2^Z$ is a set-valued map with* gr (G) *a Borel-measurable set, then u^* given by $u^*(y) = \inf_{z \in G(y)} u(y, z)$ is lower semianalytic.*
(f) *Let $u : Y \times Z \to \mathbb{R}$ be lower semianalytic and q a substochastic kernel from (Y, σ_Y) to (Z, σ_Z).*
If $u'(y) := \int_Z u(y, z) q(dz \mid y)$ exists for every $y \in Y$, then u' is lower semianalytic.

The following selection theorem is shown in [16].

Lemma 4.2. *Let (Y, σ_Y) and (Z, σ_Z) be standard Borel spaces. Let $G : Y \to 2^Z$ a set-valued map with the property that* gr $(G) \subseteq Y \times Z$ *is an analytic set, and let $u : gr(G) \to \mathbb{R}$ a lower semianalytic function with $u^*(y) := \inf_{z \in G(y)} u(y, z) > -\infty$ for all $y \in Y$. Then, for every $\varepsilon > 0$, there is a universally measurable function $g : Y \to Z$ with $g(y) \in G(y)$, and*

$$u(y, g(y)) \leq u^*(y) + \varepsilon,$$

for all $y \in Y$.

Obviously, $\mathcal{W} \subseteq \mathcal{V}$ and $\mathcal{W}$ is complete.

Lemma 4.3.

(a) *$LUTw \in \mathcal{W}$ for $w \in \mathcal{W}$.*
(b) *For every $w \in \mathcal{W}$ and $\varepsilon > 0$, there are Markovian deterministic decision rules $\pi^* = \delta_{e^*} \in \mathbf{E}$ and $\rho^* = \delta_{f^*} \in \mathbf{F}$ with*

$$\pi^* \rho T w - \varepsilon \leq LUTw \leq \pi \rho^* T w + \varepsilon,$$

for all $\pi \in \mathbf{E}$ and $\rho \in \mathbf{F}$.

Proof.

(a) The statement follows by Lemma 4.1 (d), (e), and (f).
(b) From Lemma 4.1 we get that Tw and UTw are lower semianalytic. Since $\mathbf{B}$ is countable, there is a universally measurable function $f^* : \mathbf{X} \times \mathbf{A} \to \mathbf{B}$

with $f^*(x,a) \in \boldsymbol{B}(x)$ for all $x \in \mathbf{X}$, $a \in \mathbf{A}$, and

$$UTw(x,a) \le Tw(x,a,f^*(x,a)) + \varepsilon = \delta_{f^*}Tw(x,a)) + \varepsilon$$

for all $x \in \mathbf{X}$, $a \in \mathbf{A}$. Hence,

$$LUTw(x) \le \pi UTw(x) \le \pi\delta_{f^*}Tw(x) + \varepsilon,$$

for all $x \in \mathbf{X}$, $\pi \in \mathbf{E}$.

From Lemma 4.2 the existence of a universally measurable selector e^* of $\boldsymbol{A}$ with

$$\delta_{e^*}UTw(x) - \varepsilon = UTw(x, e^*(x)) - \varepsilon \le LUTw(x)$$

follows for all $x \in \mathbf{X}$. Therefore,

$$\delta_{e^*}\rho Tw(x) - \varepsilon \le LUTw(x),$$

for all $x \in \mathbf{X}$, $\rho \in \mathbf{F}$. □

Lemma 4.4. *Let $l, N \in \mathbb{N}, l \le N$, and $w \in \mathcal{W}$. Then for each $\varepsilon > 0$ there are strategies $\Pi^* = (\pi_n^*) \in \mathbf{E}^\infty$ and $P^* = (\rho_n^*) \in \mathbf{F}^\infty$ such that for all $\Pi = (\pi_n) \in \mathbf{E}^\infty$, $P = (\rho_n) \in \mathbf{F}^\infty$ it holds that*

$$\begin{aligned}\pi_l^*\rho_l T\cdots\pi_N^*\rho_N Tw - (N-l+1)\varepsilon &\le (LUT)^{N-l+1}w \\ &\le \pi_l\rho_l^* T\cdots\pi_N\rho_N^* Tw + (N-l+1)\varepsilon. \end{aligned} \tag{15}$$

Furthermore,

$$\inf_{\Pi\in\mathbf{E}^\infty}\sup_{P\in\mathbf{F}^\infty}\pi_l\rho_l T\pi_{l+1}\rho_{l+1}T\cdots\pi_N\rho_N Tw = (LUT)^{N+1-l}w.$$

Especially,

$$\inf_{\Pi\in\mathbf{E}^\infty}\sup_{P\in\mathbf{F}^\infty} V_{\Pi P}^N = (LUT)^{N+1}0.$$

Proof. Let $w \in \mathcal{W}$. We set $(LUT)^0 w := w$. From Lemma 4.3 it follows that there are deterministic $\pi_n^* \in \mathbf{E}$, $\rho_n^* \in \mathbf{F}$, with $\pi_n^*\rho_n T(LUT)^{N-n}w - \varepsilon \le (LUT)^{N-n+1}w \le \pi_n\rho_n^* T(LUT)^{N-n}w + \varepsilon$ for all $\pi_n \in \mathbf{E}$, $\rho_n \in \mathbf{F}$, $n = l,\ldots,N$. We have especially

$$\pi_N^*\rho_N Tw - \varepsilon \le LUTw \le \pi_N\rho_N^* Tw + \varepsilon.$$

From

$$\begin{aligned}\pi_{n+1}^*\rho_{n+1}T\cdots\pi_N^*\rho_N Tw - (N-n)\varepsilon &\le (LUT)^{N-n}w \\ &\le \pi_{n+1}\rho_{n+1}^* T\cdots\pi_N\rho_N^* Tw + (N-n)\varepsilon,\end{aligned}$$

it follows that

$$\begin{aligned}&\pi_n^*\rho_n T\cdots\pi_N^*\rho_N Tw - (N-n+1)\varepsilon \\ &\quad\le \pi_n^*\rho_n T(LUT)^{N-n}w - \varepsilon \le (LUT)^{N-n+1}w \le \pi_n\rho_n^* T(LUT)^{N-n}w + \varepsilon \\ &\quad\le \pi_n\rho_n^* T\cdots\pi_N\rho_N^* Tw + (N-n+1)\varepsilon.\end{aligned}$$

Hence, we get (15) by mathematical induction.

It follows that

$$\inf_{\Pi}\sup_{P}\pi_l\rho_l T\cdots\pi_N\rho_N Tw-(N-l+1)\varepsilon\le(LUT)^{N-l+1}w$$
$$\le\sup_{P}\inf_{\Pi}\pi_l\rho_l T\cdots\pi_N\rho_N Tw+(N-l+1)\varepsilon$$
$$\le\inf_{\Pi}\sup_{P}\pi_l\rho_l T\cdots\pi_N\rho_N Tw+(N-l+1)\varepsilon.$$

For $\varepsilon\to 0$, we get the rest of the statement. □

Lemma 4.5. *Let* $\mathbf{D}_0=\emptyset$.

(a) *There is a solution* (u^*,ϕ^*), $u^*\in\mathcal{W}$, $\phi^*\in\mathcal{W}$, *of*

$$u+\kappa\phi=(LUT)^{\kappa}u, \tag{16}$$

with

$$\pi_1\rho_1 p\cdots\pi_\kappa\rho_\kappa p\phi=\phi, \tag{17}$$

for all $\pi_i\in\mathbf{E}$, $\rho_i\in\mathbf{F}$, $i=1,\ldots,\kappa$.

(b) *For all solutions* $(\tilde{u},\tilde{\phi})$, $\tilde{u}\in\mathcal{W}$, $\tilde{\phi}\in\mathcal{W}$, *of* (16) *and* (17) *it holds that*

$$\tilde{\phi}=\phi^*=\sup_{P}\inf_{\Pi}\Phi_{\Pi P}=\inf_{\Pi}\sup_{P}\Phi_{\Pi P} \tag{18}$$

and

$$\tilde{u}(x)=u^*(x)+\text{const},$$

for all $x\in\mathbf{D}_i$, $i=1,\ldots,m$.

Proof.

(a) Let S be the operator $S=(LUT)^{\kappa}-\mu$. Then $S\mathcal{W}\subseteq\mathcal{W}$. From Lemmas 3.3 and 4.4 we conclude that

$$Su_1\le Su_2+\alpha\|u_1-u_2\|_v v,$$

for all $u_1,u_2\in\mathcal{W}$. It follows

$$\|Su_1-Su_2\|_v\le\alpha\|u_1-u_2\|_v.$$

Hence, S contracts on $\mathcal{W}$ and has a unique fixed point u^* in $\mathcal{W}$. Also, u^* and $\phi^*=(1/\kappa)\mu u^*$ are a solution to (16) and (17).

(b) From Lemma 4.4 it follows that for an arbitrary $\varepsilon>0$, there are decision rules $\pi_j^*\in\mathbf{E}$, $\rho_j^*\in\mathbf{F}$, $j=0,\cdots,\kappa-1$, such that for all $(\pi_n)\in\mathbf{E}^{\infty}$, $P=(\rho_n)\in\mathbf{F}^{\infty}$, it holds that

$$\pi_0^*\rho_{n\kappa}T\cdots\pi_{\kappa-1}^*\rho_{(n+1)\kappa-1}T\tilde{u}-\kappa\varepsilon\le(LUT)^{\kappa}\tilde{u}$$
$$=\tilde{u}+\kappa\tilde{\phi}\le\pi_{n\kappa}\rho_0^*T\cdots\pi_{(n+1)\kappa-1}\rho_{\kappa-1}^*T\tilde{u}+\kappa\varepsilon. \tag{19}$$

Using (19) repeatedly, we get

$$\begin{aligned}
N\kappa\tilde{\phi} &\leq \pi_0\rho_0^* T \cdots \pi_{\kappa-1}\rho_{\kappa-1}^* T \pi_\kappa \rho_0^* T \cdots \pi_{N\kappa-1}\rho_{\kappa-1}^* T\tilde{u} - \tilde{u} + N\kappa\varepsilon \\
&= V_{\Pi P^*}^{N\kappa} + \pi_0\rho_0^* p \cdots \pi_{N\kappa-1}\rho_{\kappa-1}^* p\tilde{u} - \tilde{u} + N\kappa\varepsilon \\
&\leq V_{\Pi P^*}^{N\kappa} + \|\tilde{u}\|_v \left(\alpha^N + 1 + \frac{1}{1-\alpha}\right) v + N\kappa\varepsilon,
\end{aligned}$$

for $P^* = (\rho_0^*, \cdots, \rho_{\kappa-1}^*, \rho_0^*, \cdots, \rho_{\kappa-1}^*, \rho_0^*, \cdots)$. Analogously,

$$N\kappa\tilde{\phi} \geq V_{\Pi^* P}^{N\kappa} - \|\tilde{u}\|_v \left(\alpha^N + 1 + \frac{1}{1-\alpha}\right) v - N\kappa\varepsilon,$$

for $\Pi^* = (\pi_0^*, \cdots, \pi_{\kappa-1}^*, \pi_0^*, \cdots, \pi_{\kappa-1}^*, \pi_0^*, \cdots)$. By means of Lemma 3.4, we get

$$\begin{aligned}
\Phi_{\Pi P^*} - \varepsilon = \liminf_{N\to\infty} \frac{1}{N\kappa} V_{\Pi^* P}^{N\kappa} - \varepsilon \\
\leq \tilde{\phi} \leq \liminf_{N\to\infty} \frac{1}{N\kappa} V_{\Pi P^*}^{N\kappa} + \varepsilon = \Phi_{\Pi P^*} + \varepsilon.
\end{aligned}$$

Hence, it follows that

$$\tilde{\phi} = \sup_P \inf_\Pi \Phi_{\Pi P} = \inf_\Pi \sup_P \Phi_{\Pi P}.$$

The rest of the proof is analogous, as in the proof of Lemma 3.6. □

Theorem 4.1. *Let* $\mathbf{D}_0 = \emptyset$.

(a) *There is a solution* u^{**}, ϕ^{**} *of the equation*

$$u + \phi = LUTu, \tag{20}$$

with $\pi\rho p\phi^{**} = \phi^{**}$ *for all* $\pi \in \mathbf{E}$, $\rho \in \mathbf{F}$. *For each such solution, it holds that*

$$\phi^{**} = \inf_\Pi \sup_P \Phi_{\Pi P}.$$

(b) *For every* $\varepsilon > 0$, *both players have* ε*-optimal stationary strategies* $\pi^\infty \in \mathbf{E}^\infty$, $\rho^\infty \in \mathbf{F}^\infty$. *Furthermore, there is a constant* C *with*

$$\frac{1}{n\kappa}(V_{\pi^{*\infty} P}^{n\kappa} - n\kappa\varepsilon - Cv) \leq \phi^{**} \leq \frac{1}{n\kappa}(V_{\Pi\rho^{*\infty}}^{n\kappa} + n\kappa\varepsilon + Cv), \tag{21}$$

for all $n \in \mathbb{N}$, $\Pi \in \mathbf{E}^\infty$, *and* $P \in \mathbf{F}^\infty$.

Proof. Part (a) can be shown analogously to Theorem 3.1. The proof of part (b) is similar to the proof of Lemma 4.5. Particularly, if we use (20) in this proof, we can choose $\Pi^* = \pi^{*\infty}$ and $P^* = \rho^{*\infty}$ with $\pi^* = \pi_{\kappa-1}^*$ and $\rho^* = \rho_{\kappa-1}^*$. Then, we get also (21). □

We will now consider the case $\mathbf{D}_0 \neq \emptyset$. We remark that the above results continue to hold for states from $\overline{\mathbf{D}}_0$.

Theorem 4.2.

(a) *There is a unique solution* $\psi^* \in \mathcal{W}$ *of the functional equation*

$$\psi = LUp\psi, \tag{22}$$

with the properties that ψ^* *is bounded on* $\mathbf{D}_0$ *and*

$$\psi^*(x) = \phi^{**}(x) = \inf_{\Pi} \sup_{P} \Phi_{\Pi P}(x),$$

for all $x \in \overline{\mathbf{D}}_0$.

(b) *For every* $\varepsilon > 0$, *both players have* ε-*optimal stationary strategies* $\pi^{*\infty} \in \mathbf{E}^\infty$ *and* $\rho^{*\infty} \in \mathbf{F}^\infty$.

Proof. If $\psi(x) = \phi^{**}(x)$ for $x \in \overline{\mathbf{D}}_0$, then (22) is equivalent to

$$\psi(x) = LUp(\mathbf{I}_{\overline{\mathbf{D}}_0} \cdot \phi^{**} + \mathbf{I}_{\mathbf{D}_0} \cdot \psi)(x), \tag{23}$$

for all $x \in \mathbf{D}_0$. Let $v' \colon \mathbf{D}_0 \to \mathbb{R}$, with $v'(x) = v(x)$ for all $x \in \mathbf{D}_0$. We denote by $\mathcal{W}'$ the set of all lower semianalytic functions w' on $\mathbf{D}_0$ with $\|w'\|_{v'} := \sup_{x \in \mathbf{D}_0}(|w'(x)|/v'(x)) < \infty$. We consider on $\mathcal{W}'$ the metric given by $\| \cdot \|_{v'}$. Then $\mathcal{W}'$ is complete. Let

$$T'\psi(x,a,b) := \int_{\overline{\mathbf{D}}_0} p(d\xi \mid x,a,b)\phi^{**}(\xi) + \int_{\mathbf{D}_0} p(d\xi \mid x,a,b)\psi(\xi)$$

and

$$p'\psi(x,a,b) := \int_{\mathbf{D}_0} p(d\xi \mid x,a,b)\psi(\xi),$$

for arbitrary $\psi \in \mathcal{W}'$, $x \in \mathbf{D}_0$, $a \in \mathbf{A}$, $b \in \mathbf{B}$. Then (3) is equivalent to

$$\psi = LUT'\psi. \tag{24}$$

The operator T' has a similar structure as T. By Lemma 4.1 we get $LUT'\mathcal{W}' \subseteq \mathcal{W}'$. Analogously as in Lemma 4.4, it follows that

$$\inf_{\Pi \in \mathbf{E}^\infty} \sup_{P \in \mathbf{F}^\infty} \pi_0\rho_0 T' \cdots \pi_{\kappa-1}\rho_{\kappa-1} T' w' = (LUT')^\kappa w',$$

for every $w' \in \mathcal{W}'$. From Assumption 3.2 we get $\pi_0\rho_0 p' \cdots \pi_{\kappa-1}\rho_{\kappa-1} p' v' \le \alpha v'$ for all $\Pi = (\pi_n) \in \mathbf{E}^\infty$, $P = (\rho_n) \in \mathbf{F}^\infty$. Let w', $w'' \in \mathcal{W}'$. Then

$$\begin{aligned}
(LUT')^\kappa w' &= \inf_{\Pi \in \mathbf{E}^\infty} \sup_{P \in \mathbf{F}^\infty} \pi_0\rho_0 T' \cdots \pi_{\kappa-1}\rho_{\kappa-1} T' w' \\
&\le \inf_{\Pi \in \mathbf{E}^\infty} \sup_{P \in \mathbf{F}^\infty} \pi_0\rho_0 T' \cdots \pi_{\kappa\ 1}\rho_{\kappa-1} T'(w'' + \|w'' - w'\|_{v'} v') \\
&= \inf_{\Pi \in \mathbf{E}^\infty} \sup_{P \in \mathbf{F}^\infty} (\pi_0\rho_0 T' \cdots \pi_{\kappa-1}\rho_{\kappa-1} T' w'' \\
&\qquad + \|w'' - w'\|_{v'} \pi_0\rho_0 p' \cdots \pi_{\kappa-1}\rho_{\kappa-1} p' v') \\
&\le \inf_{\Pi \in \mathbf{E}^\infty} \sup_{P \in \mathbf{F}^\infty} \pi_0\rho_0 T' \cdots \pi_{\kappa-1}\rho_{\kappa-1} T' w'' + \alpha \|w'' - w'\|) \\
&= LUT' w'' + \alpha \|w'' - w'\|_{v'} v'.
\end{aligned}$$

Because w' and w'' can be permuted, it follows that

$$\|LUT'w' - LUT'w''\|_{v'} \leq \alpha \|w' - w''\|_{v'}.$$

Hence, LUT' is κ-stage contracting and has a unique fixed point. Thus, (a) follows.

There is a constant $\tilde{C}$ with $\psi^*(x) \leq (1/n\kappa)V_{\Pi P}^{n\kappa} + \tilde{C}v$ for all $n \in \mathbb{N}$, $\Pi = (\pi_n) \in \mathbf{E}^\infty$, $P = (\rho_n) \in \mathbf{F}^\infty$ (see Lemma 3.4). Let $\varepsilon > 0$ be arbitrary. There are $l \in \mathbb{N}$ and $\varepsilon' > 0$ with $\alpha^l < \varepsilon/2\tilde{C}$, $\varepsilon' < \varepsilon/2(l\kappa + 1)$. Lemma 4.3 and Theorem 4.1 embrace the fact that stationary strategies $\Pi^* = \pi^{*\infty} \in \mathbf{E}^\infty$ and $P^* = \rho^{*\infty} \in \mathbf{F}^\infty$ exist that fulfill (21) and

$$\pi^* \rho p \psi^*(x) - \varepsilon' \leq \psi^*(x) \leq \pi \rho^* p \psi^*(x) + \varepsilon',$$

for $x \in \mathbf{D}_0$.

Let $\Pi_l := (\pi_{n+l\kappa})$. Then

$$\begin{aligned}
\psi^* &\leq \pi_0 \rho^* p \cdots \pi_{l\kappa-1} \rho^* p \psi^* + l\kappa\varepsilon' \\
&= \pi_0 \rho^* p \cdots \pi_{l\kappa-1} \rho^* p \mathbf{I}_{\mathbf{D}_0} \cdot \psi^* + \pi_0 \rho^* p \cdots \pi_{l\kappa-1} \rho^* p \mathbf{I}_{\overline{\mathbf{D}}_0} \cdot \phi^{**} + l\kappa\varepsilon' \\
&\leq \pi_0 \rho^* p \cdots \pi_{l\kappa-1} \rho^* p \mathbf{I}_{\mathbf{D}_0} \tilde{C} \cdot v + \pi_0 \rho^* p \cdots \pi_{l\kappa-1} \rho^* p \mathbf{I}_{\mathbf{D}_0} \frac{1}{n\kappa} V_{\Pi_l \rho^{*\infty}} \\
&\quad + \pi_0 \rho^* p \cdots \pi_{l\kappa-1} \rho^* p \mathbf{I}_{\overline{\mathbf{D}}_0} \cdot \phi^{**} + l\kappa\varepsilon' \\
&\leq \pi_0\, \rho^* p \cdots \pi_{l\kappa-1} \rho^* p \mathbf{I}_{\mathbf{D}_0} \tilde{C} \cdot v + \pi_0 \rho^* p \cdots \pi_{l\kappa-1} \rho^* p \frac{1}{n\kappa} V_{\Pi_l \rho^{*\infty}} \\
&\quad + (l\kappa + 1)\varepsilon' + \frac{C}{n\kappa} v = \psi \text{ [see (21)]}.
\end{aligned}$$

For $n \to \infty$, we have

$$\psi^* \leq \Phi_{\Pi P^*} + \varepsilon,$$

for all $\Pi \in \mathbf{E}^\infty$. Analogously,

$$\psi^* \geq \Phi_{\Pi^* P} - \varepsilon,$$

for all $P \in \mathbf{F}^\infty$. From these both inequalities, we can conclude the statement. □

5 Concluding Remark

In this paper we have investigated only zero-sum Markov games. But ε-optimal strategy pairs can be used to construct ε-intimidation strategy pairs for non zero-sum Markov games (see [6]). Under certain conditions it can be shown that the class of these strategy pairs contains an ε-equilibrium strategy pair. Thus, many of the results in [6] concerning the unichain case are also true in the special multichain case considered in this paper.

Acknowledgments

The author expresses his thanks to the associate editor and the referees for their careful reading of the original manuscript and their many helpful suggestions.

REFERENCES

[1] Altman, E., Hordijk, A., and Spieksma, F. M. Contraction conditions for average and α-discount optimality in countable state Markov games with unbounded rewards, *Math. Oper. Res.*, **22**(3): pp. 588–618, 1997.

[2] Blackwell, D., Freedman, D., and Orkin, M. The optimal reward operator in dynamic programming. *Ann. Prob.* **2**, pp. 926–941, 1974.

[3] Gubenko, L. G. On multistage stochastic games (in Russian. English summary), *Teor. Verojatnost. i Mat. Stat.*. **8**, pp. 35–49, 1973.

[4] Hinderer, K. Foundations of non-stationary dynamic programming with discrete time parameter. *Lecture Notes in Operations Research and Mathematical Systems*, **33**, Springer-Verlag, Berlin, Heidelberg, New York, 1970.

[5] Küenle, H.-U. Stochastische Spiele und Entscheidungsmodelle, *Teubner-Texte zur Mathematik*, **89**. Teubner-Verlag, Leipzig, 1986.

[6] Küenle, H.-U. Stochastic games with complete information and average cost criterion, Advances in Dynamic Games and Applications. J.A. Filar, B. Gaitsgory, K. Mizukamil, eds. (Annals of the International Society of Dynamic Games, Vol. 5) Birkhäuser, Boston, pp. 325–338, 2000.

[7] Küenle, H.-U. Equilibrium strategies in stochastic games with additive cost and transition structure. *International Game Theory*, Review **1**, pp. 131–147, 1999.

[8] Kurano, M. The existence of a minimum pair of state and policy for Markov decision processes under the hypothesis of Doeblin, *SIAM J. Cont. Opt.*, **27**, pp. 296–307, 1989.

[9] Kurano, M. Average cost Markov decision processes under the hypothesis of Doeblin, *Ann. Oper. Res.*, **29**, pp. 375–386, 1991.

[10] Maitra, A. and Sudderth, W. Borel stochastic games with limsup payoff, *Ann. Prob.*, **21**, pp. 861–885, 1993.

[11] Maitra, A. and Sudderth, W. Finitely additive and measurable stochastic games, *Intl. J. Game Th.*, **22**, pp. 201–223, (1993).

[12] Maitra, A. and Sudderth, W. Finitely additive stochastic games with Borel measurable payoffs, *Intl. J. Game Th.*, **27**, pp. 257–267, 1998.

[13] Nowak, A. S. Zero-sum average payoff stochastic games with general state space, *Games Econ. Behavior*, **7**, pp. 221–232, (1994).

[14] Nowak, A. S. Optimal strategies in a class of zero-sum ergodic stochastic games. *Math. Meth. Oper. Res.* **50**, pp. 300–419, 1999.

[15] Rieder, U. Average optimality in Markov games with general state space, *Proc. 3rd Intl. Conf. Approx. Opt.*, Puebla, 1995.

[16] Shreve, S. E. and Bertsekas, D. P. Alternative theoretical frame-works for finite horizon discrete-time stochastic optimal control, *SIAM J. Cont. Opt.*, **16**, pp. 953–978, 1978.

[17] Thiemann, J. G. F. Analytic spaces and dynamic programming, *CWI Tract*, **14**, Centrum voor Wiskunde en Informatica, Amsterdam, 1985.

[18] Yushkevich, A. A. Blackwell optimal policies in a Markov decision process with a Borel state space, *ZOR—Math. Meth. Oper. Res.*, **40**, pp. 253–288, 1994.

Perturbed Zero-Sum Games with Applications to Stochastic and Repeated Games

Eitan Altman
INRIA
Sophia Antipolis, France

Eugene A. Feinberg
Department of Applied Mathematics
SUNY at Stony Brook
Stony Brook, NY, USA

Jerzy Filar
School of Mathematics
University of South Australia
The Levels, 5095, Australia

Vladimir A. Gaitsgory
School of Mathematics
University of South Australia
The Levels, 5095, Australia

Abstract

This paper deals with perturbed matrix games. The main result is that for a large class of perturbations, the sets of solutions of perturbed games converge to subsets of solutions of appropriate lexicographic games. We consider applications of these results to dynamic games with finite numbers of states. In particular, we apply our main result to the following three models: (i) infinite-horizon stochastic games with perturbed transition probabilities and rewards when only one player controls transition probabilities; (ii) finite-horizon stochastic games with perturbed transition probabilities and rewards when both players control transition probabilities; and (iii) repeated games with weighted discounted criteria.

1 Introduction

This paper deals with perturbed matrix games and their applications to stochastic and repeated games. For matrix games with continuously perturbed matrices, the values of perturbed games converge to the value of the original game; see [24] or [8], Proposition G.5. For continuous perturbations, the upper limits of the sets of optimal solutions of matrix games belong to the sets of optimal solutions of the original game; see the same references.

In this paper we study limiting properties of the sets of optimal policies of each player for nonlinear perturbation. Under some conditions, we show that the limits of these sets exist and belong to the sets of solutions of the described lexicographic game (Theorem 3.1). For the simplest case of a linear perturbation, Example 1 ahead shows that the limit of solution sets can be strictly smaller than the set of solutions for the appropriate lexicographic game. For perturbations of higher orders, it is possible to consider lexicographic games when the players sequentially play the games with perturbation matrices of higher orders on the polytopes of optimal solutions for previous lexicographic optimal policies. Example 2 shows that, in the case of a quadratic perturbation, the limiting solutions of the perturbed game may have no common points with the optimal solutions of the lexicographic game, which the players play on the sets of lexicographic solutions for the linear perturbed games with the payoff matrix being the matrix of the second-order perturbations. Our proofs are based on the analysis of asymptotic solutions of linear programs (see Jeroslow [13]) for perturbed matrix games.

We consider applications of these results to the following three models of stochastic and repeated games with finite state and action sets: (i) infinite-horizon stochastic games when only one player controls transition probabilities and with perturbed transition probabilities and rewards; (ii) finite-horizon stochastic games with finite state and action sets and with perturbed transition probabilities and reward functions; (iii) repeated games with objective functions equal to the sums of the total discounted rewards with different discount factors. In particular, for repeated games with several discount factors, we show that any limit of optimal solutions is optimal for the following lexicographic matrix game: (a) play the game with the payoffs that correspond to the biggest discount factor; then (b) play the game with the payoffs that correspond to the second-biggest discount factor on the sets of optimal policies for the first game.

2 Definitions and Preliminary Results

Let G_ϵ be a family of matrices of size $\mathbf{m} \times \mathbf{n}$ and $\epsilon \in [0, \epsilon^*]$, where ϵ^* is a positive number. We denote $G = G_0$.

This paper studies asymptotic properties of matrix games with payoff matrices G_ϵ as $\epsilon \to 0$. Player 1 maximizes over the policies U, which are probabilities over the $\mathbf{m}$ rows, and player 2 minimizes over the policies V, which are the probability measures over the $\mathbf{n}$ columns. We assume elements of G_ϵ are continuous in ϵ. The zero-sum game with matrix G_ϵ is called a *perturbed game*, and we are interested in characterizing the limit of the values and optimal policies of G_ϵ.

We consider G_ϵ that can be represented as ratios of two irrational polynomials. We say that a function $P(\epsilon)$ is an *irrational polynomial* if either it is a constant or for some $L = 1, 2, \ldots,$

$$P(\epsilon) = a_0 + \sum_{l=1}^{L} a_l \epsilon^{\alpha_l}, \tag{1}$$

where $\alpha_l > 0$, $l = 1, \ldots, L$. Without loss of generality, we assume that $\alpha_1 < \alpha_2 < \ldots < \alpha_L$.

We denote by $\mathcal{P}$ the set of all irrational polynomials and let $\mathcal{P}' = \{P \in \mathcal{P} | P(0) \neq 0\}$. Let $\mathcal{M}$ be the set of all functions $f(\epsilon) = P(\epsilon)/Q(\epsilon)$, where $P \in \mathcal{P}$ and $Q \in \mathcal{P}'$.

We observe that if $R_1, R_2 \in \mathcal{M}$, then, with the usual rules for addition and multiplication, $R_1 R_2 \in \mathcal{M}$ and $c_1 R_1 + c_2 R_2 \in \mathcal{M}$ for any two constants c_1 and c_2. In addition, if $(R_1/R_2)(\epsilon)$ is bounded on some interval $]0, \epsilon_0]$, then $R_1/R_2 \in \mathcal{M}$.

We consider perturbations of the following form:

- (P0) All entries of G_ϵ are functions from $\mathcal{M}$.

  We list examples when (P0) holds:

- (P1) $G_\epsilon = G + \epsilon F$, where G and F be two real matrices of size $\mathbf{m} \times \mathbf{n}$;
- (P2) $G_\epsilon = G + \epsilon F + \epsilon^2 F_1 + \ldots + \epsilon^{L+1} F_L$, where G, F, and $F_l, l = 0, 1, \ldots, L$, are real matrices of size $\mathbf{m} \times \mathbf{n}$;
- (P3) all entries of G_ϵ are rational functions of ϵ (the division of two polynomials of ϵ with real coefficients). We assume that G_ϵ has no poles at $\epsilon = 0$.
- (P4) $G_\epsilon = G + \epsilon F + \epsilon^{\alpha_1} F_1 + \ldots + \epsilon^{\alpha_L} F_L$, where F, G, and $F_l, l = 1, \ldots, L$, are real matrices of size $\mathbf{m} \times \mathbf{n}$ and $1 < \alpha_1 < \ldots < \alpha_L < \infty$;
- (P5) $G_\epsilon = G + \epsilon^{\alpha_1} F_1 + \ldots + \epsilon^{\alpha_L} F_L$, where G and $F_l, l = 1, \ldots, L$, are real matrices of size $\mathbf{m} \times \mathbf{n}$ and $0 < \alpha_1 < \ldots < \alpha_L < \infty$;
- (P6) $G_\epsilon = \sum_{i=1}^{N} f_i(\epsilon) F_i$ for some $N = 1, 2, \ldots,$ where $f_i(\epsilon) \in \mathcal{M}$ and F_i are $\mathbf{m} \times \mathbf{n}$ real matrices, $i = 1, \ldots, N$.

We observe that (P1)⇒(P2)⇒(P3)⇒(P6)⇒(P0) and (P2)⇒(P4)⇒(P5)⇒(P6), where "⇒" substitutes for the phrase "is a special case of."

For a matrix game with a matrix H, we denote by U_H and V_H the sets of optimal policies for players 1 and 2, respectively. These sets are convex polytopes whose extreme points can be computed via well-known linear programming (LP) formulation; see, e.g., Appendix G.3 in [8]. We consider the LP for player 1 that is the dual LP from Appendix G.3 in [8]. We say that a policy is *basic* if it corresponds to a basis in this LP. For $H = G_\epsilon$, we sometimes write U_ϵ and V_ϵ instead of U_{G_ϵ} and V_{G_ϵ}, respectively.

Condition 2.1. For player 1 there exist $\epsilon_0 > 0$ and a finite number, say K, of sets of basic variables in the LP such that for each $\epsilon \in I =]0, \epsilon_0]$, a basic policy u is optimal if and only if $u = u_\epsilon^k$, where u_ϵ^k is a basic policy for player 1 that corresponds to the kth set of basic variables, $k = 1, \ldots, K$, in game G_ϵ. A similar condition holds for player 2 for which we use notations v, L, and v_ϵ^l, $l = 1, \ldots, L$, $\epsilon \in I$.

We notice that, since the number of players is finite (2), we can use the same ϵ_0 for players 1 and 2. The following proposition describes the structure of the value and optimal basic solutions for the corresponding LP. This proposition is related to the minimax theorem for ordered fields, see [26] and [16], Theorem 3.6 in Chapter I. Here we provide the direct proof.

Proposition 2.1. *For perturbation* (P0), *there exists* $\epsilon_0 > 0$ *such that*

(i) *condition* 2.1 *holds for* $\epsilon \in I =]0, \epsilon_0]$,
(ii) *there exists* $f \in \mathcal{M}$ *such that* $\mathbf{val}\, G_\epsilon = f(\epsilon)$ *when* $\epsilon \in I$,
(iii) *for each coordinate of vectors* u^k, *say* $u^k(i)$, $i = 1, \ldots, \mathbf{m}$, $k = 1, \ldots, K$, *there exists a function* $f \in \mathcal{M}$ *such that* $u_\epsilon^k(i) = f(\epsilon)$ *for all* $\epsilon \in I$; *the similar statement holds for player* 2.

Proof. We prove (i) and (iii) for player 1. For player 2 the result is correct because of the symmetry in matrix games. Without loss of generality, we assume that $G(i, j) > 0, i = 1, \ldots, \mathbf{m}, j = 1, \ldots, \mathbf{n}$. We select $\epsilon_1 > 0$ such that all entries of G_ϵ are positive and bounded when $\epsilon \in]0, \epsilon_1]$. Similarly to the observations made by Jeroslow [13], there is a finite number of basic sets in the LP related to player 1; see, e.g., the dual LP in Appendix G3 in [8]. For a detailed discussion of the relationship between matrix games and LPs, see [9].

Let $\mathcal{N}$ be the set of all functions $f = P/Q$ for some $P, Q \in \mathcal{P}$. If $f \in \mathcal{N}$, then $f(\epsilon) = \epsilon^\alpha(a + o(1))$ for some α and a. In addition, if $f(\epsilon) \neq 0$ for some ϵ, then α and a can be selected in a way that $a \neq 0$. If $f \in \mathcal{N}$ is bounded on some interval $]0, \epsilon_1]$, then $f \in \mathcal{M}$.

We consider the dual LP from Appendix G.3 in [8]. We consider only the bases with the following property: for any $\epsilon > 0$, there exists $\epsilon' \in]0, \epsilon[$ such that this set is a basis for the LP for game G_ϵ. If a set of variables can be basic for a basis with this property, then (P0) implies that the elements of the corresponding basic solutions are in $\mathcal{N}$; see [13] for details where just rational functions were considered. For small, positive ϵ, any such basic solution of this LP is a vector of size $(\mathbf{m} + 1)$, where the first $\mathbf{m}$ coordinates form a basic policy and the last coordinate is the value of the objective function for this basic solution. All these coordinates are bounded functions of ϵ. The first $\mathbf{m}$ coordinates belong to $[0, 1]$ and the last coordinate is limited by the largest and smallest elements of G_ϵ taken for all $\epsilon \in]0, \epsilon_1]$. Therefore, all elements of this vector are in $\mathcal{M}$.

Since the number of possible basic sets is finite, there are $\epsilon_2 > 0$ and a finite subset $\mathcal{F}$, of functions from $\mathcal{M}$ such that for each $\epsilon \in]0, \epsilon_1]$, there is $f \in \mathcal{F}$ such that $\mathbf{val}\, G_\epsilon = f(\epsilon)$. Since the difference between two functions from $\mathcal{M}$ is also in $\mathcal{M}$, there exist $\epsilon_0 \in]0, \epsilon_2]$ and $f \in \mathcal{F}$ such that, for each $\epsilon \in I =]0, \epsilon_0]$, the equality $\mathbf{val}\, G_\epsilon = f(\epsilon)$ holds. Thus, (ii) is proved.

We have that $\mathbf{val}\, G_\epsilon = f(\epsilon)$ for all $\epsilon \in I$ and for some $f \in \mathcal{F}$. This f relates to at least one set of basic variables. Therefore, there is at most a finite number (K)

of distinct sets of basic variables such that the corresponding u_ϵ^k, $k = 1, \ldots, K$, is the optimal basic solution for all $\epsilon \in I$. So, (i) is proved. Since the elements of u_ϵ^k are in $\mathcal{N}$ and bounded on I (they are probabilities), (iii) is proved. □

Let $\mathcal{P}_1$ be the set of all irrational polynomials such that $\alpha_1 = 1$. We denote $\mathcal{P}_1' = \mathcal{P}_1 \cap \mathcal{P}'$. Let $\mathcal{M}_1$ be the set of all functions $f = P/Q$, where $P \in \mathcal{P}_1$ and $Q \in \mathcal{P}_1'$.

We consider a weaker version of (P0):

- (P0*) All entries of G_ϵ are functions from $\mathcal{M}_1$.

We notice that (P1)⇒ (P2)⇒ (P4)⇒ (P0*).

Remark 2.1. Every perturbation of the form (P0) can be reduced to the perturbation of the form (P0*) by replacing the variable ϵ with $\epsilon^* = \epsilon^\gamma$, where γ is the lowest power of ϵ in all entries of G_ϵ.

Condition 2.2. $G_\epsilon = G + \epsilon F + o(\epsilon)$ for some matrices G and F.

Condition 2.3. Let Condition 2.1 hold, and let all functions u_ϵ^k described in Condition 2.1 have the form

$$u_\epsilon^k = u_0^k + \sum_{n=1}^{N} u_{\beta_n}^k \epsilon^{\beta_n} + u_1^k \epsilon + o(\epsilon),$$

where $k = 1, \ldots, K$, for some $N = 0, 1, \ldots$, and for $0 < \beta_1 < \ldots < \beta_N < 1$. A similar condition holds for player 2 and functions v^l, $l = 1, \ldots, L$. Since we allow $u_{\beta_n}^k = 0$ and $v_{\beta_n}^l = 0$, we shall use the same β_n and N for all u^k, $k = 1, \ldots, K$, and for all v^l, $l = 1, \ldots, L$.

Conditions 2.2 and 2.3 imply that

$$\begin{aligned}\mathbf{val}\, G_\epsilon &= u_0^k G v_0^l + \sum_{n=1}^{N} \epsilon^{\beta_n} (u_0^k G v_{\beta_n}^l + u_{\beta n}^k G v_0^l) + \sum_{(n,m)\in Q(k,l)} \epsilon^{\beta_n+\beta_m} u_{\beta_n}^k G v_{\beta_m}^l \\ &\quad + \epsilon (u_0^k G v_1^l + u_1^k G v_0^l + u_0^k F v_0^l) + o(\epsilon), \qquad \epsilon \in I, \end{aligned} \tag{2}$$

where $k = 1, \ldots, K$, $l = 1, \ldots, L$, and $Q(k,l) = \{(n,m) : 1 \le n, m \le N, \beta_n + \beta_m \le 1\}$.

Proposition 2.2.

(i) *If $f \in \mathcal{M}$, then there exists $\epsilon_0 > 0$ such that for $\epsilon \in]0, \epsilon_0]$,*

$$f(\epsilon) = \sum_{i=0}^{\infty} \epsilon^{\beta_i} d_i, \tag{3}$$

where $\beta_0 \ge 0$, $\beta_{i+1} > \beta_i$, $i = 0, 1, \ldots$, and $\beta_i \to \infty$ as $i \to \infty$.

(ii) *If $f \in \mathcal{M}$, then there exists a unique representation of the form* (3)*, which may have a finite number or, infinite number of summands, with the following properties: $\beta_0 = 0$ and $d_i \ne 0$ when $i > 0$. If $f \in \mathcal{M}_1$, then there is a unique* *representation of this form with the following properties: $\beta_0 = 0$, $\beta_1 = 1$, and $d_i \neq 0$ when $i > 1$.*

We remark that (3) holds if there is a finite number of summands in its right-hand side, because we can add zero summands.

Proof. We observe that if we multiply sum in (3) by ϵ^α for some $\alpha \geq 0$, we get a similar sum. The sum of expressions of this form is also an expression of this form. Therefore, it is sufficient to prove (3) for $f = 1/Q$, where $Q(\epsilon) = a_0 + \sum_{i=1}^{N} a_i \epsilon^{\alpha_i}$, $\alpha_1 > 0$, and $\alpha_{i+1} > \alpha_i$, $i = 1, \ldots, N-1$. We divide explicitly 1 by Q and we have that there is a finite or countable subset T of positive real numbers such that

$$f(\epsilon) = \sum_{t \in T} d_t \epsilon^t$$

and $d_t \neq 0$ for all $t \in T$. Suppose (3) does not hold. This implies that there is an accumulation point in T. Let t_0 be the minimal accumulation point (the infinum of all accumulation points). Then we can present

$$f(\epsilon) = a_0^{-1}[1 - (Q(\epsilon) - a_0)/Q(\epsilon)],$$

and we get that the minimal accumulation point is $(t_0 + \alpha_1)$. After (i) is proved, (ii) is obvious. □

Proposition 2.3. *Conditions* 2.2 *and* 2.3 *hold for perturbation* (P0*).

Proof. Condition 2.2 follows from the second statement of Proposition 2.2 (ii). Propositions 2.1 (iii) and 2.2(i) applied to functions u_k imply Condition 2.3. □

We recall that $U_G \subset U$ and $V_G \subset V$ are the polytopes of policies that are optimal for players 1 and 2, respectively, in the game G; see Proposition G.5 in [8]. Consider the game F over the restricted set of policies U_G and V_G, and denote by U_{GF} and V_{GF} the corresponding sets of optimal policies. We call this a *lexicographic game* and we call these sets the *sets of lexicographic optimal policies*. The value of this game is denoted by **val** (GF).

3 Main Results

Consider matrix games G and G_ϵ. If $\lim_{\epsilon \to 0} G_\epsilon = G$, then

$$\lim_{\epsilon \to 0} \mathbf{val}\, G_\epsilon = \mathbf{val}\, G; \tag{4}$$

see Theorem 2.1 in [24] or Proposition G.5 in [8]. If $\epsilon(l) \to 0$, $(u_\epsilon, v_\epsilon) \in (U_\epsilon, V_\epsilon)$, and $(u_{\epsilon(l)}, v_{\epsilon(l)}) \to (u, v)$, then $(u, v) \in (U_G, V_G)$; see the same references. Under Conditions 2.2 and 2.3, the following stronger result holds.

Theorem 3.1. *Consider the perturbed game* (P0*). *Then there exist* $\lim_{\epsilon \to 0} U_\epsilon$ *and* $\lim_{\epsilon \to 0} V_\epsilon$, *and these limits are polytopes. Let* (u_ϵ, v_ϵ) *be an optimal solution* *for the perturbed game. Let $\epsilon(l) \to 0$ be any sequence along which some limits*

$$u' = \lim_{l\to\infty} u_{\epsilon(l)}, \qquad v' = \lim_{l\to\infty} v_{\epsilon(l)}$$

exist. Then $u' \in U_{GF}$ and $v' \in V_{GF}$. Therefore, $\lim_{\epsilon\to 0} U_\epsilon \subseteq U_{GF}$ and $\lim_{\epsilon\to 0} V_\epsilon \subseteq V_{GF}$.

Proof. Proposition 2.3 implies that Conditions 2.1, 2.2, and 2.3 hold. Since $G_\epsilon \to G = G_0$ as $\epsilon \to 0$, we have $u' \in U_G$ and $v' \in V_G$ by Proposition G.5 in [8]. Condition 2.3 implies that it is sufficient to establish the result for the case when the limits are basic policies. This means that it is sufficient to show that $u_0^k \in U_{GF}$, $k = 1, \dots, K$, and $v_0^l \in V_{GF}$, $l = 1, \dots, L$.

Let U^0 be the convex hull of $\{u_0^k | k = 1, \dots, K\}$, and let V^0 be the convex hull of $\{v_0^l | l = 1, \dots, L\}$. Condition 2.3 implies that $U^0 = \lim_{\epsilon\to 0} U_\epsilon$ and $V^0 = \lim_{\epsilon\to 0} V_\epsilon$. We have $U^0 \subseteq U_G$ and $V^0 \subseteq V_G$.

We take an arbitrary $u \in U_G$. Then

$$(uG)_j \geq \mathbf{val}G, \qquad j = 1, \dots, \mathbf{n}. \tag{5}$$

Indeed, if (5) is not true for some j, then the optimal response of player 2 to u yields a value strictly smaller than $\mathbf{val}\, G$, which contradicts the fact that $u \in U_G$.

Let $(uG)_j > \mathbf{val}\, G$. If $v \in V_G$, then $v(j) = 0$ because otherwise player 1 achieves more than $\mathbf{val}\, G$ by playing u against v.

Let $u \in U^0$. Condition 2.3 implies that, for each $\epsilon \in I$, there exists $u_\epsilon \in U_\epsilon$ such that $u_\epsilon \to u$ as $\epsilon \to 0$. Since $G_\epsilon \to G$ and $\mathbf{val}\, G_\epsilon \to \mathbf{val}\, G$, we have $(u_\epsilon G_\epsilon)_j > \mathbf{val}\, G_\epsilon$, $\epsilon \in]0, \epsilon_1]$, for some $\epsilon_1 > 0$. This implies that $v_\epsilon(j) = 0$ for any optimal policy v_ϵ for player 2 in game G_ϵ. If this is not true, then $u_\epsilon G_\epsilon v_\epsilon > \mathbf{val}\, G_\epsilon$, which contradicts the optimality of v_ϵ. So $v_\epsilon(j) - v(j) = 0$ if $(uG)_j > \mathbf{val}\, G$. Since $\sum_j v_\epsilon(j) = \sum_j v(j) = 1$, the last observation and (5) imply that, for all $\epsilon \in [0, \epsilon_1]$,

$$uG(v_\epsilon - v) = 0, \qquad u \in U^0,\ v \in V_G,\ v_\epsilon \in V_\epsilon. \tag{6}$$

We apply (6) to $v = v_0^l$ and $v_\epsilon = v_\epsilon^l$, $l = 1, \dots, L$. Then Condition 2.3 implies that, for any $u \in U^0$,

$$uGv_{\beta_n}^l = 0, \qquad n = 1, \dots, N,\ l = 1, \dots, L, \tag{7}$$

and

$$uGv_1^l = 0, \qquad l = 1, \dots, L. \tag{8}$$

Similarly, for any $v \in V^0$,

$$u_{\beta_n}^k Gv = 0, \qquad n = 1, \dots, N,\ k = 1, \dots, K, \tag{9}$$

and

$$u_1^l Gv = 0, \qquad l = 1, \dots, L. \tag{10}$$

We also observe that $(u_\epsilon G_\epsilon)_j \geq \mathbf{val}\, G_\epsilon$, $j = 1, \dots, \mathbf{n}$, for $u_\epsilon \in U_\epsilon$. We select $u_\epsilon = u_\epsilon^k$ for some $k = 1, \dots, K$. Proposition 1 implies that $f_j(\epsilon) = (u_\epsilon^k)_j - \mathbf{val}\, G_\epsilon$

are nonnegative functions from $\mathcal{M}$. As shown in the proof of Proposition 1, either $f_j(\epsilon)$ is identical to 0 or $f_j(\epsilon) > 0$ when $\epsilon \in I =]0, \epsilon_1[$, for some $\epsilon_1 > 0$. Since n and K are finite integers, the same I can be selected for all j and k.

Let J^k be the set of indexes $j = 1, \ldots, \mathbf{n}$ for which $f_j(\epsilon) > 0$ for all $\epsilon \in I$. For $j \in J^k$ and $\epsilon \in I$, we have $(u_\epsilon^k G_\epsilon)_j > \mathbf{val}\, G_\epsilon$. Therefore, $v_\epsilon(j) = 0$ for all $v_\epsilon \in V_\epsilon$, $j \in J^k$. Let $v \in V_0$. Since $v(j) = \lim_{\epsilon \to 0} v_\epsilon(j)$ for some $v_\epsilon(j)$ and for all $j = 1, \ldots, \mathbf{n}$, we have $v(j) = 0$ for all $j \in J^k$.

Thus, v_ϵ and v are probability vectors concentrated on $\bar{J}^k = \{1, \ldots, \mathbf{n}\} \setminus J^k$, and $(u_\epsilon^k G_\epsilon)_j = \mathbf{val}\, G_\epsilon$ when $j \in \bar{J}^k$. This implies that

$$u_\epsilon^k G_\epsilon (v_\epsilon - v) = 0.$$

We select $v^\epsilon = v_\epsilon^l$ for some $l = 1, \ldots, L$. We have

$$u_\epsilon^k G_\epsilon (v_\epsilon^l - v_0^l), \qquad \epsilon \in I. \tag{11}$$

Since the left-hand side of (11) belongs to $\mathcal{M}$, this equality holds for all $\epsilon \geq 0$. From (11) and Condition 2.2 we have

$$u_\epsilon^k G (v_\epsilon^l - v_0^l) + \epsilon u_\epsilon^k F (v_\epsilon^l - v_0^l) + o(\epsilon) = 0. \tag{12}$$

This formula and Condition 2.3 imply that

$$(u_0^k + \sum_{n=1}^{N} u_{\beta_n}^k \epsilon^{\beta_n}) G (\sum_{n=1}^{N} v_{\beta_n}^l \epsilon^{\beta_n} + v_1^l \epsilon) + o(\epsilon) = 0. \tag{13}$$

In view of (7) and (8),

$$\sum_{(n,m) \in Q(k,l)} \epsilon^{\beta_n + \beta_m} u_{\beta_n}^k G v_{\beta_m}^l + o(1) = 0.$$

This identity implies that

$$\sum_{(n,m) \in Q(k,l)} \epsilon^{\beta_n + \beta_m} u_{\beta_n}^k G v_{\beta_m}^l = 0. \tag{14}$$

From (2), (7)–(10), and (14), for $k = 1, \ldots, K$ and $l = 1, \ldots, L$,

$$\mathbf{val}\, G_\epsilon = u_0^k G v_0^l + \epsilon u_0^k F v_0^l + o(\epsilon) = \mathbf{val}\, G + \epsilon u_0^k F v_0^l + o(\epsilon), \tag{15}$$

where the second equality follows from $u_0^k \in U_G$, $v_0^l \in V_G$.

We fix arbitrary $k = 1, \ldots, K$ and $l = 1, \ldots, L$. We recall that $u_0^k \in U_G$ and $v_0^l \in V_G$. Suppose that either $u_0^k \notin U_{GF}$ or $v_0^l \notin V_{GF}$. Since $u_0^k \in U_G$ and $v_0^l \in V_G$, then either there exists $u \in U_G$ such that

$$u F v_0^l > u_0^k F v_0^l, \tag{16}$$

or there exists $v \in V_G$ such that $u_0^k F v_0^l > u_0^k F v$.

Without loss of generality, we consider the first situation when (16) holds. We have

$$u G_\epsilon v_\epsilon^l = u G v_\epsilon^l + \epsilon u F v_\epsilon^l + o(\epsilon) \geq \mathbf{val}\, G + \epsilon u F v_0^l + o(\epsilon), \tag{17}$$

where the equality follows from Condition 2.2 and the inequality follows from $u \in U_G$ and Condition 2.3. Formulas (15)–(17) imply that there is some $\epsilon_2 > 0$ such that for $\epsilon \in]0, \epsilon_2]$

$$uG_\epsilon v_\epsilon^l > \mathbf{val}\, G_\epsilon.$$

This contradicts $v_\epsilon^l \in V_\epsilon$. Therefore, $u_0^k \in U_{GF}$. Similarly, $u_0^l \in V_{GF}$. □

We notice that Theorem 3.1 and (15) imply that $\mathbf{val}\, G_\epsilon = \mathbf{val}\, G + \epsilon\, \mathbf{val}\,(GF) + o(\epsilon)$. We remark that in the case of linear perturbations (P1), this was obtained by Mills [17], whereas more general cases were handled on p. 12 in [16] and in Proposition 6 in [19].

Theorem 3.2. *For the perturbed game* (P0), *there exist* $\lim_{\epsilon\to 0} U_\epsilon \subseteq U_{GF}$ *and* $\lim_{\epsilon\to 0} V_\epsilon \subseteq V_{GF}$, *and these limits are polytopes, where the matrix F is constructed by the following procedures: (i) consider functions $G_\epsilon(i, j)$, $i = 1, \dots, \mathbf{m}$, $j = 1, \dots, \mathbf{n}$; (ii) select $\alpha = \min\{\beta_{i,j} | i = 1, \dots, \mathbf{m}, j = 1, \dots, \mathbf{n}\}$, where $\beta_{i,j}$ is the smallest positive power in the unique version (all coefficients before ϵ_i^{β} are not equal to 0 when $\beta_i > 0$) of representation* (3) *for $G_\epsilon(i, j)$ and $\beta_{i,j} = 0$ if $G_\epsilon(i, j) = \text{const}$; (iii) select $F(i, j)$ equal to the coefficient before the smallest power with nonzero multiplier in* (3) *for $G_\epsilon(i, j)$ if this power is equal to α and select $F(i, j) = 0$ otherwise.*

Proof. It follows from Theorem 3.1, Proposition 2.2, and Remark 2.1. □

In the particular cases of perturbations (P1), (P2), and (P4), the matrix F is defined explicitly. For (P5), $F = F_1$.

A natural question is whether $\lim_{\epsilon\to 0} U_\epsilon = U_{GF}$ and $\lim_{\epsilon\to 0} V_\epsilon = V_{GF}$ when $G_\epsilon = G + \epsilon F$. The following example provides the negative answer.

Example 3.1. Let

$$G = \begin{pmatrix} 1 & 1 & 1 & 0 \\ 0 & 0 & 0 & 1 \end{pmatrix}, \qquad F = \begin{pmatrix} 2 & 1 & 2 & 0 \\ 1 & 2 & 2 & 0 \end{pmatrix}.$$

Player 2 has three equivalent policies in game G. It is easy to see that $U_G = \{(0.5, 0.5)^t\}$, $\mathbf{val}\, G = 0.5$, and

$$\begin{aligned} V_G = \{&p_1(0.5, 0, 0, 0.5) + p_2(0, 0.5, 0, 0.5) \\ &+ p_3(0, 0, 0.5, 0.5) | p_1 + p_2 + p_3 = 1, p_i \geq 0, i = 1, 2, 3\}. \end{aligned}$$

The game GF is equivalent to a 1×3 matrix game with the payoff matrix $(3/4, 3/4, 1)$. Therefore,

$$V_{GF} = \{p_1(0.5, 0, 0, 0.5) + p_2(0, 0.5, 0, 0.5) |\ p_1 + p_2 = 1, p_i \geq 0, i = 1, 2\}.$$

For $\epsilon > 0$ we consider a matrix game $G + \epsilon F$. This 2×4 game can be solved explicitly. We have $U_\epsilon = \{(1 - \epsilon)/(2 + \epsilon), (1 + 2\epsilon)/(2 + \epsilon)^t\}$, $\mathbf{val}\,(G + \epsilon F) = (1+2\epsilon)/(2+\epsilon)$, and only policies 1 and 4 of player 2 are active. Therefore, policies

2 and 3 of player 2 can be excluded. We delete columns 2 and 3 of matrix G_ϵ and solve the appropriate 2×2 game. From this solution we get

$$V_\epsilon = \left\{ \left(\frac{1}{2+\epsilon}, 0, 0, \frac{1+\epsilon}{2+\epsilon} \right) \right\}.$$

We have $\lim_{\epsilon \to 0} V_\epsilon = \{(0.5, 0, 0, 0.5)\} \neq V_{GF}$.

We have defined lexicographic games for two matrices G and F. However, it is possible to define a lexicographic game for any finite sequence of $\mathbf{m} \times \mathbf{n}$ matrices $F_1, F_2, \ldots, F_k$. If $k = 1$, then the lexicographic game is F_1 and the sets of optimal solutions for players 1 and 2 are polytopes. If for some $i = 1, \ldots, k-1$, the lexicographic game $F_1 F_2 \ldots F_i$ is defined and the sets of optimal solutions for players 1 and 2 are polytopes, the lexicographic game $F_1 F_2 \ldots F_i F_{i+1}$ is a game on these polytopes with the payoff matrix $F_{(i+1)}$. Then the set of optimal solutions of this game are polytopes, too; see [2] for details.

Let G_ϵ satisfy (P2). In view of Theorem 1, a natural question is whether $\lim_{\epsilon \to 0} U_\epsilon \subseteq U_{GFF_1 \ldots F_L}$. The following example gives the negative answer to this question.

Example 3.2. Let $G_\epsilon = G + \epsilon F + \epsilon^2 F_1$, where matrices G and F are defined in Example 1 and

$$F_1 = \begin{pmatrix} 1 & 0 & 0 & 0 \\ 0 & 0 & 0 & 0 \end{pmatrix}.$$

We have from Example 1 that $U_{GF} = \{(0.5, 0.5)^t\}$ and V_{GF} is a convex combination of vectors $(0.5, 0, 0, 0.5)$ and $(0, 0.5, 0, 0.5)$. Since the second player minimizes the payoff, we have $V_{GFF_1} = \{(0, 0.5, 0, 0.5)\}$.

Now we explicitly solve the 2×4 matrix game G_ϵ. We have

$$U_\epsilon = \left\{ \left(\frac{1-\epsilon}{2+\epsilon+\epsilon^2}, \frac{(1+\epsilon)^2}{2+\epsilon+\epsilon^2} \right) \right\}, \qquad \mathbf{val}\, G_\epsilon = \frac{(1+\epsilon)^2}{2+\epsilon+\epsilon^2},$$

and only actions 1 and 4 are active for player 2 when ϵ is small. We delete columns 2 and 3 from G_ϵ and solve the 2×2 game. We have

$$V_\epsilon = \left\{ \left(\frac{1}{2+\epsilon+\epsilon^2}, 0, 0, \frac{1+\epsilon+\epsilon^2}{2+\epsilon+\epsilon^2} \right) \right\}.$$

Thus,

$$\lim_{\epsilon \to 0} V_\epsilon = \{(0.5, 0, 0, 0.5)\} \neq \{(0, 0.5, 0, 0.5)\} = V_{GFF_1}.$$

4 Application to Perturbed Stochastic Games

In this section we consider two applications of the results of the previous section to stochastic games. The first application deals with an infinite-horizon problem when only one player controls the transition probabilities. The second application deals with a finite-horizon problem when both players control transition probabilities.

Consider a perturbed Markov game with a finite state space $\mathbf{X}$ and finite *action spaces* $\mathbf{A}$ and $\mathbf{B}$. We assume without loss of generality that $\mathbf{A}$ and $\mathbf{B}$ are the same for all states. We assume that the transitions are controlled only by player 2, that is, the probability to go from state x to y in one step is only a function p^{ϵ}_{xby} of action b of player 2. It is given by the transition probability

$$p^{\epsilon}_{xby} := \sum_{l=0}^{L} \epsilon^l p_{xby}(l), \tag{18}$$

where L is some integer. We assume that $p_{xb\cdot}(0)$ is a probability measure for every x and b. We further assume that $p^{\epsilon}_{xb\cdot}$ is a probability in a neighborhood of $\epsilon = 0$ for all x and b. Clearly, for all $l > 0$, $x \in \mathbf{X}$ and $b \in \mathbf{B}$,

$$\sum_{y \in \mathbf{X}} p_{xby}(l) = 0$$

($p_{xb\cdot}(l)$ are not probability measures).

A behavioral policy u for player 1 in the *policy space* U is described as $u = \{u_1, u_2, \ldots, \}$ where the decision rule u_t, applied at time epoch t, is a probability measure over $\mathbf{A}$ conditioned on the whole history of actions of both players and states prior to t, as well as on the state at time t. A behavioral policy u is called *randomized stationary* if $u_t = u_s$ for all $s, t = 1, 2 \ldots$. A behavioral policy for which each measure u_t is concentrated on a single action is called a *pure behavioral policy*. A pure behavioral policy u is called *pure stationary* if $u_t = u_s$ for all $s, t = 1, 2 \ldots$.

The set of behavioral policies V for player 2 are defined in the same way. We further define a mixed stationary policy u of player 1 (and similarly for player 2) to be a probability measure of the set of pure stationary policies of player 1.

Given an initial distribution μ on $\mathbf{X}$, each policy pair (u, v) induces a probability measure denoted by $P^{u,v}_{\mu,\epsilon}$ on the space of sample paths of states and actions (which serves as the canonical sample space Ω). The corresponding expectation operator is denoted by $E^{u,v}_{\mu,\epsilon}$. On this probability space the state and action processes, x_t, a_t, b_t, $t = 1, 2, \ldots$, are defined, as well as the history process $h_t = (x_1, a_1, b_1, \ldots, x_t)$.

Let $r_l : \mathbf{X} \times \mathbf{A} \times \mathbf{B} \to \mathbb{R}$, be (real-valued) reward functions, $l = 0, \ldots, L$, and consider the total expected reward function

$$R^{\epsilon}_{\mu}(u, v) = \liminf_{N \to \infty} \frac{1}{N} E^{u,v}_{\mu,\epsilon} \sum_{s=1}^{N} r^{\epsilon}(x_s, a_s, b_s), \tag{19}$$

where $\epsilon > 0$, and

$$r^{\epsilon}(x, a, b) = \sum_{l=0}^{L} \epsilon^l r_l(x, a, b). \tag{20}$$

For a given fixed μ, the objective of player 1 is to maximize $R_\mu^\epsilon(u, v)$ over U, whereas player 2 wishes to minimize this quantity over V. We shall allow, however, both players also to use mixed pure stationary policies. We consider the following assumption.

According to [20], a subset Y of states is called a *strongly communicating class* if (i) Y is a recurrent class for a Markov chain defined by some randomized stationary policy of player 2 on the state space $\mathbf{X}$; (ii) Y is not a proper subset of some set Y' for which (i) holds true. Let $\{Y_1, \ldots, Y_j\}$ be the collection of strongly communicating classes. Let T be the (possibly empty) set of states that are transient under all randomized stationary policies of player 2. Then $\{Y_1, \ldots, Y_j, T\}$ forms a partition of the state space $\mathbf{X}$; see Proposition 1 in [20].

We say that a subset Y of $\mathbf{X}$ is *closed* if $p_{xby} = 0$ for all $x \in Y$, for all $b \in \mathbf{B}$, and for all $y \notin Y$. A closed subset Y of $\mathbf{X}$ can be interpreted as a state space of the game with this state space Y. We call this new game a *subgame*.

We observe that if any pure stationary policy of player 2 defines a Markov chain with the same number of ergodic classes, each subset Y_i, $i = 1, \ldots, j$, possesses the following properties: (a) it is closed; (b) any randomized stationary policy of player 2 defines on Y_i a Markov chain with one ergodic class.

Assumption 4.1. There exist some $\epsilon_0 > 0$ such that the number of ergodic classes of Markov chains defined by all pure stationary strategies v of player 2 is constant in v and ϵ when $\epsilon \in [0, \epsilon_0]$.

For the average reward games controlled by one player, Filar and Raghavan [6] have proved that the game has a saddle point that can be achieved either within the sets of randomized stationary policies or within the sets of mixed stationary policies. The results for stationary policies appeared also in [11], [12], [14], and [25]. Extensions to the nonzero sum case can be found in [18], which also considered the discounted cost. Singularly perturbed average reward stochastic games were considered in [1]. We denote the value for the saddle point by **val** R_μ^ϵ.

The set of pure stationary policies is finite. Thus, when considering the game over the set of pure stationary policies (allowing for mixed strategies), we are in the framework of (finite) matrix games described in Section 2: the matrix game G_ϵ is the one for which the (u, v)-entry is $R_\mu^\epsilon(u, v)$, where u, v are pure stationary policies, and a mixed strategy in this matrix game has the trivial interpretation as the corresponding mixed pure stationary strategy. Assumption 4.1 ensures that the G_ϵ are continuous in ϵ (see [15], p. 410, or [22], [5], [21]).

The game is of the form of (P3). Indeed, let $u(x)$ and $v(x)$ correspond to the actions chosen by pure stationary policies u and v at state x. Then the reward corresponding to the pair (u, v) is given by

$$\sum_{x \in \mathbf{X}} r^\epsilon(x, u(x), v(x)) \xi_x(v),$$

where $\xi(v)$ is the expected state frequencies of the Markov chain induced by v; see [10]. Indeed, consider any solution $(\xi_x(v), \rho_x(v), x \in \mathbf{X})$ of the set of linear equations and inequalities:

$$\xi_x(v) - \sum_{y\in\mathbf{X}} \xi_y(v) p^{\epsilon}_{y,v(y),x} = 0,$$

$$\xi_x(v) + \rho_x(v) - \sum_{y\in\mathbf{X}} \rho_y(v) p^{\epsilon}_{y,v(y),x} = \mu(x),$$

$$\xi(v) \geq 0, \qquad \rho(v) \geq 0.$$

It is well known [10] that for any v, $\xi(v)$ is uniquely determined by the above equalities. Since p^{ϵ} is polynomial in ϵ, $\xi(v)$ is clearly rational in ϵ and hence also $R^{\epsilon}_{\mu}(u, v)$.

Thus, (P3) holds. We conclude that (P0*) and Condition 2.1 hold, too. Indeed, G and F in Condition 2.2 are obtained as follows:

$$G(u, v) = \sum_{x\in\mathbf{X}} r_0(x, u(x), v(x))\xi^0_x,$$

where $\xi^0(v)$ is uniquely determined by

$$\xi^0_x(v) - \sum_{y\in\mathbf{X}} \xi^0_y(v) p_{y,v(y),x}(0) = 0,$$

$$\xi^0_x(v) + \rho^0_x(v) - \sum_{y\in\mathbf{X}} \rho^0_y(v) p_{y,v(y),x}(0) = \mu(x),$$

$$\xi^0(v) \geq 0, \qquad \rho^0(v) \geq 0.$$

For fixed pure stationary v for player 2, let $P(v)$ be the matrix whose (xy)-entry equals $p_{x,v(x),y}(0)$, and let

$$P^*(v) = \lim_{N\to\infty} N^{-1} \sum_{n=0}^{N-1} P^n(v).$$

The ergodic theorem for finite Markov chains implies the existence of this limit. Let

$$Z(v) = (I - P(v) + P^*(v))^{-1}$$

be the fundamental matrix, which is well defined [4]. Let $H(v) = Z(v) - P^*(v)$. Then $H(v)P^*(v) = P^*(v)H(v) = 0$.

We denote by $P'(v)$ be the matrix whose (xy)-entries equal $p_{x,v(x),y}(1)$. Assumption 4.1 here and Theorem 3.2 in [15] imply that

$$\xi(v) = \xi^0(v) + \epsilon D(v) + o(\epsilon),$$

where $D(v) = P^*(v)P'(v)H(v) + H(v)P'(v)P^*(v)$. We conclude that

$$R^{\epsilon}_{\mu}(u, v) = G(u, v) + \epsilon F(u, v) + o(\epsilon),$$

where

$$F(u, v) = \sum_{x\in\mathbf{X}} \left[r_1(x, u(x), v(x))\xi^0_x(v) + \sum_{y\in\mathbf{X}} D_{xy}(u, v) r_0(y, u(y), v(y)) \right].$$

Thus, Theorem 3.1 implies the following result.

Theorem 4.1. *Consider the perturbed stochastic game satisfying Assumption 4.1. Then* (i) *the values of R^ϵ_μ converge to that of R^0_μ, and*

$$\lim_{\epsilon\to 0} \frac{\mathbf{val}\, R^\epsilon_\mu - \mathbf{val}\, R^0_\mu}{\epsilon} = \mathbf{val}\,(GF),$$

(ii) *let (u_ϵ, v_ϵ) be optimal (randomized stationary or mixed pure stationary) for the ϵ stochastic game. Let $\epsilon(l) \to 0$ be any sequence along which some limits*

$$u' = \lim_{l\to\infty} u_{\epsilon(l)}, \qquad v' = \lim_{l\to\infty} v_{\epsilon(l)}$$

exist. Then $u' \in U_{GF}$ and $v' \in V_{GF}$.

Note: it is only for the mixed policies that we have the representation of the stochastic game as a matrix game. In order to obtain the results for the behavioral policies, we have to use the equivalence between the behavioral and mixed policies as well as the fact that we can select a continuous mapping between them. If $\{u_{\epsilon(l)}\}_l$ and u are behavioral, then the corresponding mixed policies also converge. Since the limit of the mixed policies is in U_{GF}, it follows that so is u'.

Theorem 4.1 extends results in [23] that show that $\lim_{\epsilon\to 0} \mathbf{val}\, R^\epsilon_\mu = \mathbf{val}\, R^0_\mu$ and that $u' \in U_G$ and $v' \in V_G$. We also remark that the same type of results can be obtained for a discounted cost and for the total expected cost criterion since the stochastic games are still equivalent to matrix games (see again [18], [11], [12], [14]).

Finally, consider the more general case when both the costs $r^\epsilon_t(x, a, b)$ as well as the transition probabilities p^ϵ_{txaby} may depend on the time t and on the actions of both players, and consider a finite horizon of N steps. One can now study the convergence of perturbed stochastic games with finite horizon through the dynamic programming equation

$$v_t(x) = \mathbf{val}_{ab} \left[r^\epsilon_t(x, a, b) + \sum_y p^\epsilon_{txaby} v_{t-1}(y) \right], \qquad t > 0, \tag{21}$$

$$v_0(x) = \mathbf{val}_{ab} r_0(x, a, b).$$

A saddle point (u, v) can be obtained for both players in Markov policies by using at time $N - t$ the argument that achieves the value in the tth equation above (as a function of x). At each stage t we are faced with a perturbed matrix game.

Let $r^\epsilon_t(x, a, b)$ and p^ϵ_{txaby} be continuous in ϵ on some interval $[0, \epsilon^*]$, and assume they are in $\mathcal{M}$ for all possible fixed t, x, a, b, and y. If $t = 0$, then Theorem 3.1 implies that $v_0(x)$ is in $\mathcal{M}$, $x \in \mathbf{X}$, when $\epsilon \in [0, \epsilon_0[$ for some $\epsilon_0 > 0$. We have by induction that $v_t \in \mathcal{M}$, and Theorem 3.1 can be applied at each step $t = 0, 1, \ldots, N$.

5 Application to Repeated Zero-Sum Games and to Repeated Zero-Sum Games with Weighted Discounted Payoffs

Let us consider a zero-sum repeated game with finite action sets $\{1, \dots, \mathbf{m}\}$ and $\{1, \dots, \mathbf{n}\}$ for players 1 and 2, respectively. For each $i = 1, \dots, \mathbf{m}$ and each $j = 1, \dots, \mathbf{n}$, there is a function $G_\epsilon(i, j)$ from $\mathcal{M}$. These functions define the matrix function G_ϵ. Let $G = G_0$, and let F be the matrix constructed in Theorem 3.2.

Consider $\epsilon_0 > 0$ such that all functions $G_\epsilon(i, j)$ are bounded on $[0, \epsilon_0]$. Let a monotone sequence of positive numbers $\epsilon_n \to 0$ with $\epsilon_1 \le \epsilon_0$ and a sequence of positive numbers γ_n, such that the row $\gamma_1 + \gamma_2 + \dots$ converges, be given. We define a repeated game with the payoff

$$G(u, v) = E^{u,v} \sum_{n=0}^{\infty} \gamma_n G_{\epsilon_n}(u_n, v_n).$$

It is easy to see that if players play optimal strategies at each step $n = 1, 2, \dots,$ then these strategies are optimal. Therefore, optimal randomized Markov policies exist. Finding an optimal randomized Markov policy for this game is equivalent to finding optimal policies for the sequence of games G_{ϵ_n}. Let U_n (V_n) be the sets of optimal policies for player 1 (2) in game G_{ϵ_n}. Theorem 3.2 implies that there exist $U_0 = \lim_{n\to\infty} U_n$ and $V_0 = \lim_{n\to\infty} V_n$. In addition, this theorem implies that $U_0 \in U_{GF}$ and $V_0 \in V_{GF}$.

Now we consider weighted discounted zero-sum repeated games. In this case, there are L payoff matrices $F_1, F_2, \dots, F_L$ and L discount factors $\beta_1, \beta_2, \dots, \beta_L$, where $1 > \beta_1 > \beta_2 \dots \beta_L > 0$. The payoff function is

$$G(u, v) = E^{u,v} \sum_{l=1}^{L} \sum_{n=0}^{\infty} \beta_l^n F_l(a_n, b_n).$$

Finding an optimal randomized Markov policy for this game is equivalent to finding optimal policies for the sequence of games

$$G_{\epsilon_n} = F_1 + \epsilon_n F_2 + \sum_{l=3}^{L} (\epsilon_n)^{\alpha_l} F_l,$$

with $\epsilon_n = (\beta_2/\beta_1)^n$, $n = 0, 1, \dots,$ and $\alpha_l = \log_{\beta_2/\beta_1} \beta_l/\beta_1$. Theorem 3.2 implies that if U_n and V_n are the sets of optimal strategies for players 1 and 2, respectively, in game G_{ϵ_n}, then there exist $U_0 = \lim_{n\to\infty} U_n$ and $V_0 = \lim_{n\to\infty} V_n$. In addition, this theorem implies that $U_0 \subseteq U_{F_1F_2}$ and $V_0 \subseteq V_{F_1F_2}$.

Filar and Vrieze [7] proved the existence of ϵ-optimal ultimately stationary policies for zero-sum stochastic games with weighted discounted payoffs and with finite sets of states and actions. Altman, et al. [2] proved the existence of optimal policies, which are Markov and ultimately stationary, for such games with perfect information. Example 6.1 in [2] shows that optimal ultimately stationary policies may not exist for general Markov games when there is no perfect information assumption. The statement in the previous paragraph implies that, for repeated games with several different discount factors, all limits of optimal actions as the time parameter tends to ∞ are optimal for the lexicographic game defined by payoff matrices corresponding to the two largest discount factors. An interesting open question is whether this result holds for stochastic games with weighted discounted payoffs and with finite state and action sets; see [2].

We remark that for stochastic games with perfect information, the sets of optimal solutions coincide from some step N onward with the set of optimal solutions for the lexicographic game; see [2]. Example 1 implies that, without the perfect information assumption, the limiting sets of optimal solutions can be strictly smaller than the sets of appropriate solutions for the lexicographic game even in the case of two discount factors. Example 2 demonstrates that, in the case of three or more different discount factors, the limiting sets of optimal solutions may have no common points with the sets of optimal solutions of the lexicographic game defined by the payoff matrices corresponding to the three largest discount factors.

Acknowledgments

The research of the second author was partially supported by NSF grants DMI-9500746 and DMI-9908258. The research of the third and fourth authors was supported by ARC grant no. A69703141.

REFERENCES

[1] Abbad, M. and Filar, J.A. Singularly pertured limiting average stochastic game problems. *Lecture Notes Econ. Math. Syst.*, **389**, pp. 69–97, 1992.

[2] Altman, E., Feinberg, E. A., and Shwartz, A. Weighted discounted stochastic games with perfect information, *Ann. Int. Soc. Dynamic Games*, **5**, pp. 303–323, 2000.

[3] Bather, J. Optimal decision procedures for finite Markov chains. Part II: Communicating systems, *Adv. Appl. Prob.*, **5**, pp. 521–540, 1973.

[4] Blackwell, D. Discrete dynamic programming, *Ann. Math. Stat.*, **33**, pp. 719–726, 1962.

[5] Feinberg, E. A. On ergodic conditions for finite controlled Markov chains with arbitrary decision sets, *Trans. Moscow Inst. Transport Eng.*, **640**, pp. 51–59, 1979 (in Russian).

[6] Filar, J. A. and Raghavan, T. E. S. A matrix game solution of the single-controller stochastic game, *Math. Oper. Res.*, **9**(3):pp. 356–362, 1984.

[7] Filar, J. A. and Vrieze, O. J. Weighted reward criteria in competitive Markov decision processes, *ZOR*, **36**, pp. 343–358, 1992.

[8] Filar, J. and Vrieze, K. *Competitive Markov Decision Processes*, Springer, New York, 1996.

[9] Goldman, A. J. and Tucker, A. W. Theory of linear programming, *Ann. Math. Stud.* **38**, pp. 53–97, 1956.

[10] Hordijk, A. and Kallenberg, L. C. M. Linear programming and Markov decision chains, *Mgmt. Sci.*, **25**, pp. 352–362, 1979.

[11] Hordijk, A. and Kallenberg, L. C. M. Linear programming and Markov games I, *Game Th. Math. Econ.*, North-Holland, pp. 291–305, 1981.

[12] Hordijk, A. and Kallenberg, L. C. M. Linear programming and Markov games II, *Game Th. Math. Econ.*, North-Holland, pp. 307–320, 1981.

[13] Jeroslow, R. G. Asymptotic linear programming, *Oper. Res.*, **21**, pp. 1128–1141, 1973.

[14] Kallenberg, L. C. M. *Linear Programming and Finite Markovian Control Problems*, Math. Centre Tracts **51**, Amsterdam, 1983.

[15] Lasserre, J. B. Exact formula for sensitivity analysis of Markov chains, *JOTA*, **71**(2): pp. 407–413, 1991.

[16] Mertens, J. F., Sorin, S., and Zamir, S. *Repeated Games*, part A, CORE discussion paper 9420, Center for Operations Research and Economics, Université Catolique de Louvain, Belgium, April 1994.

[17] Mills, H. D. Marginal values of matrix games and linear programs, *Ann. Math. Stud.*, **38**, pp. 183–193, 1956.

[18] Nowak, A. S. and Raghavan, T. E. S. A finite step algorithm via a bimatrix game to a single controller non-zero sum stochastic game, *Math. Prog.*, **59**, pp. 249–259, 1993.

[19] Rosenberg, D. and Sorin, S. An operator approach to zero-sum repeated games, report no. 494, Ecole Polytechnique, Laboratorie d'econometrie, Paris 1999.

[20] Ross, K. W. and Varadarajan, R. Multichain Markov decision processes with a sample path constraint: A decomposition approach, *Math. Oper. Res.* **16**, pp. 195–207, 1991.

[21] Schäl, M. Markovian decision models with bounded finite-stage rewards, *Oper. Res. Proc. 1983*, Springer, Berlin, pp. 470–473, 1984.

[22] Schweitzer, P. J. Perturbation theory and finite Markov chains, *J. Appl. Prob.*, **5**, pp. 401–413, 1968.

[23] Tidball, M. and Altman, E. Approximations in dynamic zero-sum games, I, *SIAM J. Cont. Opt.*, **34**, pp. 311–328, 1996.

[24] Tijs, S. H. and Vrieze, O. J. Perturbation theory for games in normal form and stochastic games, *JOTA*, **30**, pp. 549–567, 1980.

[25] Vrieze, O. J. *Linear Programming and Undiscounted Stochastic Games in which One player Controls the Transitions*. *OR Spektrum*, **3**, pp. 29–35, 1981.

[26] Weyl, H. Elementary proof of a minimax theorem due to von Neuman, *Ann. Math. Stud.*, **24**, 1950.

PART III
Solution Methods of Dynamic Games

Construction of Singular Surfaces in Linear Differential Games

Sergey S. Kumkov
and
Valery S. Patsko
Intstitute of Mathematics and Mechanics
Ural Branch of RAS
Ekaterinburg, 620219, Russia

Abstract

The paper deals with an algorithm of construction and classification of singular surfaces in linear differential games with fixed terminal time and convex payoff function depending on two components of the phase vector. The main procedures of the algorithm are imbedded into the backward algorithm for building level sets of the value function. Examples of numerically calculated singular surfaces are represented.

1 Introduction

In the theory of differential games, sets of the game space where the optimal motions have some peculiarities (dispersion, refraction, junction, etc.) are named as singular surfaces. The classification of the singular surfaces was suggested in [6]. Necessary conditions, which characterize different types of singularity, were studied by Bernhard [2] and Melikyan [12]. In many papers (see, for example, [11], [13], [14], and [20]), the analysis of the singular surfaces appearing in concrete differential games was carried out.

In this paper, a computer algorithm for construction and classification of singular surfaces is shortly described.

We consider a linear antagonistic differential game

$$\dot{x} = A(t)x + B(t)u + C(t)v, \tag{1}$$
$$x \in R^n, \quad u \in P,\ v \in Q,\ \varphi(x_i(T), x_j(T)),$$

with fixed terminal time T and convex payoff function φ, which depends on two coordinates x_i, x_j of the phase vector. The first (second) player governs the control u (v), choosing it from the convex compact set P (Q), and minimizes (maximizes) the value of the function φ at the instant T.

It is known (see, for example, [8], p. 354, and [10], pp. 89–91) that the substitution $y(t) = X_{i,j}(T,t)x(t)$, where $X_{i,j}(T,t)$ is a matrix combined of two corresponding rows of the fundamental Cauchy matrix, provides the transformation to the equivalent differential game of the second order.

At the beginning of 1980s, the backward constructions were elaborated ([3], [4], [22], and [23]) for building level sets of the value function in the linear differential game (1). The works of Pontryagin [16], Krasovskii and Subbotin [9] and [10], and Pschenichnyi [18] and [19] were the theoretical basis for corresponding algorithms. Recently, software for interactive investigation of level sets was created [1].

The algorithm for constructing the singular surfaces described in this paper is imbedded into the algorithm from [7]. The calculations are carried out by the recurrent backward procedure. The section of a level set of the value function at the next instant is calculated using the section at the previous one. The data obtained during these calculations are used for detection and classification of the singular points in the newly constructed section. Unrolled through the time, these singular points give the singular lines, which go along the surface of the level set. The singular lines from the different level sets are joined in the singular surfaces in the three-dimensional space t, y_1, y_2 of the equivalent game. If the original game (1) is of the second order on its phase variable, then the reverse transition to the original phase coordinates t, x_1, x_2 is possible.

The main idea of [9] and [10], was a consideration of level sets of the value function (maximal stable bridges) as a base for the solution of differential games. In this paper, the idea is adapted for detection and classification of singular surfaces. As close to this approach, [21] can be noted. It is devoted to an investigation of the singular surfaces in one concrete game of the type (1) with the elliptical constraints on the players' controls. Special analysis is essentially used for detecting the singular surfaces. The analysis is based on description of level sets of the value function via the Hamiltonian of the game and its integral.

While the level sets can be built by the algorithm [7] for arbitrary polyhedra P and Q, now the algorithm for constructing the singular surfaces has been elaborated only for the case of scalar controls of the first and second players. Hence, calculating the singular surfaces, it is supposed that the sets P and Q are segments.

2 Constructing Level Sets of the Value Function

Below we give the algorithm from [7] for constructing level sets of the value function. This consideration is necessary for further understanding the procedure of constructing and classifying singular surfaces.

2.1 Backward Procedure

Assume that the transfer from the game (1) with the payoff function φ depending on two coordinates of the phase vector to the equivalent game

$$\dot{y} = D(t)u + E(t)v, \quad y \in R^2,\ u \in P,\ v \in Q,\ \varphi(y(T)), \tag{2}$$

$$D(t) = X_{i,j}(T, t)B(t), \quad E(t) = X_{i,j}(T, t)C(t)$$

is already done. On the interval $[0, T]$, let the sequence of instants $t_i : t_N = T, \ldots, t_i = t_{i+1} - \Delta, \ldots, t_0 = 0$ dividing the interval with the step Δ be given. The interest is to find the time sections $W_c(t_i) = \{y \in R^2 : V(t_i, y) \le c\}$ of the level set $W_c = \{(t, y) \in [0, T] \times R^2 : V(t, y) \le c\}$ of the value function V for the given value of the parameter c.

Replace the dynamics (2) by the piecewise-constant dynamics

$$\dot{y} = \mathbf{D}(t)u + \mathbf{E}(t)v, \qquad \mathbf{D}(t) = D(t_i), \qquad \mathbf{E}(t) = E(t_i), \qquad t \in [t_i, t_{i+1}). \quad (3)$$

Instead of the sets P and Q, let us consider their polyhedral approximations $\mathbf{P}$ and $\mathbf{Q}$. Let $\hat{\varphi}$ be the approximating payoff function. For any c, its level set $\mathbf{M}_c = \{y : \hat{\varphi}(y) \le c\}$ is a convex polygon.

The approximating game (3) is taken so that, for each step $[t_i, t_{i+1}]$ of the backward procedure, we deal with the game with simple motions [6] and polyhedral convex control constraints. The game solvability set $\mathbf{W}_c(t_{N-1})$ of the first player for the game with the target set $\mathbf{W}_c(t_N) = \mathbf{M}_c$ can be computed. Further, starting from $\mathbf{W}_c(t_{N-1})$, the set $\mathbf{W}_c(t_{N-2})$ can be built in the same way, and so on. As a result, the collection of convex polygons is obtained, which approximate [3], [15] sections $W_c(t_i)$ of the level set W_c of the value function in the game (2) in the Hausdorff metrics. The algorithm for a-posteriori estimation of the error of numerically constructed level sets of the value function is considered in [5].

Let $\mathcal{P}(t_i) = -D(t_i)\mathbf{P}$ and $\mathcal{Q}(t_i) = E(t_i)\mathbf{Q}$. The support function $l \to \rho(l, \mathbf{W}_c(t_i))$ of the polygon $\mathbf{W}_c(t_i)$ is the convex hull [19] of the function

$$\gamma(l, t_i) = \rho(l, \mathbf{W}_c(t_{i+1})) + \Delta\rho(l, \mathcal{P}(t_i)) - \Delta\rho(l, \mathcal{Q}(t_i)).$$

The function $\gamma(\cdot, t_i)$ is positively homogeneous and piecewise-linear. The property of local convexity of this function can be violated only at the boundary of the linearity cones of the function $\rho(\cdot, \mathcal{Q}(t_i))$, that is, at the boundary of the cones generated by the normals to the edges of the polygon $\mathcal{Q}(t_i)$, which have the common vertex.

2.2 Algorithm of Convex Hull Construction

Let's agree to omit the argument t_i in the notation of the function γ.

The linearity cones of γ are determined by the normals to the convex polygons $\mathbf{W}_c(t_{i+1})$, $\mathcal{P}(t_i)$, and $\mathcal{Q}(t_i)$. Gathering the outer normals of these sets and ordering them clockwise, we obtain the collection L of the vectors. The collection of values $\gamma(l)$ of the function γ on the vectors $l \in L$ is denoted by Φ. The collections L and Φ completely describe the function γ.

The collection of the normals to $\mathcal{Q}(t_i)$ ordered clockwise is denoted by S. The collection S is called the collection of "suspicious" vectors. This name is connected with the fact that the function γ is locally convex on the cones, whose interior does not contain the normals of the set $\mathcal{Q}(t_i)$. The violation of the local convexity can appear only on the cones, whose interior contains at least one normal of the polygon $\mathcal{Q}(t_i)$.

Let $L^{(1)} = L$, $\Phi^{(1)} = \Phi$, and $S^{(1)} = S$. The $(k+1)$-step of the multistep convexification process consists of replacing the collections $L^{(k)}$ and $\Phi^{(k)}$ by the collections $L^{(k+1)} \subset L^{(k)}$ and $\Phi^{(k)} \subset \Phi^{(k+1)}$, respectively. The collection $S^{(k)}$ is also replaced by the new one $S^{(k+1)}$.

Let's now describe one step of the convexification process. Suppose that the angle between two neighbor vectors from the collection $L^{(k)}$ counted clockwise is less than π. Let $l \to \gamma^{(k)}(l)$ be the piecewise-linear function determined by the collections $L^{(k)}$ and $\Phi^{(k)}$. Since $L^{(k)} \subset L^{(k-1)} \subset \ldots \subset L^{(1)}$ and $\Phi^{(k)} \subset \Phi^{(k-1)} \subset \ldots \subset \Phi^{(1)}$, then for any vector $\bar{l} \in L^{(k)}$ the value $\gamma^{(k)}(\bar{l})$ is equal to $\gamma(\bar{l})$.

Take some vector $l_* \in S^{(k)}$ and check the local convexity of the function $\gamma^{(k)}$ on the cone generated by the vector l_* and two its neighbor vectors l_- and l_+ selected counterclockwise and clockwise from the collection $L^{(k)}$. In other words, check whether the inequality $l_*^{'} y \leq \gamma(l_*)$ is active in the triple of the inequalities $l_-^{\prime} y \leq \gamma(l_-)$, $l_*^{\prime} y \leq \gamma(l_*)$, and $l_+^{\prime} y \leq \gamma(l_+)$. If the system of three inequalities is compatible, then (by virtue of the ordering of the vectors l_-, l_*, l_+) only the middle one can be inactive.

The algorithm of verification finds the intersection point y_* of the straight lines $l_-^{\prime} y = \gamma(l_-)$ and $l_*^{\prime} y = \gamma(l_*)$, and then check the inequality $l_+^{\prime} y_* < \gamma(l_+)$. If it holds, the local convexity takes place (the middle inequality is active). In the opposite case, the local convexity is absent (the middle inequality is inactive).

In the first case, the vector l_* is taken away from the collection $S^{(k)}$, and the remaining set is denoted by $S^{(k+1)}$. Let $L^{(k+1)} = L^{(k)}$ and $\Phi^{(k+1)} = \Phi^{(k)}$.

In the second case, two situations are distinguished. Let α be the angle counted clockwise from l_- to l_+:

- $\alpha < \pi$. The vector l_* is taken away from the collection $S^{(k)}$, and, simultaneously, the vectors l_- and l_+ are included into the collection (one of them, or even both, could already be presented in the collection $S^{(k)}$). Denote the new collection of the "suspicious" vectors by $S^{(k+1)}$. The difference of the new collection $L^{(k+1)}$ from the collection $L^{(k)}$ is that the vector l_* is absent in $L^{(k+1)}$. When processing $\Phi^{(k)}$ to $\Phi^{(k+1)}$, the values $\gamma^{(k)}(l_*) = \gamma(l_*)$ are taken away.
- $\alpha \geq \pi$. The construction is ceased.

One step of the convexification algorithm has been described.

The algorithm finishes at the step with the number j when for the first time $S^{(j)} = \emptyset$, that is, when the collection of the "suspicious" vectors becomes empty. It means that the function $\gamma^{(j)}$, which corresponds to the collections $L^{(j)}$ and $\Phi^{(j)}$, is locally convex everywhere. Thus, the function $\gamma^{(j)}$ is the convex hull of the function γ. In this case, let us denote the final collections $L^{(j)}$ and $\Phi^{(j)}$ as $\overline{L}$ and $\overline{\Phi}$, respectively. These collections determine the polygon $\mathbf{W}_c(t_i)$.

The second variant of the algorithm termination is the following: the angle α between the vectors l_- and l_+ becomes greater than or equal to π after elimination of the checked vector l_* from the collection of the "suspicious" vectors at some step. It means that the convex hull of the function γ does not exist, i.e., $\mathbf{W}_c(t_i) = \emptyset$. (If $\alpha = \pi$, it is possible that $\mathbf{W}_c(t_i)$ is a degenerate polygon, i.e., $\mathbf{W}_c(t_i)$ is a point or a segment. We cease further constructions in this case also.)

The algorithm described is close to Graham's algorithm [17] for constructing the convex hull of a collection of points in the plane. The main feature of our algorithm is that it uses the specific properties of the problem. Namely, at the beginning of the convexification process, the places of possible violation of local convexity are known. These places are determined by the normals from the polygon $\mathcal{Q}(t_i)$ of the second player constraint, that is, by the vectors from the collection S. The information about the violation of the local convexity continues to be known through iterations of the procedure and is stored in the collections $S^{(k)}$. Therefore, the summary number j of iterations is not too large: usually, three or five iterations are necessary to stop the procedure.

3 Singular Surfaces

In this section, we suppose that the controls of both players are scalar, that is, the sets P and Q in system (1) are segments. So the sets $\mathcal{P}(t_i)$ and $\mathcal{Q}(t_i)$ are also segments in the space y_1, y_2 for any instant t_i.

Let's suppose in addition that the segments $\mathcal{P}(t_i)$ and $\mathcal{Q}(t_i)$ are not parallel and more than that, each of them is not parallel to any edge of the polygon $\mathbf{W}_c(t_{i+1})$. Let's call this supposition "the nonparallelism condition."

Now we shall describe the algorithm for constructing singular surfaces. Similarly to [6], the behavior of the optimal motions is taken as the base of consideration. In Figure 1, the situations of dispersion, equivocalness, and the situation of switching are shown. As for the dispersal surface, two optimal motions go out from any point. One motion goes to one side from the surface, and the second one goes to the other side. When a motion reaches an equivocal surface, it splits into two motions. One of these motions leaves the surface, and the second one goes upon it. The latter motion also splits in the same way at each future instant. Switching surfaces gives fracture to optimal motions, but without splitting. There are two types of switching surfaces: with leaving and without it.

For the game (1), each optimal motion goes along the boundary of the corresponding level set of the value function. So, the basis of our consideration is the investigation of collection (or bundle) of optimal motions taken from the boundary of a level set. Points where such motions have some peculiarities are called *singular*. Figure 2 schematically shows equivocal and dispersal situations.

Making numerical computations, we analyze optimal motions appearing in the approximating differential game (3). Detection and classification of singular points are made by means of information accumulated during the process of convexification of the function γ.

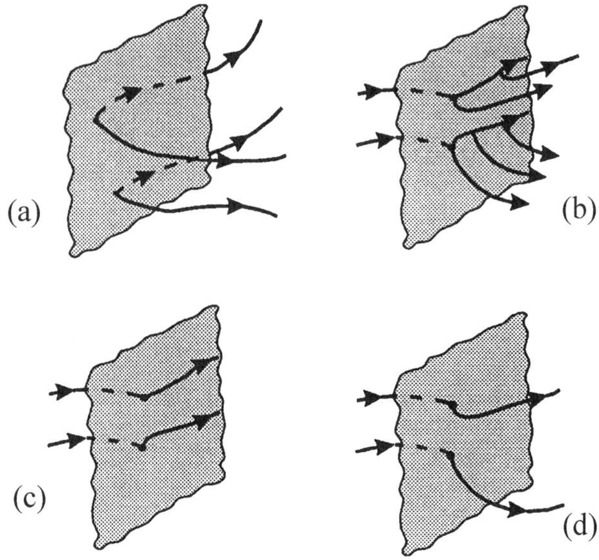

Figure 1: Behavior of the optimal motions near singular surfaces of different type: (a) dispersal surface; (b) equivocal surface; (c) switching surface without leaving; (d) switching surface with leaving.

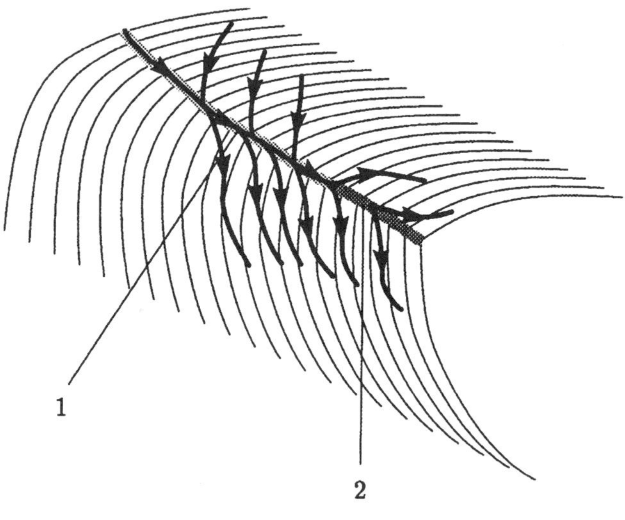

Figure 2: Scheme of a fragment of a level set boundary with optimal motion bundle. *Notations*: 1—equivocal line; 2—dispersal line.

Singular surfaces in the three-dimensional space are constructed on the base of computed singular points as a result of an additional special procedure.

3.1 Optimal Motions and Singular Points

For each time interval $[t_i, t_{i+1}]$, we have the game (3) with simple motions as the approximating game. The first player tries to transfer the system from the polygon $\mathbf{W}_c(t_i)$ onto the polygon $\mathbf{W}_c(t_{i+1})$; the second player hinders to prevent it. Entering the discrimination of the second player, determine the optimal motions.

Let's fix some arbitrary point y_0 on the boundary of the polygon $\mathbf{W}_c(t_i)$. The second player's control (constant on the interval $[t_i, t_{i+1})$) is called the *optimal* one if the first player cannot direct the resulting motion from the point y_0 into the interior of the polygon $\mathbf{W}_c(t_{i+1})$. For fixed optimal control of the second player, the first player's control (also constant on the same interval) is called *optimal parrying* if the corresponding motion comes to the boundary of the set $\mathbf{W}_c(t_{i+1})$. The motion generated by the optimal players controls is called *optimal*.

The control v^* satisfying the condition of maximum $v^* = \arg\max\{l'v : v \in \mathcal{Q}(t_i)\}$ is called the *extremal* control of the second player on the vector l. Similarly, the *extremal* control of the first player on the vector l is the control u^*, which satisfies the condition of minimum $u^* = \arg\min\{l'u : u \in \mathcal{P}(t_i)\}$.

It is easy to see that if y_0 is an internal point of some edge of the polygon $\mathbf{W}_c(t_i)$, then the constant control of the second player is optimal if and only if it is extremal on the normal vector to $\mathbf{W}_c(t_i)$ at the given point. The optimal parrying control of the first player is extremal on the same vector.

If y_0 is a vertex of the polygon $\mathbf{W}_c(t_i)$, then two normals correspond to the vertex. The structure of the optimal controls is described by the following statements.

Proposition 3.1. *For any vertex of the polygon $\mathbf{W}_c(t_i)$, the second player's control, being extremal at least on one of two normal vectors at this vertex, is optimal. With that, the optimal parrying control of the first player is extremal on the same vector.*

A proof of this statement does not use the supposition of a scalar character of players' controls. Hence, the condition of nonparallelism is not used also. In the following statement, these suppositions are essential.

Proposition 3.2. *Let the players' controls be scalar, and the condition of nonparallelism is satisfied. Then, for any vertex of the polygon $\mathbf{W}_c(t_i)$, the totality of the optimal controls of the second player consists only of the extremal controls on two normal vectors of the polygon $\mathbf{W}_c(t_i)$ at this vertex.*

The point y_0 is called *regular* if the optimal motion emanating from this point is unique and generated by the extreme players' controls. A point, which is not regular, is called the *singular* one. Here, "extreme" means that the control value is constant and is a boundary point of the segment $\mathcal{P}(t_i)$ (or $\mathcal{Q}(t_i)$).

3.2 Classification of Singularities

The classification of the singular points of the polygon $\mathbf{W}_c(t_i)$ is described below. It is based on the analysis of the character of the optimal motions emanated from these points. For the classification, we use two marks. These marks are attached to normals participated in the process of convex hull construction of the function $\gamma(\cdot, t_i)$. The mark FS ("former suspicious") is added to normals denoted as "suspicious" during the convexification process but, after the process, remaining in the final collection $\overline{L}(t_i)$. The mark $N\mathcal{P}$ is added to the normals, which were taken from the set $\mathcal{P}(t_i)$. The classification is represented in Table 1.

Table 1 deals with the normals from the final collection $\overline{L}$. Individual normals or pairs of neighbors are analyzed. The number in the first column describes the number of normals considered in some combination of normals. In the second column, the marks are represented whose presence is checked out on the considered normals. With that, if only one mark is shown for some concrete vector, then the second mark is considered to be absent. The third column contains the additional condition. Marks in the second column, together with the satisfaction of the additional condition, determine in whole the singularity type of the object from the fourth column. Such an object can be the polygon $\mathbf{W}_c(t_i)$ vertex, which is incident for edges determined by normals (row 1), either the edge corresponding to the considered normal (row 3) or one of two shown normals (row 2). Inside the case of row 1, a subdivision exists, determined by the condition from the fifth column. In the last column, the type of singularity is indicated, which is finally prescribed to the mentioned object. A type of singularity is determined by the character of the optimal motions emanating from the marked point or points of the marked edge. The names of singularities are coordinated with ones used in the Isaacs' book.

The term "$\mathcal{P}$-normal" ("$\mathcal{Q}$-normal") means a normal taken from the set $\mathcal{P}(t_i)$ [$\mathcal{Q}(t_i)$].

Table 1: Table of classification of singularities

<table>
<tr><th>Number of normals</th><th>Flags</th><th>Additional condition</th><th>Object for marking</th><th>Secondary condition</th><th>Type of singularity</th></tr>
<tr><td rowspan="2">2</td><td rowspan="2">FS, FS</td><td rowspan="2">$\mathcal{Q}$-normal is strictly between</td><td rowspan="2">Vertex</td><td>$\mathcal{P}$-normal is not between</td><td>Dispersal for 2nd</td></tr>
<tr><td>$\mathcal{P}$-normal is between</td><td>Dispersal</td></tr>
<tr><td>2</td><td>FS, $N\mathcal{P}+FS$</td><td>$\mathcal{Q}$-normal is strictly between</td><td>$\mathcal{P}$-normal edge</td><td></td><td>Equivocal</td></tr>
<tr><td>1</td><td>$N\mathcal{P}$</td><td></td><td>Edge</td><td></td><td>Switching for 1st</td></tr>
</table>

The term "dispersal" corresponds to the situation when two optimal motions are emanated from the given point. One of these motions is generated by players' controls, which are extremal on one of the regarded vectors with the mark FS. The second motion corresponds to the players' controls, which are extremal on the second vector with the same mark. With that, all controls are extreme in its segments.

The term "equivocal" corresponds to the situation when, from the vertex, determined by the normals with marks FS and $\mathcal{NP}$+FS, two optimal motions go out. One motion is generated by players' controls, which are extremal on the vector with the mark FS. The second one is generated by extremal controls on the vector with the mark $\mathcal{NP}$+FS. With that, both players' controls get their extreme values on the first motion. But on the second motion, only the second player's control is extreme. (It is namely the difference of the equivocal case from the dispersal one.) The latter motion comes to such a vertex of the polygon $\mathbf{W}_c(t_{i+1})$, where the cone generated by two normals contains the considered normal with the mark $\mathcal{NP}$+FS. Every other point of the marked edge emanates only one motion, which comes to the mentioned vertex.

The term "switching for the first player" corresponds to the case when only one optimal motion emanates from each internal point of the marked edge. It is generated by an extreme control of the second player and a nonextreme control of the first player. The difference from the equivocal case is in the fact that only one motion emanates from both vertices of the edge. With that, the first player's control gets one extreme value at one vertex of the edge and the opposite value at another vertex.

On different sides of the equivocal edge, both players' controls change. In the dispersal situation, either only the second player can change its control or both players change their controls together. So, in the dispersal case, a special subdivision appears. In the switching situation, only the first player changes its control.

It can be proved that any point on the boundary of the polygon $\mathbf{W}_c(t_i)$ is the regular one if it is not a marked vertex or is not included to any marked edge. And vice versa, any point, which is a marked vertex or included to a marked edge, is singular, except maybe the case when it is an endpoint of a marked edge.

The proof of these facts is not too complicated but is quite large and has a combinatory character. The main idea of the proof contains consideration of all possible variants of local behavior of optimal motions in the game with simple motions on the interval $[t_i, t_{i+1}]$ and a polygonal target set $\mathbf{W}_c(t_{i+1})$. The polygon $\mathbf{W}_c(t_i)$ is the solvability set for this problem. Optimal motions go from the boundary of $\mathbf{W}_c(t_i)$ to the boundary of $\mathbf{W}_c(t_{i+1})$. Regarding the nonparallelism condition, the whole number of variants of optimal motion behavior is not too large, and they are clearly connected to the Isaacs' classification. (It is important here that the players' controls are scalar, because otherwise the number of variants increases and, consequently, the number of singularity types also grows.) The problem of the analysis of optimal motion behavior can be considered separately.

But its solving becomes simpler by using the information accumulated during the process of convex hull construction for the function $\gamma(\cdot, t_i)$, that is, during construction of the polygon $\mathbf{W}_c(t_i)$. Different singularity types give their own ways for the convexification process. Thus, the terms used during this process (normals marks) are put to the base of singularity classification.

The authors plan to give the full proof in a separate paper.

3.3 Constructing Singular Surfaces

The data for the algorithm for classification of singularities are accumulated during the process of the convex hull constructing, and after that the algorithm of classification begins to work. Its work is actually reduced to the verification of the presence of situations described by rows of Table 1. As a result, the collection of the points and intervals with a description of the type of their singularity is obtained for the current backward-time section of the level set.

For a graphical presentation of the singular surfaces, a collection of the level sets is calculated on the given grid of magnitudes of the value function, and, simultaneously, the singular points and edges are detected on each section of each level set. Constructing the singular surfaces on the basis of the singular points and intervals is carried out in the program of visualization.

Note that the construction of singular surfaces on the basis of separate singular points is a reasonable euristic algorithm. Moreover, authors think that it is possible to invent examples where the "ideal" singular surfaces taken in a part of the game space consist of arbitrary small pieces with different types of singularity. For such "exotic" examples, the singularity classification itself is an open problem, especially in the case of numerical constructions. But when we investigate a "natural" problem, the configuration of singular surfaces is also natural, and it is relevant to speak about separate singular surfaces and their concatenation.

The validity of the elaborated algorithm was verified on the test example from [14]. In that work, the singular surfaces had been investigated analytically in detail. Results of our calculations coincide well with the results of the mentioned paper.

4 Examples

In this section, the examples of constructing the level sets of the value function and singular surfaces are represented for two differential games. The level sets ("tubes") are constructed with the time step $\Delta = 0.05$. The Gouraud shading is used for visualization of a tube surface built from separate sections $\mathbf{W}_c(t_i)$ by triangulation. The surface is illuminated by a dot radiant, whose position can be changed by the user. In contrast with the visualization of tubes, the visualization of singular surfaces is implemented now by the simplest methods.

Example 4.1. The system is a conflict-controlled oscillator

$$\dot{x}_1 = x_2 + v, \quad \dot{x}_2 = -x_1 + u, \qquad |u| \leq 1, \; |v| \leq 0.9, \; \varphi(x_1, x_2) = x_1^2 + x_2^2. \quad (4)$$

Figure 3, shows three level sets. They were calculated for the interval [0, 8] of the backward time τ. The sets correspond to the values $c = 1.05, 1.4, 2.7$ of the value function.

The internal tube terminates. The outer tubes are visualized nontransparently. To see the internal arrangement of the collection, two outer tubes in Figure 3(a) are dissected by a plane parallel to the coordinate axes x_1, τ. In Figure 3(b), all three tubes are dissected.

In Figure 4, the singular surfaces are shown for two viewpoints: Figure 4(a) corresponds to the view from the x_1-axis; Figure 4(b) is the view from the τ-axis. For constructing the singular surfaces, 24 level sets were calculated.

The surface near the τ-axis is dispersal. Some distance from the τ-axis, the equivocal surface is situated. It comes into the switching surface for the first player and dispersal one for the second player. The "empty" parts around the time axis can be filled by calculations with a smaller time step Δ of the backward procedure.

Example 4.2. Let the dynamics be described by relations

$$\dot{x}_1 = x_1 + 2x_2 + v, \quad \dot{x}_2 = x_2 + u, \qquad |u| \le 1,\ |v| \le 0.9,\ \varphi(x_1, x_2) = x_1^2 + x_2^2. \tag{5}$$

This example has no any mechanical interpretation. But the level sets are beautiful: they look like a drawing-pin.

In Figure 5(a), the level set for $c = 7.0$ is represented. A termination of the tube takes place at $\tau = 6.9$. One of two fracture lines is evidently visible on the tube surface. A sharp narrowing of the tube in time is explained by the fact that system (5) is asymptotically stable in the backward time.

In Figure 5(b), two tubes are shown for $c = 4.0, 7.0$. On each of them, the contours of sections are represented. The contours are polygons calculated by the computer program with the time step 5Δ.

The dependence of the singular surfaces on the second player's constraint Q is demonstrated in Figure 6. Let

$$\begin{aligned} \dot{x}_1 &= x_1 + 2x_2 + v_1, \qquad & |u| \le 1,\ (v_1, v_2)' \in Q, \\ \dot{x}_2 &= x_2 + u + v_2, & \varphi(x_1, x_2) = x_1^2 + x_2^2. \end{aligned} \tag{6}$$

Four variants of the set (segment) Q are considered:

(a) $Q = \{(v_1, v_2)' : |v_1| \le 1.1,\ v_2 = 0\}$;
(b) $Q = \{(v_1, v_2)' : |v_1| \le 1.1/\sqrt{2},\ v_2 = v_1\}$;
(c) $Q = \{(v_1, v_2)' : |v_1| \le 0.9,\ v_2 = 0\}$;
(d) $Q = \{(v_1, v_2)' : |v_1| \le 0.9/\sqrt{2},\ v_2 = v_1\}$.

Cases (a) and (b) correspond to the situation when the second player is "stronger" than the first one (the length of Q is greater than the length of P). When the segment Q goes to P, the dispersal part near the τ-axis increases while the equivocal part decreases [compare Figures 6(a) and 6(b)]. Approaching the outer dispersal surface of the second player to the switching surface of the first player happens.

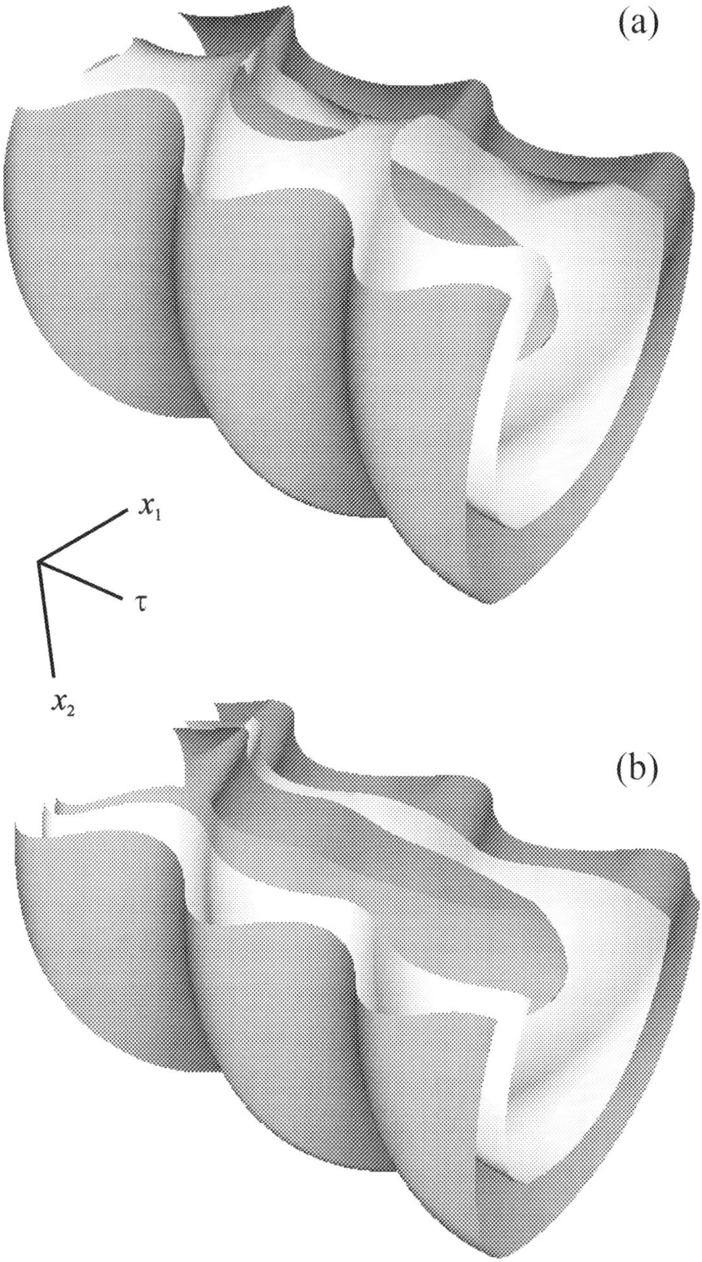

Figure 3: Three level sets of the value function for game (4).

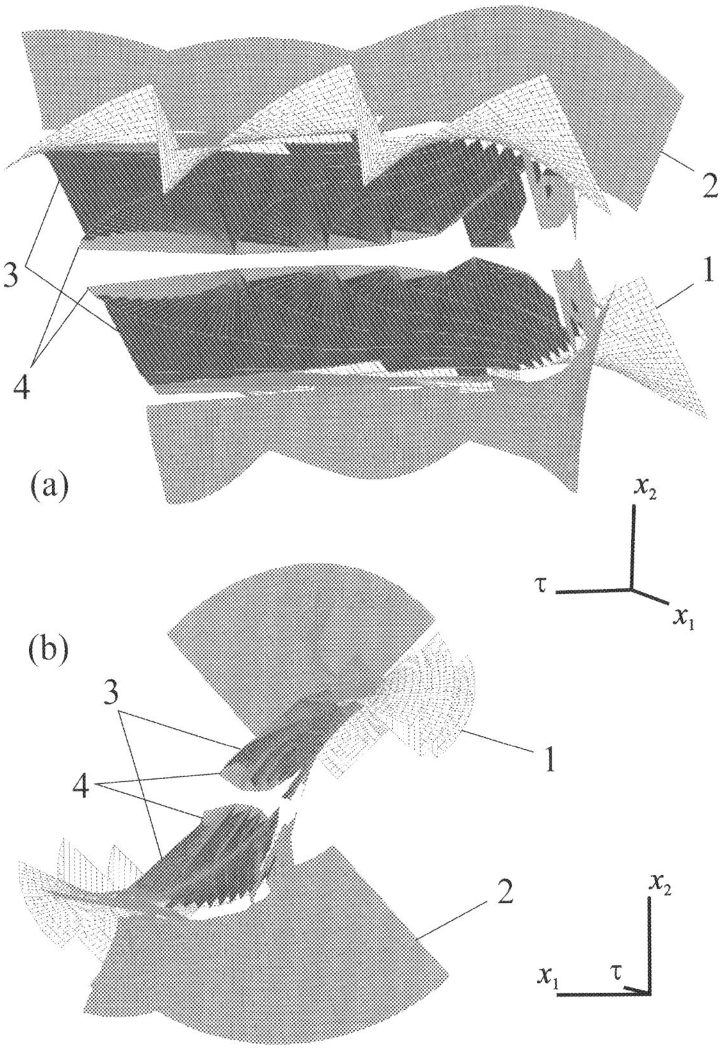

Figure 4: Singular surfaces for game (4). (a) View from the x_1-axis. (b) View from the τ-axis. *Notations*: 1—switching surface for the first player; 2—dispersal surface for the second player; 3—equivocal surface; 4—dispersal surface.

Cases (c) and (d) correspond to the opposite situation: the second player is "weaker" than the first one. When the segment Q goes to P, the equivocal part increases [see Figures 6(c) and 6(d)].

For construction of the singular surfaces 14 level sets were calculated.

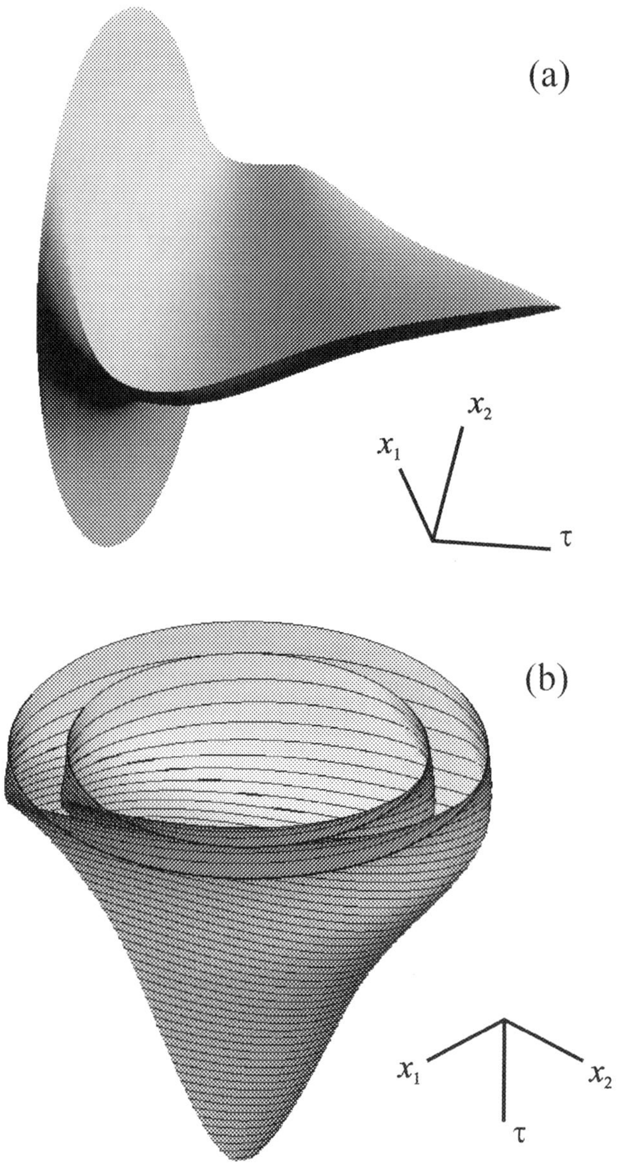

Figure 5: Level sets for game (5) from different points of view.

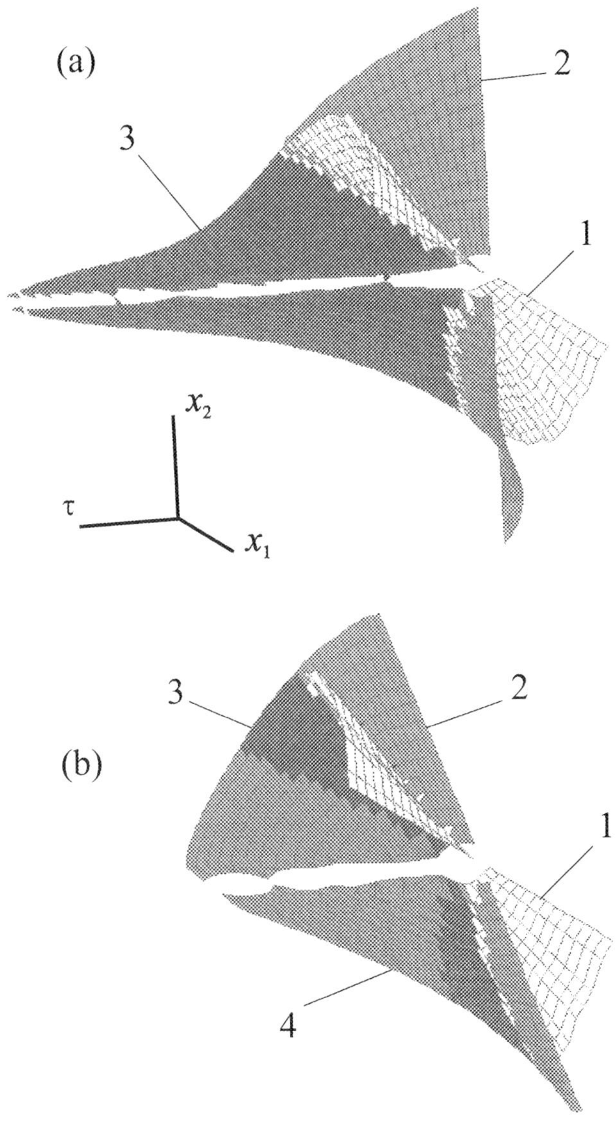

Figure 6: (a), (b). Singular surfaces for game (6). The case of a strong second player. *Notations*: 1—switching surface for the first player; 2—dispersal surface for the second player; 3—equivocal surface; 4—dispersal surface.

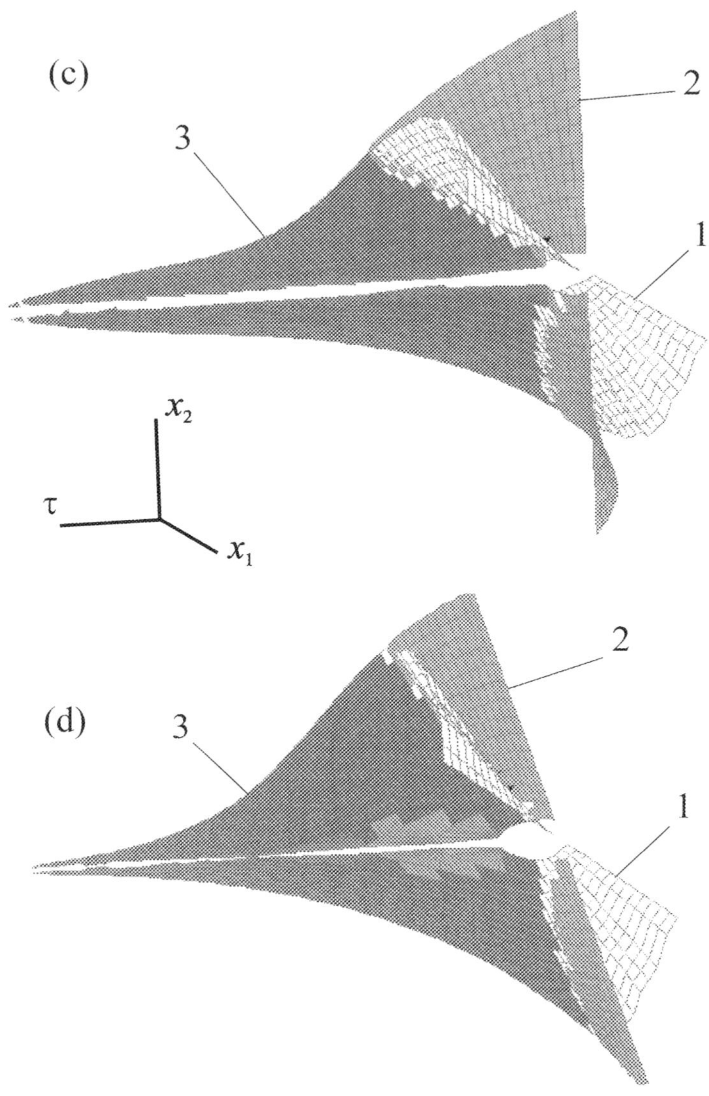

Figure 6: (cont.) (c), (d). Singular surfaces for game (6). The case of a weak second player. *Notations*: 1—switching surface for the first player; 2—dispersal surface for the second player; 3—equivocal surface.

Acknowledgments

This research was supported by the Russian Foundation for Basic Research under Grants no. 97-01-00458 and 00-01-00348.

REFERENCES

[1] Averbukh, V. L., Kumkov, S. S., Patsko, V. S., Pykhteev, O. A., and Yurtaev, D. A. Specialized visualization systems for differential games, pp. 301–309, in *Progress in Simulation, Modeling, Analysis and Synthesis of Modern Electrical and Electronic Devices and Systems*, N. E. Mastorakis, ed., World Scientific and Engineering Society Press, 1999.

[2] Bernhard, P. Singular surfaces in differential games, pp. 1–33, in *Lecture Notes in Control and Information Sciences*, Vol. 3, P. Hagedorn, H. W. Knobloch and G. J. Olsder, eds., Springer-Verlag, Berlin, 1977.

[3] Botkin, N. D. Evaluation of numerical construction error in differential game with fixed terminal time, *Problems of Control and Information Theory*, no. 11, pp. 283–295, 1982.

[4] Botkin, N. D., Kein, V. M., and Patsko, V. S. The model problem of controlling the lateral motion of an aircraft during landing, *J. Appl. Math. Mech.*, **48**(4):395–400, 1984.

[5] Botkin, N. D. and Zarkh, M. A. Estimation of error for numerical constructing absorption set in linear differential game, pp. 39–80, in: Ref. [22] (in Russian).

[6] Isaacs, R. *Differential Games*, p. 384, John Wiley and Sons, New York, 1965.

[7] Isakova, E. A., Logunova, G. V. and Patsko, V. S. Computation of stable bridges for linear differential games with fixed time of termination, pp. 125–158, in Ref. [22] (in Russian).

[8] Krasovskii, N. N. *Games Problems about Contact of Motions*, p. 420, Nauka, Moscow (in Russian), 1970.

[9] Krasovskii, N. N. and Subbotin, A. I. *Positional Differential Games*, p. 455, Nauka, Moscow (in Russian), 1974.

[10] Krasovskii, N. N. and Subbotin, A. I. *Game-theoretical Control Problems*, p. 517, Springer-Verlag, New York, 1988.

[11] Lewin, J. and Olsder, G. J. Conic surveillance evasion, *J. Opt. Th. Appl.*, **27**(1):107–125, 1979.

[12] Melikyan, A.A. *Generalized Characteristics of First Order PDEs: Application in Optimal Control and Differential Games*, p. 310, Birkhauser, Boston, 1998.

[13] Merz, A. W. *The Homicidal Chauffeur—a Differential Game*, Ph.D. dissertation, Stanford University, 1971.

[14] Patsko, V. S. and Tarasova, S. I. Nonregular differential game of approaching, *Eng. Cybernetics*, **22**(5):59–67, 1985.

[15] Ponomarev, A. P. and Rozov, N. H. Stability and the convergence of the Pontryagin's sums, *Vestnik Moskov. Univ., Ser. Vyčisl. Mat. Kibernet.*, no. 1, pp. 82–90 (in Russian), 1978.

[16] Pontryagin, L. S. Linear differential games, 2, pp. 910–912, *Soviet Math. Dokl.*, no. 8 (in Russian), 1967.

[17] Preparata, F. P. and Shamos, M. I. *Computational Geometry. An Introduction*, p. 390, Springer-Verlag, New York, 1985.

[18] Pschenichnyi, B. N. The structure of differential games, *Soviet Math. Dokl.*, no. 10, pp. 70–72, 1969.

[19] Pschenichnyi, B. N. and Sagaidak, M. I. Differential games of prescribed duration, *Cybernetics*, **6**(2):72–83, 1970.

[20] Shinar, J., Medinah, M., and Biton, M. Singular surfaces in a linear pursuit-evasion game with elliptical vectograms, *J. Opt. Th. Appl.*, **43**(3):431–456, 1984.

[21] Shinar, J. and Zarkh, M. Pursuit of a faster evader—a linear game with elliptical vectograms, *Proc. Seventh Int. Symp. Dynamic Games*, pp. 855–868, Yokosuka, Japan, 1996.

[22] Subbotin, A. I. and Patsko, V. S. (eds.) *Algorithms and Programs for Solving Linear Differential Games*, p. 295, Institute of Mathematics and Mechanics, Sverdlovsk (in Russian), 1984.

[23] Taras'ev, A. M. and Ushakov, V. N. Algorithm for constructing stable bridge in the linear problem of approach with convex target, pp. 82–90, *Investigations of Minimax Control Problems*, A. I. Subbotin and V. S. Patsko, eds., Institute of Mathematics and Mechanics, Sverdlovsk (in Russian), 1985.

Parallel Algorithms for the Isaacs Equation

Maurizio Falcone
University of Rome "La Sapienza"
Dipartimento di Matematica
Rome, Italy

Piero Lanucara
University of Rome "La Sapienza"
Dipartimento di Matematica
Rome, Italy

Monica Marinucci
CASPUR
Rome, Italy

Abstract

In this paper we apply a domain decomposition technique to construct an approximation scheme for the Isaacs equation in $\mathbb{R}^n$. The algorithm is presented for a two-domain decomposition, and some hints are given for the case of d subdomains having crossing points. The parallel algorithm is proved to have the same fixed point as the serial algorithm so that previous results guarantee convergence to the viscosity solution of the Isaacs equation. The efficiency of the algorithm is discussed by analyzing some numerical tests.

1 Introduction

In this paper we deal with the construction of a parallel algorithm for the solution of pursuit–evasion games. We focus our attention on the major numerical difficulty of the dynamic programming approach, which consists of the solution of the Isaacs equation. In fact, once we have computed that solution (the value function if the game has a value), we can also compute a synthesis of optimal feedback controls and derive from that knowledge the optimal trajectories starting in a subset of the domain where the solution has been computed. The analysis of convergence of the controls and of the corresponding approximate optimal trajectories goes beyond the scope of this paper. For the discussion of these aspects, we refer the reader to [3], [5], [8], and [19]. Let's concentrate on the computation of the value function. The need for parallel algorithms is essentially motivated by the huge number of nodes necessary to accurately solve the Isaacs equation. It is worthwhile to note that, in general, if a game is set in $\mathbb{R}^n$, then the value function depends on $2n$ variables, that is, the Isaacs equation is set in $\mathbb{R}^{2n}$. However, when the game dynamics depend on the players' relative positions, the problem can be written introducing new variables (the relative coordinates) so that the Isaacs equation can be set in $\mathbb{R}^n$. Since we are interested in algorithms that can solve the general case and we would like to solve problems in high dimensions ($n > 3$), the above remark does not provide a feasible solution. For problems in low dimensions, it should be observed that a large number of nodes is necessary only where there are abrupt changes in the gradient of the value function. The knowledge of the location of gradient discontinuities is crucial to obtain an accurate synthesis of optimal feedback controls; to this end one could also use adaptive grid algorithms concentrating the nodes around the jumps (see [12] for an adaptive grid algorithm for a Bellman equation). Although that is an interesting problem, we defer it to a future work, limiting ourselves here to the analysis of a class of parallel algorithms on a *fixed grid* using the domain decomposition technique. Those algorithms seem to be easier to implement and, as we say, will be necessary for problems in high dimensions.

The basic idea of the domain decomposition technique is that one can divide the computation, assigning subsets of nodes to each processor of a parallel machine. The most challenging architecture is a MIMD (multiple instructions, multiple data) machine, where each processor can perform different tasks and has its own local memory (see [11] for an introduction to parallel architectures and algorithms). The processors are linked by a network or by a fast switch; this is the case of the IBM/SP2 or of a cluster of workstations, for example.

The domain decomposition usually consists of a simple geometrical splitting of the original domain Ω into d subdomains Ω_r, $r = 1, \ldots, d$. The subdomains are chosen in such a way that their boundaries (or interfaces) are as simple as possible (e.g., a rectangle is divided into rectangles). Another possibility is to make a subdivision based on the physical properties of the dynamics, for example, separating low-speed and high-speed regions, but in that case the boundaries of the subdomains could be rather complicated. In any case, to obtain a correct global solution in Ω by a computation divided on d processors, one has to introduce transmission boundary conditions on the interfaces and pass the information from one processor to the others. This requires fast connections between the processors and message passing. The crucial point in a domain decomposition parallel algorithm is to determine which conditions at the interfaces will guarantee the convergence to the correct global solution. A practical evaluation of the algorithm is also needed to check that the message-passing overhead does not destroy all the advantages of distributing the computation load on several processors.

We should mention that the serial algorithm that is the basis for our parallel version was introduced in [4]. A series of convergence results for continuous as well as discontinuous value functions has been developed in the framework of viscosity solutions in the papers [1], [2], [4], [6], and [16] (see also the recent survey paper [5] for a more complete list of references on the numerical approximation of pursuit–evasion games). Another characterization of the solution (the so-called minmax solutions) to the Isaacs equation has been proposed by Subbotin in [17] and used for numerical purposes by Tarasyev in [18] and [19]. Finally, we should mention that our parallel algorithm extends to the Isaacs equation the splitting algorithm designed for the Bellman equation related to an infinite-horizon control problem in [10] (see also [7] for a nonoverlapping domain decomposition and [9]). The first numerical experiments in this direction were presented in [14].

The outline of the paper is as follows. Section 2 is devoted to a quick start on the basic convergence results for the serial algorithm. In Section 3 we introduce the splitting algorithm based on the domain decomposition and we prove its convergence. Finally, Section 4 is devoted to the evaluation of the performances of two test cases computing the speedup and efficiency with respect to the serial algorithm for an increasing number of processors and grid points.

2 The Background for the Serial Algorithm

Let's consider the dynamical system controlled by two players

$$\begin{cases} \dot{y}(t) = f(y(t), a(t), b(t)), & t > 0, \\ y(0) = x, \end{cases} \tag{1}$$

where $\dot{y}(t) \in \mathbb{R}^n$, is the state, and the functions a and b are the controls, respectively, for the first and second players. We assume

$$\begin{array}{ll} f : \mathbb{R}^n \times A \times B \longrightarrow \mathbb{R}^n & \text{is continuous,} \\ A, B & \text{are compact metric spaces,} \end{array} \tag{2}$$

and, for simplicity, for some constant L_f

$$|f(x, a, b) - f(y, a, b)| \le L_f |x - y|, \qquad \forall x, y \in \mathbb{R}^n, \quad a \in A, \quad b \in B. \tag{3}$$

Our admissible controls are $a \in \mathcal{A}$, $b \in \mathcal{B}$, where

$$\mathcal{A} := \{a : [0, +\infty[\longrightarrow A, \text{ measurable}\}, \tag{4}$$

$$\mathcal{B} := \{b : [0, +\infty[\longrightarrow B, \text{ measurable}\}. \tag{5}$$

We denote by $y_x(\,\cdot\,) = y_x(\,\cdot\,; a, b)$ the solution of (1). By standard results on ordinary differential equations, the trajectory of the system (to be understood in the Carathèodory sense) will exist at any time for any admissible a and b.

A closed set $\mathcal{T} \subseteq \mathbb{R}^n$ is also given, and the first time of arrival of the trajectory on $\mathcal{T}$ is defined as

$$t_x(a, b) \equiv \begin{cases} \min\{t : y_x(t; a, b) \in \mathcal{T}\} & \\ +\infty & \text{if } y_x(t; a, b) \notin \mathcal{T} \text{ for all } t. \end{cases} \tag{6}$$

The game is the following: the first player, "a," wants to minimize the arrival time, and the second player, "b," wants to maximize the same cost. For computational purposes, it is convenient to rescale the time variable by the nonlinear monotone transformation

$$\Phi(r) \equiv \begin{cases} 1 - e^{-\mu r} & \text{if } r < +\infty, \\ 1 & \text{if } r = +\infty, \end{cases} \tag{7}$$

which corresponds to substituting the time $t_x(a, b)$ by the discounted cost functional

$$J(x, a, b) = \Phi(t_x(a, b)) = \int_0^{t_x} e^{-\mu s}\, ds. \tag{8}$$

Note that, after the rescaling, the new time variable belongs to the interval [0, 1].

The above game is the generalization of the classical pursuit–evasion problem, where each player controls his own dynamics:

$$\dot{y}_1 = f_1(y_1, a), \qquad \dot{y}_2 = f_2(y_2, b), \qquad y_p \in \mathbb{R}^{n/2}, \quad p = 1, 2 \tag{9}$$

(naturally n will be even in that case), and the target is defined by

$$\mathcal{T}_0 = \{(y_1, y_2) \in \mathbb{R}^n : y_1 = y_2\}.$$

However, for technical reasons we will always consider the nonempty interior target

$$\mathcal{T} = \{(y_1, y_2) \in \mathbb{R}^n : |y_1 - y_2| \leq \varepsilon\},$$

for some $\varepsilon \geq 0$. The pursuer, a, wants to minimize the capture time t_x (or $\phi(t_x)$ if we use the renormalization) and the evader b, wants to avoid the capture or at least maximize t_x. It is well known (see, e.g., [3], Ch. VIII and the references therein) that, restricting to nonanticipating strategies, one can define the lower and upper values for the game as

$$v(x) := \inf_{\alpha \in \Gamma} \sup_{b \in \mathcal{B}} J(x, \alpha[b], b), \tag{10}$$

$$u(x) := \sup_{\beta \in \Delta} \inf_{a \in \mathcal{A}} J(x, a, \beta[a]), \tag{11}$$

where Γ and Δ are the nonanticipating strategies, respectively, for the first and second players. We say that the game has a value if the lower and upper values coincide, but even when they do not coincide, one can prove a dynamic programming principle and derive from it a characterization of the lower value of the game (i.e., the minimum time of capture). Provided some local capturability assumptions are satisfied, u is Lipschitz-continuous and will be the only viscosity solution of the Isaacs equation

$$\mu w + \min_{b \in B} \max_{a \in A} \{-f(x, a, b) \cdot Dw\} - 1 = 0, \qquad \text{in } \Omega \equiv \mathbb{R}^n \setminus \mathcal{T}, \tag{12}$$

coupled with the homogeneous Dirichlet boundary condition on the target

$$w = 0, \qquad \text{in } \partial\mathcal{T}. \tag{13}$$

Note that the Isaacs equation (12) has a nonconvex Hamiltonian, and this will require an extension of the usual definition of the viscosity solution to get uniqueness (see also [17] for a different notion of solution). In the sequel we will fix $\mu = 1$. The serial algorithm to compute the solution can be obtained via a time discretization (with time step $h \equiv \Delta t$) of the dynamics and of the payoff coupled with a projection on a grid with a fixed number of nodes (see [4], [5], and [6] for details). To simplify, assume that there exists a set Ω that is invariant with respect to the dynamics [e.g., assume $\text{supp}(f) \in \Omega$] so we can actually restrict the computation to Ω and construct a finite triangulation.

Following [4], we say that a polyhedron $Q \subset \mathbb{R}^n$ is *discretized with step k* if we are given a finite family of simplices $\{S_l\}$ such that $Q = \cup_l S_l$, $\text{int}(S_i) \cap \text{int}(S_j) = \emptyset$ for $i \neq j$, $k \equiv \max_l \text{diam}(S_l)$. As usual in the approximation of partial differential equations, any open set can be approximated from the interior by a polyhedron. Let us denote by x_i the vertices of the simplices of the triangulation and by $\mathcal{G} \equiv \{x_i\}_{i \in I}$ the family of the vertices (the grid). Let us denote by N the number of nodes in $\mathcal{G}$. Any point $x \in Q$ belongs to at least one simplex S_l, and it can be written as a convex combination of the vertices of S_l, that is,

$$x = \sum_{m \in I} \lambda_m x_m, \text{ where } \lambda_m \geq 0, \qquad \sum_{m \in I} \lambda_m = 1, \qquad \lambda_m = 0 \text{ if } x_m \notin S_l. \quad (14)$$

Note that, for simplicity, we always make the sums over the whole set of indices I although the vertices really involved in the computations are only those referring to the simplex containing the point x (i.e., only $n+1$ coefficients are really needed in $\mathbb{R}^n$).

We define the map $F : \mathbb{R}^N \to \mathbb{R}^N$ componentwise as follows:

$$F_i(V) \equiv \begin{cases} \gamma \max\limits_b \min\limits_a P_i(a, b, V) + 1 - \gamma & \text{if } x_i \in \mathcal{G} \setminus \mathcal{T}, \\ 0 & \text{if } x_i \in \mathcal{T} \cap \mathcal{G}, \end{cases} \quad (15)$$

where $\gamma \equiv e^{-h}$, $P_i(a, b, V) \equiv \sum_m \lambda_{im}(a, b) V_m$, and the coefficients λ_{im} are such that $z_i(a, b) \equiv x_i + hf(x_i, a, b)$,

$$\sum_m \lambda_{im}(a, b) x_m = z_i(a, b), \qquad \lambda_{im} \in [0, 1], \qquad \sum_m \lambda_{im} = 1. \quad (16)$$

As far as the algorithm is concerned, one first has to determine the simplex that contains z_i. This is quite simple if one uses a structured grid made by rectangles or ipercubes (in $\mathbb{R}^n$). The simplices are obtained simply by dividing the rectangular cells by two. Only n divisions (in $\mathbb{R}^n$) are necessary to determine which rectangular cell contains z_i. The second step is to compute the coefficients λ_i, which is done solving a linear system of size $n+1$ (in $\mathbb{R}^n$).

It can be proved that F has a unique fixed point V^* in $[0, 1]^N$; that fixed point is our approximate solution at the nodes of the grid. By linear interpolation, we can also define $w : Q \to [0, 1]$, which is the local reconstruction of the fixed point V^*. Then the function w will satisfy

$$\begin{cases} w(x) = \sum\limits_m \lambda_m w(x_m) & \text{if } x = \sum\limits_m \lambda_m x_m, \\ w(x_i) = \gamma \max\limits_b \min\limits_a w(x_i + hf(x_i, a, b)) + 1 - \gamma & \text{if } x_i \in \mathcal{G} \setminus \mathcal{T}, \\ w(x_i) = 0 & \text{if } x_i \in \mathcal{T} \cap \mathcal{G}. \end{cases} \quad (17)$$

Theorem 2.1. *Let* (2) *and* (3) *be verified. Moreover, assume that*

$$|f(x,a,b)| \leq M_f, \qquad \text{for all } x \in \partial\mathcal{T},\ a \in A,\ \text{and } b \in B,$$

and that $\mathcal{T}$ *is the closure of an open set with a Lipschitz boundary. Let* h_n, k_n, *and* k_n/h_n *converge to* 0*; and assume that there is a bounded, continuous viscosity solution* v *of* (12) *and* (13). *Then* w_n *converges uniformly to* v *as* $n \to \infty$ *in* $\overline{\Omega}$.

Moreover, the following rate of convergence has been established in [16].

Theorem 2.2. *Let* (2) *and* (3) *be verified and let* $\mu \geq L_f$. *Moreover, let us assume that* $\mathcal{T}$ *is the closure of an open set with Lipschitz boundary and that local controlability assumptions are satisfied (see [16]). Then there is a positive constant* C *such that, for all* h *and* k, *we have*

$$\sup_{x\in\Omega} |w^{h,k}(x) - v(x)| \leq Ch^{\rho}(1 + (k/h)^2), \tag{18}$$

where ρ *depends on the regularity of the solution and on the rescaling* (7).

The above results show that we can compute an approximation of v by the fixed-point algorithm defined in (15). Note that the use of interpolation in the local reconstruction will make the algorithm converge, provided the target $\mathcal{T}$ contains at least one single node of the grid $\mathcal{G}$.

3 The Domain Decomposition Algorithm

Let's analyze how we can modify the fixed-point algorithm in order to distribute the computation on a parallel machine. To simplify the presentation, we take $\Omega \in \mathbb{R}^2$. First we construct a domain decomposition splitting Ω into d subdomains Ω_r, $r = 1, \ldots, d$, by a number of piecewise regular curves Γ_j, $j = 1, \ldots, m$. Note that in this way the domains of the decomposition cross only at the interfaces (i.e., they have an empty interior intersection). However, we can always build from there a decomposition with overlappings between the domains considering the neighborhoods centered at each interface Γ_j,

$$\Gamma_j^{\delta_j} \equiv \Gamma_j + B(0,\delta_j), \qquad \delta_j > 0. \tag{19}$$

In fact, denote by L the number of internal interfaces that have a nonempty intersection with the boundary of Ω_r. For every index j such that

$$\partial\Omega_r \cap \Gamma_j \neq \emptyset,$$

we define first

$$\mathcal{Q}_{rj} \equiv (\Omega_r + B(0,\delta_j)) \cap \Gamma_j^{\delta_j} \tag{20}$$

and, finally,

$$\mathcal{Q}_r \equiv \Omega_r \bigcup \left(\bigcup_j \mathcal{Q}_{rj} \right). \tag{21}$$

The only restriction on δ_j will be that δ_j cannot be too large since we want to keep the overlapping regions as small as possible. In particular, the first assumption we make is that

$$\overline{\Omega}_i \cap \overline{\Omega}_j = \emptyset \text{ implies } \mathcal{Q}_i \cap \mathcal{Q}_j = \emptyset, \qquad \forall i, j = 1, \ldots, d. \tag{22}$$

That condition means that we cannot have new neighboring domains if we enlarge the domains Ω_r, $r = 1, \ldots, d$, that is, the overlapping regions cannot completely include the neighboring subdomains. Let's divide the nodes of $\mathcal{G}$ taking into account the subdomains $\mathcal{Q}_r$ and the location of the points

$$z_i(a, b) = x_i + hf(x_i, a, b).$$

For every $r = 1, \ldots, d$ and $b \in B$, we define the following sets:

$$A_r(x_i, b) \equiv \{a \in A : z_i(a, b) \in \mathcal{Q}_r\}; \tag{23}$$

$$\mathcal{G}_r \equiv \{x_i \in \mathcal{G} \cap \mathcal{Q}_r : \forall b \in B \, \exists a \in A \text{ such that } z_i(a, b) \in \mathcal{Q}_r\}; \tag{24}$$

$$\mathcal{G}_r^{\text{in}} \equiv \{x_i \in \mathcal{G}_r : x_i \notin \mathcal{T}\} \text{ and } \mathcal{G}_r^{\text{out}} \equiv \{x_i \in \mathcal{G} \cap \mathcal{Q}_r : x_i \notin (\mathcal{G}_r^{\text{in}} \cup \mathcal{T})\}. \tag{25}$$

The set $\mathcal{G}_r^{\text{in}}$ is the set of nodes in $\mathcal{Q}_r$ such that, for any choice of the second player (player b), it is always possible for player a to make a choice that keeps $z_i(a, b)$ in $\mathcal{Q}_r$, that is, on those nodes we are always able to compute a value just using the information on the nodes of $\mathcal{Q}_r$. Naturally, at the nodes in $\mathcal{T}$ there is nothing to compute since we just set their value to 0. The nodes belonging to G_r^{out} would require information coming from other subdomains no matter which control player a will choose. By the above remarks, we can define a local operator that will act just on the nodes belonging to $\mathcal{Q}_r$,

$$S_r(x_i, U, b) \equiv \begin{cases} \min\limits_{a \in A_r(x_i, b)} \gamma P_i(a, b) + 1 - \gamma, & \text{for } x_i \in \mathcal{G}_r^{\text{in}}, \\ 1 & \text{for } x_i \in \mathcal{G}_r^{\text{out}}, \\ 0 & \text{for } x_i \in \mathcal{T} \cap \mathcal{Q}_r. \end{cases} \tag{26}$$

Then we can define a global operator on $\mathcal{G}$ based on the family of local operators S_r, $r = 1, \ldots, d$. First define $S : \mathcal{G} \times \mathbb{R}^N \times B \to \mathbb{R}^N$,

$$S(x_i, U, b) \equiv \begin{cases} S_r(x_i, U, b), & \text{for } x_i \in \mathcal{Q}_r, \\ \min\limits_j [S_j(x_i, U, b)], & \text{for } x_i \in \bigcap\limits_j \mathcal{Q}_j. \end{cases} \tag{27}$$

Finally, we define the ith component of the fully discrete splitting operator as

$$S(x_i, U) \equiv \max_{b \in B} \{S(x_i, U, b)\}, \qquad \text{for every } x_i \in \mathcal{G}. \tag{28}$$

It should be noted that the above definition is used here to simplify the notations, but in the real algorithm the max operator in (28) is computed locally in every subdomain (i.e., separately in every processor) and does not need message passing. The only coupling between subdomains that requires message passing appears in the definition (27).

In order to simplify the presentation, let us consider a domain decomposition based on two subdomains with overlapping, $\mathcal{Q}_1$ and $\mathcal{Q}_2$, and denote by $\mathcal{Q}_0$ their intersection.

Now let's introduce the following assumptions:

A1. $\mathcal{Q}_0 \equiv \mathcal{Q}_1 \cap \mathcal{Q}_2 \neq \emptyset$;
A2. The time step h satisfies the bounds

$$0 < h < \frac{1}{M_f} \inf_{\substack{x \in \mathcal{Q}_1 \setminus \mathcal{Q}_0 \cap \mathcal{G} \\ y \in \mathcal{Q}_2 \setminus \mathcal{Q}_0 \cap \mathcal{G}}} \operatorname{dist}(x, y); \tag{29}$$

A3. $A_r(x_i, b) \neq \emptyset$, for $x_i \in \mathcal{Q}_r$, $r = 1, 2$;
A4. The triangulation of $\mathcal{Q}$ is such that each simplex is not crossing the interface between $\mathcal{Q}_1 \setminus \mathcal{Q}_0$ and $\mathcal{Q}_0$ and the interface $\mathcal{Q}_2 \setminus \mathcal{Q}_0$ and $\mathcal{Q}_0$.

The second assumption simply guarantees that the discrete dynamics cannot cross the overlapping region passing from a node $x_i \in \mathcal{Q}_1 \setminus \mathcal{Q}_0$ to a point in $\mathcal{Q}_2 \setminus \mathcal{Q}_0$. That assumption is necessary to reduce the memory storage of the splitting algorithm since the processor computing the solution in $\mathcal{Q}_j$, $j = 1, 2$ (i.e., applying the local operator S_r), will not require the values on the nodes belonging to the subdomains $\mathcal{Q}_r$, $r \neq j$. The third assumption is a compatibility condition between the domain decomposition and the vector field (we will see later how it can be removed). It implies that the regions of the domain decomposition have to be large enough. Finally, the last assumption simply means that the interfaces can be seen in the space discretization since they are formed by the sides of the triangles of the triangulation. Let's divide the nodes $x_i \in \mathcal{G}$ into three subsets depending on the regions that contain them. Let's introduce the sets of indices

$$I_r \equiv \{i : x_i \in \mathcal{Q}_r\}, \qquad I_r^{\text{in}} \equiv \{i : x_i \in \mathcal{G}_r^{\text{in}} \setminus \mathcal{Q}_0\}, \text{ for } r = 1, 2, \tag{30}$$

$$I_0 \equiv \{i : x_i \in \mathcal{Q}_0 \setminus \mathcal{T}\}, \qquad I_{\mathcal{T}} \equiv \{i : x_i \in \mathcal{T} \cap \mathcal{G}\}. \tag{31}$$

Let N_r, $r = 1, 2$, be the number of nodes in $\mathcal{Q}_r$. We define the discrete restriction operators $R_r : \mathbb{R}^N \to \mathbb{R}^{N_r}$, which select among the N components of a vector (representing the solution on the grid) those corresponding to the nodes belonging to the subdomain $\mathcal{Q}_r \setminus \mathcal{T}$,

$$R_r(U) = \{U_i\}_{i \in I_r \cup I_0}, \qquad \text{for } r = 1, 2. \tag{32}$$

Given the vectors V^0 and W^0 in $\mathbb{R}^N$, we define by recursion the two sequences

$$V^{n+1} = S(V^n), \qquad W^{n+1} = F(W^n), \tag{33}$$

where F is the the global operator introduced in Section 2 and S is the splitting operator defined in (27) and (28).

Theorem 3.1. *Let Assumptions* (A1)–(A4) *be satisfied. Moreover, let S and F be defined as in* (27) *and* (28) *and* (15) *and $V^0 = W^0$. Then $V^n = W^n$ for any $n \in \mathbb{N}$.*

Proof. Denote by $\Lambda_i(a,b)$ the ith row of the $N \times N$ matrix $\Lambda(a,b) = (\lambda_{ij}(a,b)) = (\lambda_j(x_i,a,b))$ defined in (16). We can write

$$F_i(W) \equiv \begin{cases} \gamma \max_b \min_a \Lambda_i(a,b) \cdot V + 1 - \gamma & \text{if } x_i \in \mathcal{G} \setminus \mathcal{T}, \\ 0 & \text{if } x_i \in \mathcal{G} \cap \mathcal{T}. \end{cases} \tag{34}$$

Let's compare the above definition with the splitting operator. First note that the nodes belonging to $\mathcal{T}$ always keep the same value, 0. Then F and S obviously coincide on $\mathcal{T}$.

Let's consider the nodes in $\mathcal{G} \setminus \mathcal{T}$. By Assumptions (A2) and (A4), the matrix $\Lambda(a,b)$ has a special structure. In fact, we can write

$$\Lambda(a,b) = \begin{pmatrix} \Lambda_{I_1 I_1}(a,b) & \Lambda_{I_1 I_0}(a,b) & \Lambda_{I_1 I_2}(a,b) \\ \Lambda_{I_0 I_1}(a,b) & \Lambda_{I_0 I_0}(a,b) & \Lambda_{I_0 I_2}(a,b) \\ \Lambda_{I_2 I_1}(a,b) & \Lambda_{I_2 I_0}(a,b) & \Lambda_{I_2 I_2}(a,b) \end{pmatrix}, \tag{35}$$

where all the elements depend on (a,b) and

$$\Lambda_{I_r I_s}(a,b) = \{\lambda_{ij}(a,b)\}, \qquad i \in I_r, \quad j \in I_s, \quad r,s = 0,1,2.$$

Assumptions (A2) and (A4) imply that

(i) $\Lambda_{I_1 I_2}(a,b) = 0$, $\Lambda_{I_2 I_1}(a,b) = 0$;

(ii) the operator S_1 will only require the coefficients belonging to the submatrix

$$\begin{pmatrix} \Lambda_{I_1 I_1}(a,b) & \Lambda_{I_1 I_0}(a,b) \\ \Lambda_{I_0 I_1}(a,b) & \Lambda_{I_0 I_0}(a,b) \end{pmatrix}, \tag{36}$$

whereas the operator S_2 will use only the coefficients in the submatrix

$$\begin{pmatrix} \Lambda_{I_0 I_0}(a,b) & \Lambda_{I_0 I_2}(a,b) \\ \Lambda_{I_2 I_0}(a,b) & \Lambda_{I_2 I_2}(a,b) \end{pmatrix}; \tag{37}$$

(iii) note that by (i) the sum of the rows of the two submatrices (36) and (37) will always give 1.

In the above notations the local operator S_r, $r = 1,2$, already defined in (27), can be written as

$$S_r(Z) \equiv \begin{cases} \min_{a \in A_r(x_i,b)} \gamma \Lambda_i(a,b) \cdot R_r(U) + 1 - \gamma, & \text{for } x_i \in \mathcal{G}_r^{\text{in}}, \\ 1 & \text{for } x_i \in \mathcal{G}_r^{\text{out}}, \\ 0 & \text{for } x_i \in \mathcal{T}, \end{cases} \tag{38}$$

where $Z \equiv (x_i, R_r(U), b)$. By definition, on the nodes belonging to the overlapping region, we will have two different values. Moreover, by (A3), $A_r(x_i,b)$ is not empty at least on the nodes belonging to $\mathcal{Q}_r$ and

$$A_1(x_i,b) \cup A_2(x_i,b) = A$$

by (23) and by the invariance condition in Ω. We conclude that, for any control $b \in B$, any vector $U \in \mathbb{R}^N$, and any $x_i \in \mathcal{G} \cap \mathcal{Q}_r \setminus \mathcal{T}$,

$$\min_{r=1,2} \min_{a \in A_r(x_i,b)} \gamma \Lambda_i(a,b) \cdot R_r(U) + 1 - \gamma = S(x_i,U,b). \tag{39}$$

The equivalence between the global and splitting operators on $\mathcal{G} \setminus \mathcal{T}$ follows by the relation

$$
\begin{aligned}
F_i(U) &= \max_{b \in B} \min_{a \in A} \gamma \Lambda_i(a,b) \cdot U + 1 - \gamma \qquad (40)\\
&= \max_{b \in B} S(x_i, U, b) = S(x_i, U). \qquad \square
\end{aligned}
$$

Corollary 3.1. *Let the assumptions of Theorem* 3.1 *be satisfied. Then the operators F and S have the same fixed point.*

It should be noted that if A3 is not satisfied, one can obtain the same result by a slight change of the definition of S in order to take into account the possibility $A_1(x_i, b)$ or $A_2(x_i, b) = \emptyset$. Say, for instance, that $A_1(x_i, b) = \emptyset$ and $A_2(x_i, b) = A$. Then we define

$$
S(x_i, U, b) \equiv \min\left(1, \min_{a \in A_2(x_i, b)} \gamma \Lambda_i(a,b) \cdot R_2(U) + 1 - \gamma\right), \qquad (41)
$$

that is, we assign the value 1 to the minimum over the empty set $A_1(x_i, b)$. However, in all our test problems assumption A3 will be satisfied.

We can now sketch the numerical *splitting algorithm* corresponding to the definition of V^n in (33) and starting at

$$
V^0 \equiv \begin{cases} 1 & \text{for } x_i \in \mathcal{G} \setminus \mathcal{T}, \\ 0 & \text{for } x_i \in \mathcal{G} \cap \mathcal{T}. \end{cases} \qquad (42)
$$

Step 0. Define $V^{1,0} = R_1(V^0) \in \mathbb{R}^{N_1}$ and $V^{2,0} = R_2(V^0) \in \mathbb{R}^{N_2}$. Set $n = 0$.
Step 1. Compute for $r = 1, 2$, $b \in B$,

$$
V_i^{r,n+1/2} = \min_{a \in A_h^r(x_i)} \gamma \sum_{j \in I_r \cup I_0} \lambda_{ij}^{(r)}(a,b) V_j^{r,n} + 1 - \gamma, \qquad i \in I_r \cup I_0.
$$

Step 2. For $r = 1, 2$, $b \in B$, compute

$$
V_i^{r,n+1} = \begin{cases} V_i^{r,n+1/2}, & i \in I_r, \\ \min\{V_i^{1,n+1/2}, V_i^{2,n+1/2}\}, & i \in I_0. \end{cases}
$$

Step 3. Compute

$$
\begin{aligned}
V_i^{n+1} &= \max_{b \in B} \{V_i^{r,n+1}\}_{r=1,2}, \quad \text{for } i = I \setminus I_{\mathcal{T}}, \\
V_i^{n+1} &= 0, \qquad \text{for } i \in I_{\mathcal{T}}.
\end{aligned}
$$

Step 4. Check a stopping criterion.
IF it is satisfied THEN STOP
ELSE
Increase n by 1 and GO TO *Step 1.*

Note that the definition of S guarantees $V_i^{1,n+1} = V_i^{2,n+1}$ for each $i \in I_0$. The above algorithm first splits the computation in each subdomain (step 1), making a link at the end of each iteration (steps 2 and 3). Its speed of convergence to the fixed point can be quite slow since the contraction-mapping coefficient is $\gamma = e^{-h}$.

Let's comment on the construction of a domain decomposition with more than two subdomains and give some hints for an efficient implementation. The first remark is that, to guarantee the convergence to the value function, one should carefully treat the internal conditions on the overlapping regions. In particular, the splitting operator should collect the values from all the processors acting on a single node x_i and compute the maximum. Moreover, for an efficient parallelization, some basic requirements must be satisfied:

1. The domain decomposition should be made by trying to maintain a balanced load between the processors. Typically, if the target is contained in just one subdomain and assigned to a single processor, it will result in an unbalanced load because there is nothing to compute on the nodes belonging to the target. The target should be divided into pieces assigned to different processors.
2. The overlapping regions should be as small as possible. Naturally, the width of the overlapping region strongly depends on the vector field and on the time–step h. A reasonable choice would be to reduce h in order to guarantee that the overlapping region is limited to a narrow band, but this will slow down the convergence of the algorithm (which is based on a fixed-point iteration). A good balance still depends on the problem.
3. The internal subdomains have more intersections with their neighboring subdomains if one compares them to the subdomains containing part of the boundary of Ω; this causes an unbalanced (and in some sense unavoidable) message-passing load. A possible remedy would be to make smaller overlapping regions between the internal subdomains to reduce the consequences on the performances of the algorithm.

4 Numerical Tests and Algorithm Performances

We present the numerical results obtained on two tests in $\mathbb{R}^2$ comparing the serial and the parallel algorithms. The parallel code has been implemented using the standard Message Passing Interface (MPI) library, both for *portability* and for *efficiency* of the program on a variety of hardware (cf. [15]).

In comparing serial and parallel code, we take into account the problem size N (the number of nodes in the grid being $N \times N$), the size of the overlapping regions, and the number of physical processors involved in the computation. The performances of the parallel code are measured in terms of two well-known parameters, the *speedup* A and the *efficiency* E. Let T_{ser} and T_{par} be the CPU times corresponding respectively to the execution of the serial and parallel algorithms (over N_P processors) for the solution of the same problem. We define,

$$A \equiv \frac{T_{\text{ser}}}{T_{\text{par}}}, \qquad E \equiv \frac{A}{N_P}. \tag{43}$$

Note that an ideal parallel algorithm without message-passing loads would have $A = N_P$ and $E = 1$, so that a parallel algorithm is considered to be efficient and have good performance as far as its values for A and E are close to the above ideal values. Interested readers could find useful information to build efficient parallel codes in [11].

The platform considered is a cluster of two Digital AlphaServers 3/400 4100 with a 4 21164 (400-MHz clock) processor each and two Gigabytes of global RAM.

The parallel code has been developed with an optimized version of MPI, particularly suitable for the digital cluster architecture, using shared memory on each machine and a (very fast) memory channel link for the communication of different processors.

4.1 Test 1: The Tag–Chase Game

Let's consider the following dynamics for the players:

$$\begin{cases} \dot{z}_1 = v_1 a, \\ \dot{z}_2 = v_2 b, \end{cases} \tag{44}$$

where $z_i \in \mathbb{R}^2$, and v_1 and v_2 are two positive real parameters representing the relative velocities of the two players. In the tag–chase game we choose $A \equiv B(0, 1)$, $B \equiv B(0, 1)$, and the (unbounded) target will be given by $\mathcal{T}_\varepsilon \equiv \{(z_1, z_2) \in \mathbb{R}^4 : d(z_1, z_2) \leq \varepsilon\}$. Studying the game in $\Omega \equiv [-1, 1]^2$ would imply that we have to restrict the target to $\mathcal{T}_\varepsilon \cap [-1, 1]^2$. However, note that Ω is not invariant with respect to the trajectories and that the time of capture just depends on the players' relative positions. The second remark suggests introducing the following change of coordinates in the plane variables:

$$y = x - z_1, \tag{45}$$

which corresponds to considering new dynamics with respect to a moving system of coordinates (see [13] for an extensive use of it in the study of several differential games). This new system has its origin at the position of the pursuer and is oriented in the direction of the evader, that is, the pursuer is fixed at the origin in the new coordinate system. The advantage of this change of coordinates is that the target $\mathcal{T}_\varepsilon$ is now a ball around the origin. It is then reasonable to impose the boundary condition $w = 1$ outside our domain of computation. This implies that if the evader exits from the domain of computation, he will never be captured. This also tells us that if we end the computation with a value $w = 1$ on a piece of the boundary of Ω, we have to enlarge the domain of computation to get the right information.

If we start the game at (z_1^0, z_2^0), the evader's optimal strategy will be to move as fast as he can along the line $\overline{z_1 z_2}$ in the opposite direction with respect to the purser's position. The pursuer's optimal strategy will be to follow the evader on the same line, moving at his maximum speed. A simple computation gives the explicit expression of T:

$$T(z_1^0, z_2^0) = [-\epsilon + |z_1^0 - z_2^0|]^+/(v_1 - v_2).$$

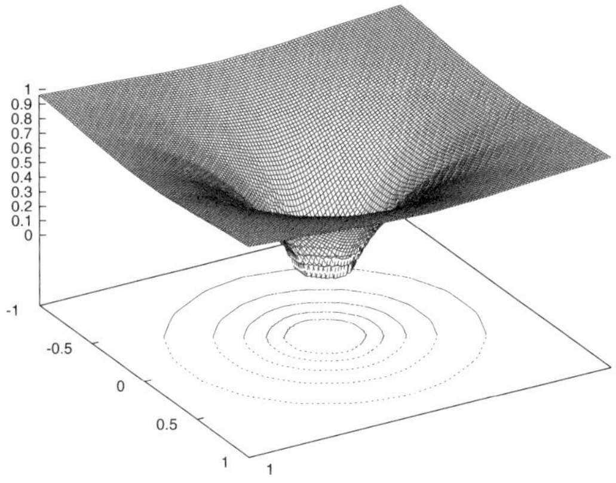

Figure 1: Test 1's value function.

Figure 1 shows the numerical approximation of the value function w (remember that w is the renormalized capture-time function). Note that the solution is differentiable outside the target.

4.2 Test 2: The Tag–Chase Game with Control Constraints

Let's consider new dynamics where the pursuer places a constraint on the directions of displacement. We keep the dynamics of the pursuer and the evader of Test 1, restricting the set of admissible controls for the pursuer to $A \equiv \{a \in B(0, 1) : a \cdot (0, 1) \geq -\sqrt{2}/2\}$. The evader can still take any direction in the unit ball, $B \equiv B(0, 1)$. We study the game in the square $\Omega \equiv [-1, 1]^4$ and set $\mathcal{T}_\varepsilon$ as in Test 1, reducing the state-space dimension by the same change of variables. Figure 2 shows the approximate value function of this game. It should be noted that the

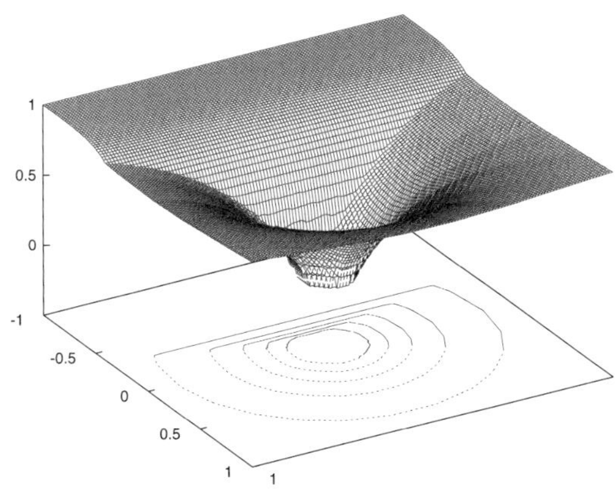

Figure 2: Test 2's value function.

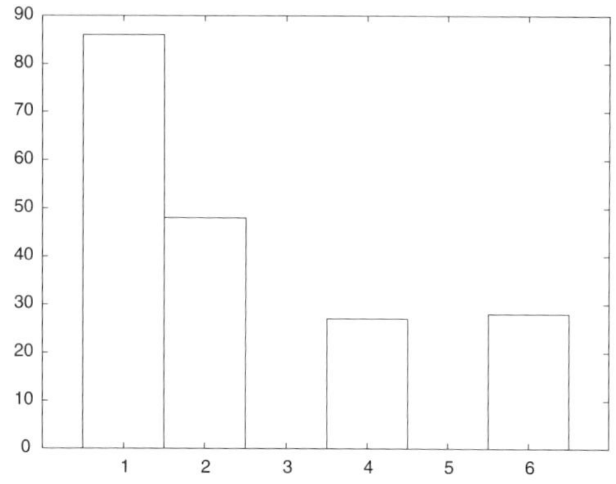

Figure 3: Test 1's elapsed time (sec.), 60×60.

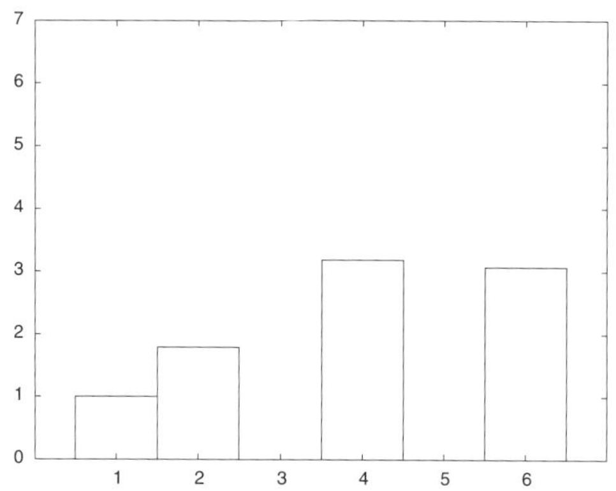

Figure 4: Test 1's speedup, 60×60.

solution here is just Lipschitz-continuous and the gradient of T jumps along two straight lines.

Let's now examine the performances of the parallel algorithm proposed in Section 3. Figures 3 to 8 contain the results for Test 1. The elapsed time (in seconds), the speedup and the efficiency are given for problems of sizes $N = 60$ and $N = 180$.

In the same way, Figures 9 to 14 contain the elapsed time (in seconds), the speedup, and the efficiency for Test 2, for problems of sizes $N = 60$ and $N = 180$.

Figures 15 and 17 give a summary of the speedup in the two tests. The speedup is plotted against the number of the processor, considering problems of increasing size ($N = 60$, 180).

Finally, Figures 16 and 18 give a summary of the efficiency in the two tests. The efficiency is plotted against the number of the processor, considering problems of increasing size ($N = 60$, 180).

We can make some general comments on the parallel implementation:

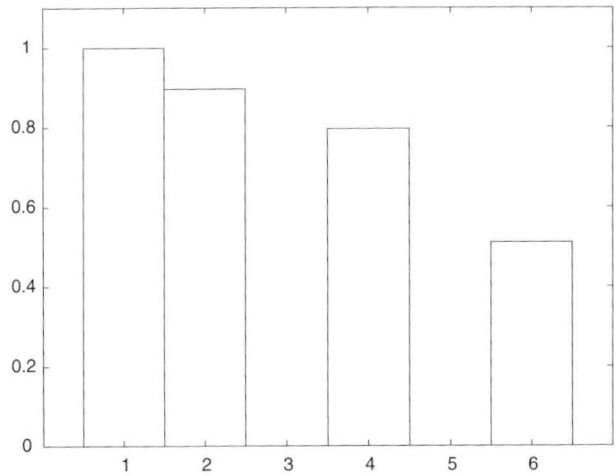

Figure 5: Test 1's efficiency, 60×60.

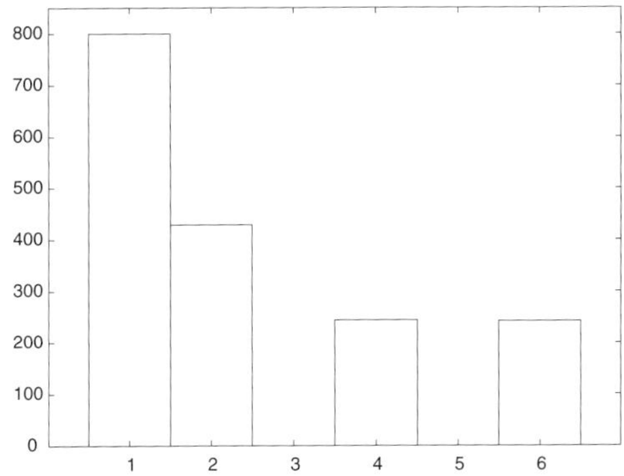

Figure 6: Test 1's elapsed time (sec.), 180×180.

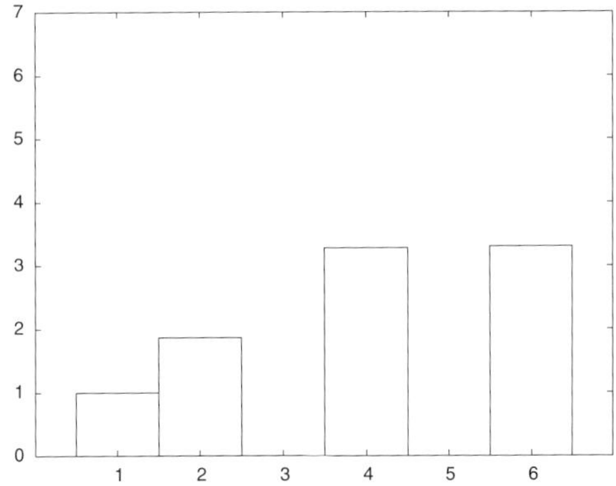

Figure 7: Test 1's speedup, 180×180.

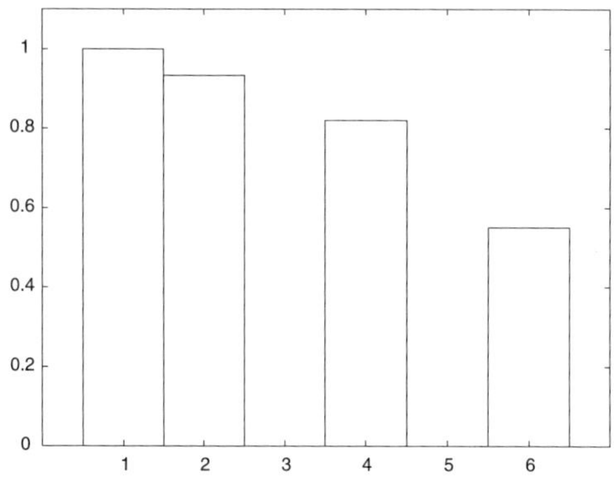

Figure 8: Test 1's efficiency, 180 × 180.

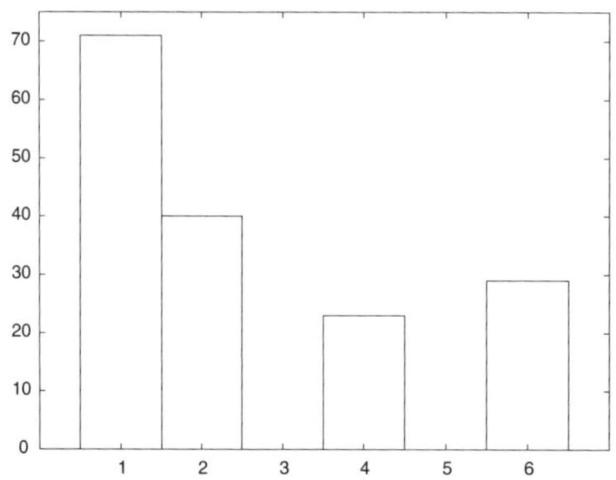

Figure 9: Test 2's elapsed time (sec.), 60 × 60.

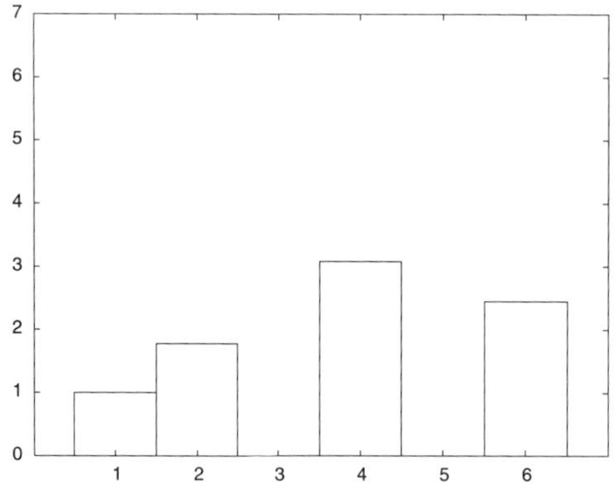

Figure 10: Test 2's speedup, 60 × 60.

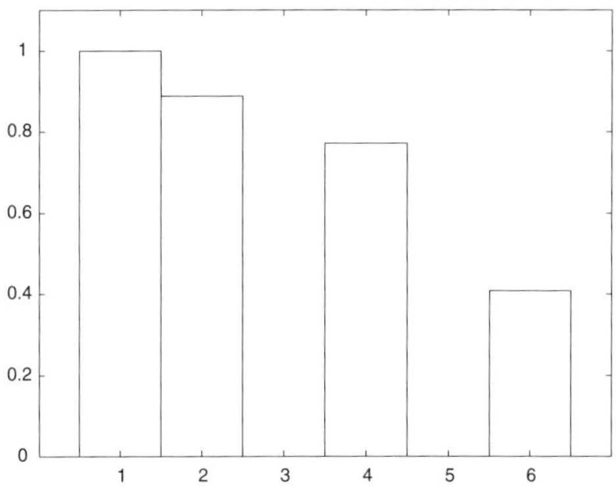

Figure 11: Test 2's efficiency, 60×60.

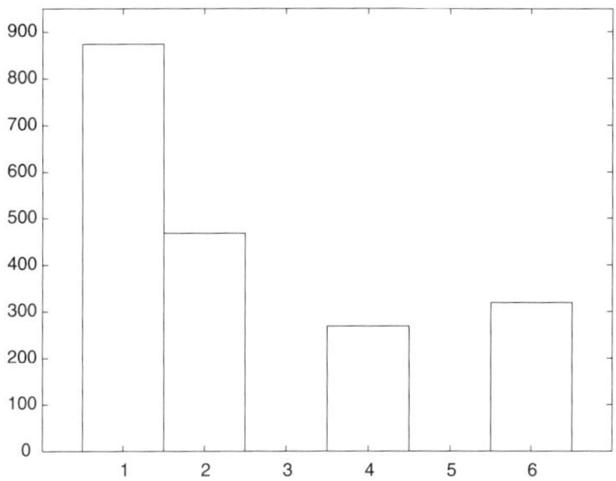

Figure 12: Test 2's elapsed time (sec.), 180×180.

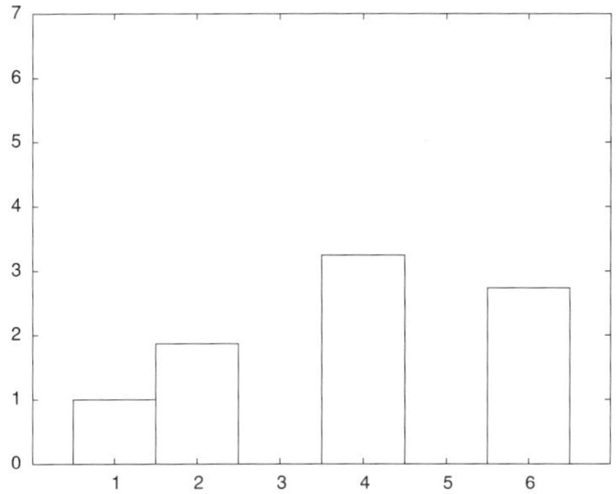

Figure 13: Test 2's speedup, 180×180.

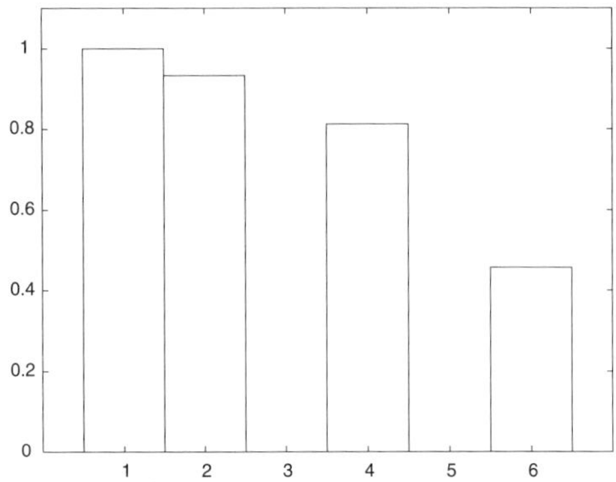

Figure 14: Test 2's efficiency, 180 × 180.

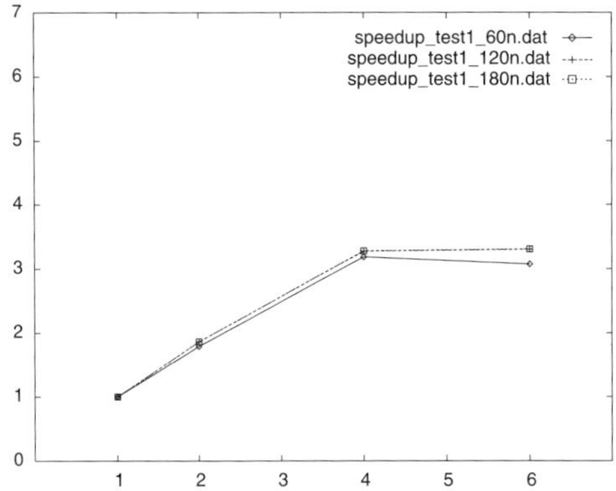

Figure 15: Test 1's global speedup.

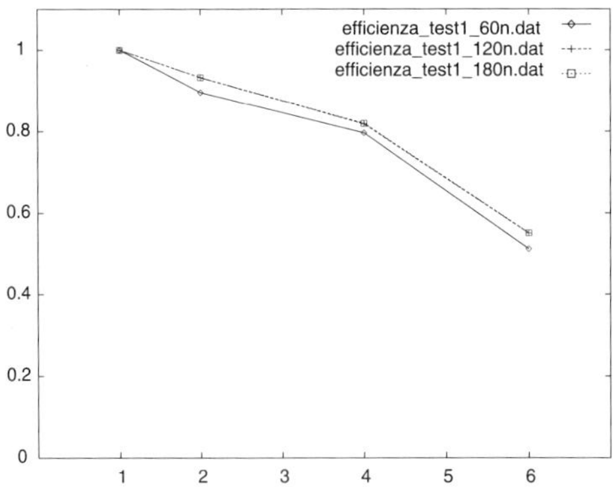

Figure 16: Test 1's global efficiency.

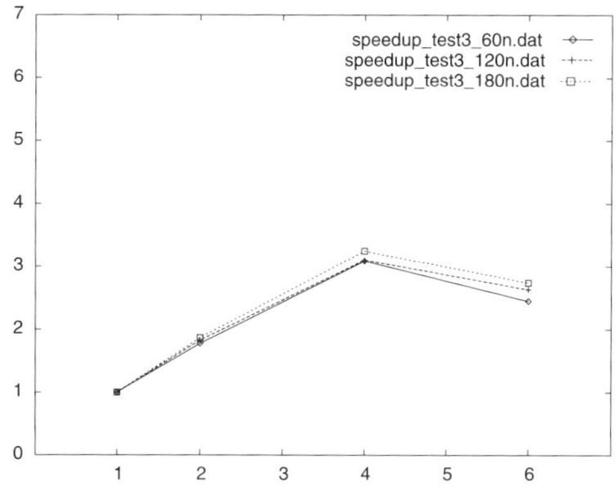

Figure 17: Test 2's global speedup.

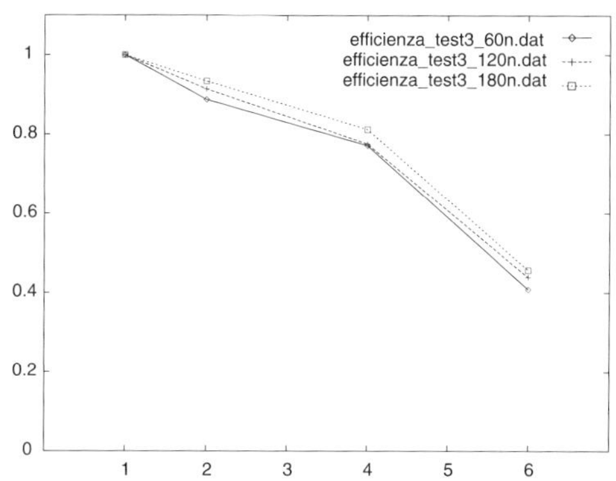

Figure 18: Test 2's global efficiency.

- The speedup is almost the same going from size $N = 60$ to 180. This fact can easily be connected to the high number of nodes in the overlapping region, which has a major impact on the ratio between communication and computation loads. For large numbers of subdomains we have the worst possible case because (for h fixed) the overlapping regions are rather large relative to the global size of the subdomains. The result is that each subdomain communicates a number of nodes almost equal to its size.
- The summary results (Figures 15 and 17) show that speedup and efficiency remain almost constant if we increase the size of the problem (the three lines almost coincide).
- The summary results (Figures 15 and 17) also show that the speedup does not increase when going from 4 to 8 processors (it slightly decreases in Test 2). This obviously implies that the efficiency decreases when going from 4 to 8 processors. However, there are no significant differences with respect to the

size of the problems (the three lines almost coincide), and the efficiency is always beyond the limit 0.4.

There is no significant difference between Tests 1 and 2 in terms of speedup and efficiency, so it seems that those parameters are not affected by the regularity of the value function. The above results suggest that although the algorithm converges properly, its efficiency can be improved. A possible remedy would be to decrease the time step in order to obtain smaller overlapping regions. However, this would also require the use of an acceleration technique in the fixed-point iteration to avoid a dramatic increase in the number of iterations. An extension to the Isaacs equation of the parallel version of the acceleration algorithm studied in [10] is rather complicated due to the lack of convexity of the Hamiltonian, we will try to overcome this difficulty in a forthcoming paper.

Acknowledgments

This work has been supported by M.U.R.S.T. funds (project Matematica Computazionale e Analisi Numerica) and by INTAS Network Optimal Control and Differential Games (contract 93-2622). We gratefully acknowledge the technical support given by CASPUR for the development of this research.

REFERENCES

[1] Alziary de Roquefort, B. Jeux différentiels et approximation numérique de fonctions valeur, 1re partie: Étude théorique; 2e partie: Étude numérique, *RAIRO Math. Model. Numer. Anal.*, **25**, pp. 517–560, 1991.

[2] Bardi, M., Bottacin, S., and Falcone, M. Convergence of discrete schemes for discontinuous value functions of pursuit–evasion games, pp. 273–304, in G. J. Olsder, ed., *New Trends in Dynamic Games and Applications*, Birkhäuser, Boston, 1995.

[3] Bardi, M. and Capuzzo Dolcetta, I. *Optimal Control and Viscosity Solutions of Hamilton–Jacobi–Bellman Equations*, Birkhäuser, Boston, 1997.

[4] Bardi, M., Falcone, M., and Soravia, P. Fully discrete schemes for the value function of pursuit–evasion games, pp. 89–105, in T. Basar and A. Haurie, eds., *Advances in Dynamic Games and Applications*, Birkhäuser, Boston, 1994.

[5] Bardi, M., Falcone, M., and Soravia, P. Numerical methods for pursuit–evasion games via viscosity solutions, pp. 289–303, in M. Bardi, T. Parthasarathy, and T. E. S. Raghavan, eds., *Stochastic and Differential Games: Theory and Numerical Methods*, Annals of the I.S.D.G., **4**, Birkhäuser, Boston, 1999.

[6] Bardi, M. and Soravia, P. Approximation of differential games of pursuit–evasion by discrete-time games, pp. 131–143, in *Differential Games—Developments in Modeling and Computation*, R. P. Hamalainen and H. K. Ethamo, eds., Lecture Notes in Control and Information Sciences, **156**, Springer-Verlag, New-York, 1991.

[7] Camilli, F., Falcone, M., Lanucara, P., and Seghini, A. *A domain decomposition method for Bellman equations*, pp. 477–483, in D. E. Keyes and J. Xu, eds., *Domain Decomposition Methods in Scientific and Engineering Computing*, Contemporary Mathematics no. 180, AMS, Providence, RI, 1994.

[8] Falcone, M., Numerical solution of Dynamic Programming equations, Appendix A in M. Bardi and I. Capuzzo Dolcetta, *Optimal Control and Viscosity Solutions of Hamilton-Jacobi-Bellman Equations*, Birkhäuser, Boston, 1997.

[9] Falcone, M. and Lanucara, P. Parallel algorithms for Hamilton-Jacobi equations, ICIAM/GAMMJ95 Special Issue of *Zeitschrift für Angewandte Mathematik und Mechanik* (ZAMM), **3**, O. Mahrenholtz, K. Marti, and R. Mennicken, eds., *Applied Stochastics and Optimization*, pp. 355–359.

[10] Falcone, M., Lanucara, P., and Seghini, A. A splitting algorithm for Hamilton–Jacobi–Bellman equations, *Appl. Num. Math.*, **15**, pp. 207–218, 1994.

[11] Foster, I., *Designing and Building Parallel Programs*, Addison-Wesley, New York, 1995.

[12] Grüne, L. An adaptive grid scheme for the discrete Hamilton–Jacobi–Bellman equation, *Numerische Mathematik*, **75**, pp. 319–337, 1997.

[13] Isaacs, R. *Differential Games*, John Wiley & Sons, New York, 1965.

[14] Marinucci, M. *Giochi Differenziali: Metodi Numerici Recenti e Loro Parallelizzazione*, Tesi di Laurea, Roma, 1995.

[15] Message Passing Interface Forum, *MPI: A Message-Passing Interface Standard*, `http://www.mpi-forum.org/docs/mpi-11-html/mpi-report.html`, 1995.

[16] Soravia, P. Estimates of convergence of fully discrete schemes for the Isaacs equation of pursuit–evasion games via maximum principle, *Siam J. Cont. Opt.*, **36**, pp. 1–11, 1998.

[17] Subbotin, A. *Generalized Solutions of First Order PDEs: The Dynamic Optimization Perspective*, Birkhäuser, Boston, 1995.

[18] Tarasyev, A. M. Approximation schemes for constructing minimax solutions of Hamilton–Jacobi equations, *J. Appl. Math. Mechs.* **58**, pp. 207–221, 1994.

[19] Tarasyev, A. M. Control synthesis in grid schemes for Hamilton–Jacobi equations, *Ann. Oper. Res.* **88**, pp. 337–359, 1999.

Computation of S-adapted Equilibria in Piecewise Deterministic Games via Stochastic Programming Methods

Alain Haurie
and
Francesco Moresino
University of Geneva
LOGILAB/HEC
Geneva, Switzerland

Abstract

This paper explores a class of equilibria for piecewise deterministic differential games, with an information structure permitting the players to adapt to the jump Markov disturbance sample path but not to the continuous state trajectory. This information structure is therefore halfway between the open-loop and feedback ones. The paper shows that these equilibria can be approximated through a sequence of stochastic programming–variational inequalities problems for which efficient computational techniques exist.

1 Introduction

The aim of this paper is to propose a numerical technique for the approximation of a class of equilibria in a stochastic game of oligopoly. These equilibria, called "S-adapted" in [13], correspond to an information structure where the players adapt their actions to an observation of the realization of the random disturbances affecting the game dynamics. These disturbances are supposed to take the form of an uncontrolled jump process. Recently this class of problems has received a renewed attention from researchers in mathematical programming circles (see, e.g., [8] and [7]) who extended the numerical experiments reported in [13]. The present paper complements these previous works in the following way:

1. the oligopoly model is formulated in continuous time as in [11];
2. the S-adapted information structure is compared with the piecewise open-loop (POL) information structure used in [11];
3. an approximation to the S-adapted equilibrium is obtained through the solution of a sequence of variational inequality (VI) problems defined via a discretization over time of the game dynamics and perturbing jump process;
4. the approximating S-adapted equilibrium is proved to be unique under strict diagonal concavity of the total reward function;

5. convergence results are proved for the approximating S-adapted equilibrium;
6. a numerical example, consistent with the one given in [11], is fully detailed and shows the proximity of the equilibria under S-adapted and piecewise open-loop information structure.

The paper is organized as follows. In Section 2 we formulate a differential game of oligopoly with an open-loop information structure and we show that the Nash equilibrium can be approximated via a VI solution using mathematical programming techniques. This provides another efficient way to approximate an open-loop equilibrium in a differential game of oligopoly. In Section 3 a piecewise deterministic game version of the same oligopoly model is proposed, and the concept of an S-adapted information structure is discussed. S-adapted equilibria are compared with POL equilibria, and one conjectures that these equilibria could coincide in many cases. An approximation of the S-adapted equilibria through a sequence of VI solutions is proposed. In Section 5 some numerical experiments are reported, and a comparison with the POL information structure is made on the basis of the numerical solutions obtained, which tends to confirm the conjecture.

2 The Deterministic Dynamic Oligopoly

In this section we consider a deterministic differential game model of oligopoly, propose an approximating discrete-time model, show that the open-loop equilibrium in the discrete-time model can be computed via the solution of a variational inequality, and prove that the equilibrium of this approximating game converges to the equilibrium of the initial game as the discretizations step tends to 0.

2.1 The Formulation of the Oligopoly

The model is similar to those studied in [4], [10], and [11]. There are J competing firms (also called players) supplying a market for a homogeneous good. Let $\mathcal{J} = \{1, 2, \dots, J\}$ be the set of all players. The control variables are such that $u_j \in [u_j^{\min}, u_j^{\max}]$, $j \in \mathcal{J}$; they represent the investment in production capacity by each firm. The state variables are $x_j \in \mathbb{R}^+$, $j \in \mathcal{J}$; they represent the accumulated production capacity of each firm.

The state equation for player j is given by[1]

$$\dot{x}_j(t) = u_j(t) - \mu_j x_j(t), \qquad j \in \mathcal{J}, \tag{1}$$

$$x_j(0) = x_j^0. \tag{2}$$

With $u_j(\cdot)$ a measurable function over $[0, T]$, bounded above by $u_j^{\max}$ and below by $u_j^{\min}$, such that the generated trajectory $x_j(t)$ is not negative. μ_j is the capacity

[1]This is the same dynamic equation that was analyzed in Section 3 of [1] for a constrained optimization situation.

depreciation rate for firm j. Given the initial state $x_j(0) = x_j^0$, the solution $x_j(\cdot)$ of (1) can be expressed as a function of the control $u_j(\cdot)$:

$$x_j(t) = e^{-\mu_j t} x_j^0 + \int_0^t e^{-\mu_j (t-s)} u_j(s)\, ds. \tag{3}$$

The information structure is open-loop; hence, each player j knows the initial states $(x_1^0, \dots, x_J^0) = x^0$ and chooses a control function $u_j(\cdot) : [0, T] \to [u_j^{\min}, u_j^{\max}]$ that generates a positive trajectory. Let U_j be the set of admissible control functions and X_j the set of trajectories $x_j(\cdot)$ generated by admissible controls $u_j(\cdot)$ in U_j.

The profit functions are thus defined by

$$V_j(x^0; u_1(\cdot), \dots, u_J(\cdot)) = \int_0^T e^{-\rho_j t} L_j(x(t), u_j(t))\, dt, \tag{4}$$

where ρ_j is the discount rate for player j, $x(t) = (x_1(t), \dots, x_J(t))$, and $L_j(x, u_j)$ is a profit rate function assumed to be C^1 in x and in u_j.

Definition 2.1. Let $\epsilon \geq 0$. The J-tuple $(u_1^*(\cdot), \dots, u_J^*(\cdot))$ is an ϵ-Nash open-loop equilibrium if we have, for all j in $\mathcal{J}$ and for all $u_j(\cdot)$ in U_j,

$$V_j(x^0; u_1^*(\cdot), \dots, u_j(\cdot), \dots, u_J^*(\cdot)) \leq V_j(x^0; u_1^*(\cdot), \dots, u_J^*(\cdot)) + \epsilon.$$

If $\epsilon = 0$, we obtain a Nash equilibrium.

2.2 An Equilibrium Principle

The Hamiltonian of player j is defined as usual by

$$\mathcal{H}_j(t, x, u_j, p_j) = L_j(x, u_j) + p_j(u_j - \mu_j x_j).$$

The optimized Hamiltonian is

$$H_j(t, x, p_j) = \max_{u_j \in [u_j^{\min}, u_j^{\max}]} \mathcal{H}_j(t, x, u_j, p_j).$$

In order for $(x_j(\cdot), u_j(\cdot))_{j \in \mathcal{J}}$ to be an open-loop equilibrium, it is necessary that there exist absolutely continuous costate trajectories $(p_j(\cdot))_{j \in \mathcal{J}}$ such that

$$\dot{p}_j(t) = -\frac{\partial H_j(t, x(t), p_j(t))}{\partial x_j},$$

$$\dot{x}_j(t) = \frac{\partial H_j(t, x(t), p_j(t))}{\partial p_j},$$

and

$$u_j(t) = \arg \max_{u_j \in [u_j^{\min}, u_j^{\max}]} \mathcal{H}_j(t, x(t), u_j, p_j(t)),$$

with the transversality condition $p_j(T) = 0$.

Definition 2.2. The combined Hamiltonian $\sum_j H_j(t, x, p)$ is strictly diagonally concave in x, convex in p if, for all t, x, $\tilde{x}$, p, and $\tilde{p}$,

$$\sum_j (p_j - \tilde{p}_j)(\nabla_{p_j} H_j(t, x_1, \ldots, x_J, p_j) - \nabla_{p_j} H_j(t, \tilde{x}_1, \ldots, \tilde{x}_J, \tilde{p}_j)) \\ -(x_j - \tilde{x}_j)(\nabla_{x_j} H_j(t, x_1, \ldots, x_J, p_j) - \nabla_{x_j} H_j(t, \tilde{x}_1, \ldots, \tilde{x}_J, \tilde{p}_j)) > 0.$$

Definition 2.3. The total reward function $\sum_j L_j(x_1, \ldots, x_J, u_j)$ is strictly diagonally concave in (x, u) if, for all x, $\tilde{x}$, p, and $\tilde{p}$

$$\sum_j (u_j - \tilde{u}_j)(\nabla_{u_j} L_j(x_1, \ldots, x_J, u_j) - \nabla_{u_j} L_j(\tilde{x}_1, \ldots, \tilde{x}_J, \tilde{u}_j)) \\ +(x_j - \tilde{x}_j)(\nabla_{x_j} L_j(x_1, \ldots, x_J, u_j) - \nabla_{x_j} L_j(\tilde{x}_1, \ldots, \tilde{x}_J, \tilde{u}_j)) < 0.$$

The strict diagonal concavity in x, convexity in p of the combined Hamiltonian can be verified by applying the following lemma, which we borrow from [5].

Lemma 2.1. *Assume that $L_j(x, u_j)$ is concave in (x, u_j) and assume that the total reward function $\sum_j L_j(x, u_j)$ is strictly diagonally concave in (x, u); then the combined Hamiltonian $\sum_j H_j(t, x, p)$ is strictly diagonally concave in x, convex in p.*

The following uniqueness result can then be proved as in [5].

Theorem 2.1. *If the combined Hamiltonian is strictly diagonally concave in x, convex in p, then the open-loop equilibrium is unique.*

2.3 A Discrete-Time Approximation

We shall now explore a method for approximating the open-loop equilibrium of the duopoly game that uses a sequence of solutions of variational inequalities (VI) to get an ϵ-equilibrium of the continuous-time game.

2.3.1 A Variational Inequality Reformulation

We use a discrete-time approximation of the dynamic oligopoly model. The approximating game of order K is defined as follows: let $t_k = \delta k$, with $k = 0, \ldots, K$ and $\delta = T/K$. The discrete-time state and control variables are $\mathbf{x}_j^K = (x_j^K(k))_{k=1}^K$ and $\mathbf{u}_j^K = (u_j^K(k))_{k=1}^K$, respectively. Using a slight abuse of notation, we call $x_j^K(k)$ the discrete-time state at time t_k, and similarly for $u_j^K(k)$. The state equations are the difference equations

$$x_j^K(k) = u_j^K(k)\frac{1 - e^{-\mu_j \cdot \delta}}{\mu_j} + e^{-\mu_j \cdot \delta} x_j^K(k-1) \tag{5}$$

$$= x_j^0 e^{-\mu_j t_k} + \frac{1 - e^{-\mu_j \cdot \delta}}{\mu_j} \sum_{l=1}^{k} u_j^K(l) e^{-\mu_j (t_k - t_l)}, \tag{6}$$

and the profit functions are given by

$$V_j^K(x^0;\mathbf{u}_1^K,\dots,\mathbf{u}_J^K)=\sum_{k=1}^{K}e^{-\rho_j t_k}L_j(x_1^K(k),\dots,x_J^K(k),u_j^K(k))\delta, \tag{7}$$

where $(x_1^K(k),\dots,x_J^K(k))_{k=1,\dots,K}$ is the trajectory J-tuple emanating from x^0 and generated by the controls as shown in (5).

An admissible open-loop strategy for player j in the approximating game of order K is a vector $\mathbf{u}_j^K\in[u_j^{\min},u_j^{\max}]^K$ such that the generated trajectory $\mathbf{x}_j^K$ remains positive. Let U_j^K be the set of strategies and X_j^K the set of the corresponding trajectories for player j. We have thus defined a game where the strategies are elements of a Euclidean space. The equilibrium $\mathbf{u}^{K*}=(\mathbf{u}_1^{K*},\dots,\mathbf{u}_J^{K*})$ is a solution of the following variational inequality:

$$\langle F(\mathbf{u}^{K*}),\mathbf{u}^{K*}-\mathbf{u}^K\rangle\geq 0\quad\forall\mathbf{u}^K\in U_1^K\times\dots\times U_J^K=U^K, \tag{8}$$

where $\langle\cdot,\cdot\rangle$ denotes the scalar product and

$$\mathbf{u}^K=\begin{pmatrix}\mathbf{u}_1^K\\ \vdots\\ \mathbf{u}_J^K\end{pmatrix},$$

$$F(\mathbf{u}^K)=\begin{pmatrix}\nabla_{\mathbf{u}_1^K}V_1^K(x^0;\mathbf{u}_1^K,\dots,\mathbf{u}_J^K)\\ \vdots\\ \nabla_{\mathbf{u}_J^K}V_J^K(x^0;\mathbf{u}_1^K,\dots,\mathbf{u}_J^K)\end{pmatrix}. \tag{9}$$

The gradients of the reduced profit functions can easily be obtained from (7) once one expresses

$$\frac{\partial x_j^K(k)}{\partial u_j^K(l)}=\begin{cases}\frac{1-e^{-\mu_j\cdot\delta}}{\mu_j}e^{-\mu_j(t_k-t_l)} & \text{if } k\geq l,\\ 0 & \text{if } k<l.\end{cases}$$

2.4 Convergence Results

Let us first recall the definition of monotony.

Definition 2.4. $G(\cdot):U^K\to\left(\mathbb{R}^K\right)^J$ is a monotone operator in U^K if it satisfies

$$\langle G(\mathbf{u}^K)-G(\tilde{\mathbf{u}}^K),\mathbf{u}^K-\tilde{\mathbf{u}}^K\rangle\geq 0\qquad\forall\mathbf{u}^K,\tilde{\mathbf{u}}^K\in U^K.$$

Theorem 2.2. *Assume the total reward function $\sum_j L_j(x_1,\dots,x_J,u_j)$ is strictly diagonally concave in (x,u). Then the operator $-F(\cdot)$ defined in (9) is monotone.*

Proof. By straightforward verification. □

Theorem 2.3. *Under the assumptions of Theorem 2.2, there exists a unique equilibrium for the approximating game of order K.*

Proof. This theorem is a special case of Theorem 3.2, to be proved later on for the stochastic case. □

Assumption 2.1. We suppose the following:

- there is state and control separation in the profit rate functions, that is,

$$L_j(x_1, \ldots, x_J, u_j) = \mathcal{L}_j(x_1, \ldots, x_J) + \mathcal{M}_j(u_j),$$

  where $\mathcal{L}_j$ and $\mathcal{M}_j$ are Lipschitz-continuous functions.
- $\sum_j \mathcal{L}_j(x)$ is strictly diagonally concave in x, and $\mathcal{M}_j(u_j)$ is strictly concave in u_j.

We now address the question of approximating the solution of the continuous-time game through the solution of approximating games. Our approach is inspired from [15] and [2]. To establish a correspondence between the continuous-time game and its approximation of order K, let us define the mappings $\phi_j^K : U_j^K \to U_j$ and $\sigma_j^K : U_j \to U_j^K$ as follows:

$$\phi_j^K(\mathbf{u}_j^K)(t) = u_j^K(k) \quad \text{where } k \text{ is such that } t_k = \min_s\{t_s | t_s \geq t\},$$

$$\sigma_j^K(u_j)(k) = \frac{\mu}{1 - e^{-\mu\delta}} \int_{t_{k-1}}^{t_k} u_j(s) e^{-\mu(t_k - s)}\, ds.$$

With each control for the discrete-time game of order K, the mapping ϕ_j^K associates a piecewise-constant control for the continuous-time game. With each control for the continuous-time game, the mapping σ_j^K associates a control for the discrete-time game of order K. One can verify that these mappings satisfy the following property: $\sigma_j^K \circ \phi_j^K = \mathbf{1}$. Furthermore, the mappings preserve the property of nonnegativity of the generated trajectory.

The convergence of the discrete-time equilibrium toward the continuous-time equilibrium is stated in the following two theorems.

Theorem 2.4. *Suppose Assumption* 2.1 *holds. Let* $u_1^*, \ldots, u_J^*$ *be the equilibrium controls of the continuous-time oligopoly. Then, for all positive* ϵ, *there exists* K_ϵ *such that, for all* $K > K_\epsilon$, *the control vector* $(\sigma_1^K(u_1^*), \ldots, \sigma_J^K(u_J^*))$ *is an* ϵ*-Nash equilibrium for the discrete-time oligopoly of order* K.

Theorem 2.5. *Suppose Assumption* 2.1 *holds. Let* $\mathbf{u}_1^{K*}, \ldots, \mathbf{u}_J^{K*}$ *be the equilibrium controls of the discrete-time oligopoly of order* K. *Then, for all positive* ϵ, *there exists* K_ϵ *such that, for all* $K > K_\epsilon$, *the controls vector* $(\phi_1^K(\mathbf{u}_1^{K*}), \ldots, \phi_J^K(\mathbf{u}_J^{K*}))$ *is an* ϵ*-Nash equilibrium for the continuous-time oligopoly.*

Proof. The proofs of these two theorems can be found in Appendix B. □

3 The Stochastic Dynamic Oligopoly

A stochastic oligopoly model has been proposed, in a discrete-time setting, by Haurie, Smeers, Zaccour and Legrand in [13] and [12]. In the proposed model the random disturbances were uncontrolled. The information structure used in these papers has been called S-adapted, for "sample path adapted," and the equilibrium has been computed via the solution of a variational inequality. These papers extended, in some sense, the stochastic programming technique to the case of Nash–Cournot equilibria. Recently, this discrete-time stochastic equilibrium framework has been further studied in [8].

In [11], Haurie and Roche have studied a stochastic oligopoly model, in a continuous-time setting and with uncontrolled random jump disturbances. These authors used the information structure called *piecewise open-loop* (POL). The POL equilibrium was characterized and approximated through the solution of a discrete event dynamic programming equation.

In this section we revisit the stochastic oligopoly model presented in [11] but with the S-adapted information structure. We compare the S-adapted equilibrium with the POL equilibrium and we conjecture that, in many instances, these equilibrium solutions will coincide. We propose an approximation via a discrete-time model and show that the S-adapted equilibrium in the discrete-time model can be computed via the solution of a variational inequality. We prove that under strict diagonal concavity of the total reward function, there exists a unique S-adapted equilibrium for the approximating game. Finally, we show that the equilibria of the approximating games converge to an equilibrium of the continuous-time game.

3.1 A System with Jump Markov Disturbances

The state equations are still given by

$$\dot{x}_j(t) = u_j(t) - \mu_j x_j(t), \qquad j \in \mathcal{J},$$

with $u_j^{\min} \le u_j(t) \le u_j^{\max}$ and $x_j(0) = x_j^0$. Let $\{\xi(\cdot) : [0, T] \to \mathcal{I}\}$ be a continuous-time, discrete-state Markov chain taking values in the finite set $\mathcal{I} = \{1, \dots, I\}$ that describes random changes in the market condition. The dynamics of $\xi(t)$ are defined by the transition rate matrix $Q = [q_{ih}]$, $i, h \in \mathcal{I}$. The profit rate functions $L_j^i(x, u_j)$ now also depend on the market condition $i \in \mathcal{I}$.

3.2 The S-adapted Information Structure

Let $(\Omega, \mathcal{B}, P)$ be the probability space for the $\xi(\cdot)$ process. We call $\xi(\omega, \cdot) : [0, T) \to \mathcal{I}$, $\omega \in \Omega$, a sample path of the ξ-process and $\xi_\theta(\omega, \cdot) : [0, \theta] \to \mathcal{I}$ its history up to time θ. We assume that the players know the initial state $x^0 = (x_1^0, \dots, x_J^0)$ and, at each instant t, the ξ-process history up to time t, namely, $\xi_t(\omega, \cdot)$. The game is played as follows.

- Let $\tau_0 = 0, \tau_1, \tau_\nu, \nu = 0, 1, 2, \ldots,$ be the successive jump times of the ξ-process and $\xi_0 = \xi(0)$, $\xi_1 = \xi(\tau_1)$, $\xi_\nu = \xi(\tau_\nu)$ the visited states.
- Call $h_\nu = \{\tau_0, \xi_0, \tau_1, \xi_1, \ldots, \tau_\nu, \xi_\nu\}$ the jump process history up to jump ν.
- At any jump time τ_ν, each player $j \in \mathcal{J}$, knowing x^0 and h_ν, chooses an open-loop control $u_j^\nu(\cdot) : [\tau_\nu, T) \to [u_j^{\min}, u_j^{\max}]$ that will be used until the next jump $\tau_{\nu+1}$ occurs.
- This information is called S-adapted as the control of each player is progressively adapted to the sample path $\xi_t(\omega, \cdot)$.

Denote by Γ_j the set of the S-adapted strategies for player j. Let $\gamma_j(\xi_t(\cdot), t) : t \in [0, T]$ be the control used at time t by player j according to the S-adapted strategy γ_j. The strategic J-tuple is denoted by $\gamma \in \Gamma$. The profit functions are then defined by

$$V_j(x^0; \gamma_1, \ldots, \gamma_J) = E_\gamma \left[\int_0^T e^{-\rho_j t} L_j^{\xi(t)}(x_1(t), \ldots, x_J(t), u_j(t))\, dt \right], \qquad (10)$$

where $u_j(t) = \gamma_j(\xi_t(\cdot), t)$ and $x_j(t)$ is the resulting state trajectory.

3.3 Comparison between S-adapted and POL Information Structures

In the POL information structure studied and used, e.g., in [9] and [11], the players observe, at each jump time τ_ν, the discrete state ξ_ν and the continuous state $x^\nu = (x_1(\tau_\nu), \ldots, x_J(\tau_\nu))$ and then choose open-loop controls to be implemented until the next jump occurs. Therefore, in the POL information structure, the players are confronted with a sequential game where decisions are made at each jump time with full state information.

In the S-adapted information structure, the players cannot observe the current continuous state $x(t)$, even at jump time. We can immediately notice the following.

Remark 3.1. In the POL information structure, with perfect recall, one could as well assume that the players know, at any jump time τ_ν, the whole $(\xi(\cdot), x(\cdot))$-history $H_\nu = \{\tau_0, \xi_0, x^0, \tau_1, \xi_1, x^1 \ldots, \tau_\nu, \xi_\nu, x^\nu\}$. Therefore, clearly, the S-adapted information structure corresponds to a coarser information structure than POL with perfect recall.

It appears that the S-adapted information structure is close to the open-loop information structure, whereas the POL one is closer to the feedback information structure. It is appropriate to recall here the definition of the concepts of *time consistency* and *subgame perfectness at jump times.*

Definition 3.1. An equilibrium is time-consistent, if, at any time θ, given that the equilibrium has been played up to that time and given the state $s^*(\theta) = (\xi(\theta), x^*(\theta))$ that has been reached, the same strategies would remain an equilibrium if one restarts the game at that time θ with initial state s^*.

Definition 3.2. An equilibrium is subgame-perfect at jump times, if, at any jump time τ_ν, whatever has been played up to that time, given the state $s_\nu = (\xi_\nu, x^\nu)$ has been reached, the same strategies would remain an equilibrium if one restarts the game at that time τ_ν, with initial state s_ν.

Now the difference between the two information structures can be seen in the following remarks.

Remark 3.2. A Markov strategy in the POL information structure uses only the information available at τ_ν to update the control on the next random time interval $[\tau_\nu, \tau_{\nu+1})$. An equilibrium based on Markov strategy will indeed be time-consistent and subgame-perfect at jump times.

The time consistency and subgame perfectness at jump times is a direct consequence of the characterization of the equilibrium via a discrete event dynamic programming equation (see [11] and [9]).

Remark 3.3. In an open-loop information structure, for a deterministic game, each player knows the initial state x^0 and chooses a control $u_j(\cdot) : [0, T] \to [u_j^{\min}, u_j^{\max}]$. Given a J-tuple $u(\cdot) = (u_j(\cdot))_{j\in\mathcal{J}}$, the players know at each instant of time the state $x(t)$ that has been reached. In an open-loop equilibrium $u^*(\cdot) = (u_j^*(\cdot))_{j\in\mathcal{J}}$, the trajectory generated $x^*(\cdot)$ is therefore known to each player. Time consistency is observed along $x^*(\cdot)$.

Remark 3.4. In the S-adapted information structure, each player knows x^0 and chooses a control adapted to the history of the $\xi(\cdot)$-process. Given a J-tuple of $\xi(\cdot)$-adapted controls and the history h_ν, where $\tau_\nu = \sup\{\tau_l : \tau_l \le t\}$, each player is able to find the current state $x(t)$. In an S-adapted equilibrium, there is a set of possible trajectories, indexed over the sample space Ω of the $\xi(\cdot)$-process. Call $x_\omega^*(\cdot)$ the equilibrium trajectory associated with the sample value $\omega \in \Omega$. For any $\omega \in \Omega$, $t \in [0, T]$, the S-adapted equilibrium strategy will still be an equilibrium for the game starting at time t, with initial state $s_\omega^*(t) = (\xi_\omega(t), x_\omega^*(t))$. Let us call $X_\Omega^*(t) = \{x_\omega^*(t) : \omega \in \Omega\}$ the reachable set at time t, associated with the S-adapted equilibrium strategy J-tuple and the set of all possible sample paths of the ξ-process. Hence, we propose as conjecture that the S-adapted equilibrium is subgame-perfect at any jump time τ_ν for any x^ν contained in the set $X_\Omega^*(\tau_\nu)$.

The time consistency and subgame perfectness at jump time on the reachable set of S-adapted equilibria is a consequence of the dynamic optimality conditions.

It would be interesting to compare the set $X_\Omega^*(t)$ with the reachable set $\hat{X}_\Omega^*(t)$ associated at time t with a POL equilibrium. We conjecture that they may coincide for many games, in particular the oligopoly game considered in the present paper. If these sets coincide and if they are locally convex, then the POL and the S-adapted information structure will yield the same equilibrium values. A more precise study of this conjecture should be the object of further investigation.

3.4 A Discrete-Time Stochastic Game Approximation

We shall again proceed to an approximation of the S-adapted equilibrium of the continuous-time game through a sequence of variational inequality (VI) solutions. Each VI corresponds to the equilibrium of an associated approximating game of order K. We proceed as follows

- We discretize time. Let $t_k = \delta \cdot k$ with $k = 0 \dots K$ and $\delta = T/K$.
- We consider the discrete-time, discrete-state Markov chain $\tilde{\xi}^K(k)$ with state set $\mathcal{I}$ and transitions probabilities:

$$P[\tilde{\xi}^K(k+1) = h | \tilde{\xi}^K(k) = i] = \begin{cases} e^{q_{ii}\delta} & \text{if } i = h, \\ (1 - e^{q_{ii}\delta})\frac{-q_{ih}}{q_{ii}} & \text{otherwise.} \end{cases}$$

- We represent the set of all sample paths of the Markov chain $\tilde{\xi}^K(k)$ as an event tree. Let $\mathcal{N} = \{1, 2, \dots, N\}$ be the set of the nodes of this tree and $\mathcal{N}(k)$ the set of nodes associated with period k. In this representation, each node n_k at period k corresponds to a whole history of the Markov chain from period 1 to period k. A complete path along the event tree is also often called a *scenario*. Let $\mathcal{A}(n_k)$ denote the unique predecessor of n_k along the unique path going from n_1 to n_k; let $\mathcal{S}(n_k)$ denote the set of nodes $n_{k+1} \in \mathcal{N}(k+1)$ that can be successors of n_k along a sample path; denote also $\mathcal{U}(n_k)$ the set of all the upstream nodes w.r.t. n_k (including n_k) and $\mathcal{D}(n_k)$ the set of downstream nodes w.r.t. n_k (including n_k), respectively. The number of nodes in the event tree is given by $N = (I^K - 1)/(I - 1)$.
- We index the state and control variables over the set of nodes of the event tree:

$$\mathbf{x}_j^K = (x_j^K(n_k))_{n_k \in \mathcal{N}},$$
$$\mathbf{u}_j^K = (u_j^K(n_k))_{n_k \in \mathcal{N}},$$

and we introduce the state equations

$$x_j^K(n_k) = u_j^K(n_k)\frac{1 - e^{-\mu_j \cdot \delta}}{\mu_j} + e^{-\mu_j \cdot \delta} x_j^K(\mathcal{A}(n_k)) \tag{11}$$

$$= x_j^0 e^{-\mu_j t_k} + \frac{1 - e^{-\mu_j \cdot \delta}}{\mu_j} \sum_{l=1}^{k} u_j^K(\mathcal{A}^{k-l}(n_k)) e^{-\mu_j (t_k - t_l)}, \tag{12}$$

where $\mathcal{A}^{k-l}$ means $\mathcal{A}$ to the power $k-l$, that is, the $k-l$ step predecessor. In the S-adapted information structure, a strategy for player j in the approximating game of order K is a vector $\mathbf{u}_j^K \in [u_j^{\min}, u_j^{\max}]^N$ such that the generated trajectory $\mathbf{x}_j^K$ remains positive. Let U_j^K be the set of such strategies and X_j^K the set of the corresponding trajectories. Associated with a strategy J-tuple, we have the payoff functions[2]

$$V_j^K(x^0; \mathbf{u}_1^K, \dots, \mathbf{u}_J^K) = \sum_{k=1}^{K} \sum_{n_k \in \mathcal{N}(k)} L_j^{n_k}(x^K(n_k), u_j^K(n_k)) p(n_k) e^{-\rho_j t_k}, \quad (13)$$

where $p(n_k)$ denotes the probability of the node $n(k)$ and is given by

$$p(n_k) = \begin{cases} 1 & \text{if } k = 1 \\ p(\mathcal{A}(n_k))\mathrm{Prob}(\tilde{\xi}^K(n_k)|\tilde{\xi}^K(\mathcal{A}(n_k))) & \text{otherwise,} \end{cases}$$

while $x^K(n_k)$ is the state reached at period k, given the history summarized by n_k and the controls $u_1, \dots, u_J$ as indicated in (12).

We have thus defined a game in normal form with strategies in an Euclidean space. An equilibrium $\mathbf{u}^{K*} = (\mathbf{u}_1^{K*}, \dots, \mathbf{u}_J^{K*})$ is a solution of the following variational inequality:

$$\langle F(\mathbf{u}^{K*}), \mathbf{u}^{K*} - \mathbf{u}^K \rangle \geq 0 \quad \forall \mathbf{u}^K \in U_1^K \times \cdots \times U_J^K = U^K, \quad (14)$$

with

$$\mathbf{u}^K = \begin{pmatrix} \mathbf{u}_1^K \\ \vdots \\ \mathbf{u}_J^K \end{pmatrix},$$

$$F(\mathbf{u}^K) = \begin{pmatrix} \nabla_{\mathbf{u}_1^K} V_1^K(x^0; \mathbf{u}_1^K, \dots, \mathbf{u}_J^K) \\ \vdots \\ \nabla_{\mathbf{u}_J^K} V_J^K(x^0; \mathbf{u}_1^K, \dots, \mathbf{u}_J^K) \end{pmatrix}.$$

The partial derivatives of the reduced profit functions can be calculated from (13) once one expresses

$$\frac{\partial x_j^K(n_k)}{\partial u_j^K(n_l)} = \begin{cases} \frac{1-e^{-\mu_j \cdot \delta}}{\mu_j} e^{-\mu_j(t_k - t_l)} & \text{if } n_l \in \mathcal{U}(n_k), \\ 0 & \text{otherwise.} \end{cases}$$

Theorem 3.1. *Assume the total reward function $\sum_j L_j^i(x_1, \dots, x_J, u_j)$ is strictly diagonally concave in (x, u) for all i in $\mathcal{I}$. Then the operator $-F(\mathbf{u}^K)$ is monotone.*

Proof. By direct verification. □

Theorem 3.2. *Under the assumptions of Theorem 3.1, there exists a unique equilibrium for the approximating game of order K.*

Proof. As $-F$ is monotone, we have that $V_j^K(\mathbf{u}_1^K, \dots, \mathbf{u}_J^K)$ is concave in $\mathbf{u}_j^K$. Moreover, the set $U = U_1 \times \cdots \times U_J$ is convex, so the game is a concave game, and according to Theorem 1 in [16] we know that there exists an equilibrium.

[2] We use the notation $L_j^{n_k}$ in place of the more correct notation $L_j^{i(n_k)}$.

From the monotony of $-F$ and Theorem 2 in [16], we know that the equilibrium is unique. □

For the approximation results that we shall establish, it is convenient to have a property of continuity of the strategies w.r.t. sample paths. This requires first a definition of a distance on the space of sample paths of the $\xi(\cdot)$-process. Let Λ be the class of strictly increasing, continuous mapping of $[0, T]$ into itself.

Definition 3.3. Let $\xi(\omega_1, \cdot)$ and $\xi(\omega_2, \cdot)$ be two sample paths of the continuous-time Markov chain. Define the distance[3] $d(\xi(\omega_1, \cdot), \xi(\omega_2, \cdot))$ as the infimum of those positive D for which there exists a λ in Λ such that $\sup_t |\lambda(t) - t| \leq D$ and $\sup_t |\xi(\omega_1, t) - \xi(\omega_2, \lambda(t))| \leq D$.

Assumption 3.1. We suppose the following:

- state and control separation: $L^i_j(x, u_j) = \mathcal{L}^i_j(x) + \mathcal{M}^i_j(u_j)$ with $\mathcal{L}^i_j$ and $\mathcal{M}^i_j$ Lipschitz-continuous,
- monotonicity: $\sum_j \mathcal{L}^i_j(x)$ is strictly diagonally concave in x and $\mathcal{M}^i_j(u_j)$ strictly concave in u_j for all j in $\mathcal{J}$ and i in $\mathcal{I}$;
- stability: the admissible strategies γ verify, except perhaps on a null measure set, the following property: for all ϵ and for all realizations $\xi(\omega_1, \cdot)$ and $\xi(\omega_2, \cdot)$, there exists a η such that[4] $d(\xi(\omega_1, \cdot), \xi(\omega_2, \cdot)) < \eta \Rightarrow \|\gamma(\omega_1, \cdot) - \gamma(\omega_2, \cdot)\| < \epsilon$, where the norm is the L_1-norm: $\|v\| = \int_0^T |v(t)|\, dt$.

The last assumption says that the control must be continuous with respect to the random disturbance trajectory; it ensures a stability of the control with respect to the random process.

In order to establish convergence results, it is convenient to modify the representation of the approximating game and of the continuous-time game in order to explicit the dependence on the sample paths. In the approximating game of order K, the control, previously denoted $u^K_j(n(k))$, will now be written $u^K_j(\tilde{\omega}^K, k)$, where $\tilde{\omega}^K$ indexes the sample paths for the discrete-time Markov chain and k is the current period. The same representation is used for the state variables.

Let $\xi(\omega, \cdot)$ be a sample path of the continuous-time Markov chain and $\tilde{\xi}^K(\tilde{\omega}^K, \cdot) = (\tilde{\xi}^K(\tilde{\omega}^K, 1), \ldots \tilde{\xi}^K(\tilde{\omega}^K, K))$ be a sample path of the discrete-time Markov chain. We define the projection $\pi_K : \{\xi(\omega, \cdot) : [0, T] \to \mathcal{I}\} \to \mathcal{I}^K$ as

$$\pi_K(\xi(\omega, \cdot)) = (\xi(\omega, t_0), \ldots, \xi(\omega, t_{K-1})). \tag{15}$$

We define, as in the deterministic case, $\phi^K_j : U^K_j \to U_j$ and $\sigma^K_j : U_j \to U^K_j$ by

$$\phi^K_j(\mathbf{u}^K_j)(\omega, t) = u^K_j(\tilde{\omega}^K, k)$$

[3] Also called Skorohod topology; see [3].
[4] We prefer the shorter notation $\gamma(\omega, \cdot)$ to the notation $\gamma(\xi_t(\omega, \cdot), \cdot)$.

where k is such that $t_k = \min_s\{t_s | t_s \geq t\}$ and $\tilde{\xi}^K(\tilde{\omega}^K, \cdot) = \pi_K(\xi(\omega, \cdot))$ and

$$\sigma_j^K(\gamma_j)(\tilde{\omega}^K, k) = \frac{\mu}{1 - e^{-\mu\delta}} \int_{t_{k-1}}^{t_k} \gamma_j(\omega, s) e^{-\mu(t_k - s)}\, ds$$

where $\xi(\omega, t) = \tilde{\xi}^K(\tilde{\omega}^K, k)$ and $k = \arg\min_s\{t_s | t_s \geq t\}$, respectively.

Let $x^K_{j,\sigma_j^K(\gamma)}(\tilde{\omega}^K, \cdot)$ be the discrete-time trajectory generated by $\sigma_j^K(u_j)(\tilde{\omega}^K, \cdot)$ and $x_{j,\phi_j^K(\mathbf{u}_j^K)}(\omega, \cdot)$ the continuous-time trajectory generated by $\phi_j^K(\mathbf{u}_j^K)(\omega, \cdot)$. One can verify that we again have the property $\sigma_j^K \circ \phi_j^K = \mathbf{1}$. The mappings preserve the property of nonnegativity of the generated trajectory and the nonanticipativity with respect to the S-adapted information structure.

The convergence of the discrete-time oligopoly toward the continuous-time oligopoly is shown in the following two theorems.

Theorem 3.3. *Suppose Assumption* 3.1 *holds. Let* $\gamma_1^*, \ldots, \gamma_J^*$ *be the equilibrium strategies of the continuous-time oligopoly. Then, for all positive* ϵ*, there exists* K_ϵ *such that, for all* $K > K_\epsilon$*, the strategies vector* $(\sigma_1^K(\gamma_1^*), \ldots, \sigma_J^K(\gamma_J^*))$ *is an* ϵ*-Nash equilibrium for the discrete-time oligopoly of order* K*.*

Theorem 3.4. *Suppose Assumption* 3.1 *holds. Let* $\mathbf{u}_1^{K*}, \ldots, \mathbf{u}_J^{K*}$ *be the equilibrium controls of the discrete-time oligopoly of order* K*. Then, for all* j *in* $\mathcal{J}$*, all* γ_j *in* Γ_j*, and all positive* ϵ*, there exists* $K_\epsilon(\gamma_j)$ *such that for all* $K > K_\epsilon(\gamma_j)$*, we have*

$$\begin{aligned} &V_j(\phi_1^K(\mathbf{u}_1^{K*}), \ldots, \gamma_j, \ldots, \phi_J^K(\mathbf{u}_J^{K*})) \\ &\quad \leq V_j(\phi_1^K(\mathbf{u}_1^{K*}), \ldots, \phi_j^K(\mathbf{u}_j^{K*}), \ldots, \phi_J^K(\mathbf{u}_J^{K*})) + \epsilon. \end{aligned}$$

Proof. The proofs of these two theorems can be found in Appendix A. □

4 Numerical Experiments

In this section we recall the stochastic duopoly model presented in [11] and show that the total reward function is strictly diagonally concave. We then propose to use an algorithm due to Konnov [14] to solve the variational inequalities. Finally, the numerical results are presented and discussed.

4.1 A Stochastic Duopoly Model

We take the same duopoly model as in [11]. The depreciation rates are $\mu_1 = 0.08$ and $\mu_2 = 0.06$, respectively. Assume that the firms supply, according to their production capacity, a market characterized by an inverse demand law depending on the market condition i:

$$D^i(x_1(t) + x_2(t)) = \frac{a_i}{x_1(t) + x_2(t) + b_i} - c_i.$$

Here D^i is the market clearing price, given the total supply $x_1(t) + x_2(t)$. The continuous-time Markov chain, describing the market condition, takes three different values, corresponding to three different demand functions $D^i(x_1 + x_2)$, $i \in \mathcal{I} = \{1, 2, 3\}$. The coefficients are $a_1 = 120$, $a_2 = 100$, $a_3 = 80$, $b_i = 20$, $i = 1, 2, 3$, $c_1 = 3$, $c_2 = 2.5$, and $c_3 = 2$. The dynamics of the continuous-time Markov chain are described by the following transition rate matrix:

$$Q = \begin{pmatrix} -0.2 & 0.2 & 0.0 \\ 0.01 & -0.05 & 0.04 \\ 0.1 & 0.0 & -0.1 \end{pmatrix}.$$

We also assume that each firm has a quadratic maintenance and investment cost. So the reward functions are given by

$$L^i_j(x_1, x_2, u_j) = D^i(x_1 + x_2)x_j - (x_j)^2 - (u_j)^2.$$

The discount rate is $\rho = 0.09$ for both firms. The time horizon is $T = 10$. For the time discretization, we take the number of periods $K = 10$. So the number of nodes is given by

$$N = \frac{I^K - 1}{I - 1} = 29{,}524.$$

Applying Theorem 6 from [16], it can be checked that the total reward function $\sum_j L_j(x_1, x_2, u_j)$ is strictly diagonally concave in (x, u) for all i in $\mathcal{I}$. So according to Theorems 3.1 and 3.2, the function $-F(\mathbf{u}^K)$ is monotone and there exists a unique equilibrium for the approximating game.

4.2 Implementation

If $-F(\mathbf{u})$ is a monotone function that is Lipschitz with constant L, the solution $\mathbf{u}^*$ of the following variational inequality

$$\langle F(\mathbf{u}^*), \mathbf{u}^* - \mathbf{u} \rangle \geq 0, \qquad \forall \mathbf{u} \in U \subseteq R^{2N},$$

can be obtained via the following algorithm given by Konnov [14]:

Step 0 (initialization)
Choose λ, $0 < \theta < 1/(\lambda(1 + L))$, ε, and $\mathbf{u}(0)$.
Set $n = 0$.

Step 1 (computation of the next point and stopping criterion)

$p(n+1) = \text{proj}_U(\mathbf{u}(n) + \lambda F(\mathbf{u}(n)))$

$\mathbf{v}(n+1) = \mathbf{u}(n) + \theta(p(n) - \mathbf{u}(n))$

IF $F(\mathbf{v}(n+1)) = 0$ THEN STOP $\mathbf{u}^* = \mathbf{v}(n+1)$ $\alpha = \langle F(\mathbf{v}(n+1)), \mathbf{v}(n+1) - \mathbf{u}(n)\rangle / \| F(\mathbf{v}(n+1))\|$

$\mathbf{u}(n+1) = \text{proj}_U(\mathbf{u}(n)) + \alpha F(\mathbf{v}(n+1))$

Evaluate $g(\mathbf{u}(n+1)) = \min_{\mathbf{u}\in U} \langle F(\mathbf{u}(n+1)), \mathbf{u}(n+1) - \mathbf{u}\rangle/(2N)$

IF $g(\mathbf{u}(n+1)) \geq -\varepsilon$ THEN STOP $\mathbf{u}^* = \mathbf{u}(n+1)$

ELSE Increment n

GO TO Step 1.

We implemented this algorithm and computed the S-adapted open-loop equilibrium of the stochastic duopoly taking as stopping criterion $\varepsilon = 0.001$.

4.3 Numerical Results

All the equilibrium trajectories illustrated in this report are for the state $i = 2$. They are similar for the other states. Figure 1(a) shows the equilibrium trajectories for the initial state $x_1(0) = 0$ and $x_2(0) = 0$. As the model is nearly symmetric (the only difference between players 1 and 2 is the depreciation rate), it is not surprising to have the equilibrium trajectories nearly similar for the two players. Figures 1(b) and (c) show the equilibrium trajectories for the initial state $x_1(0) = 0$ and $x_2(0) = 1.5$. We see that they are consistent with the results in [11]. Figure 1(d) compares the equilibrium trajectories for player 1 when player 2 has two different initial states [$x_2(0) = 0$ and $x_2(0) = 1.5$]. We see that the turnpikes[5] are identical and, as expected, the first trajectory lies above the second one.

In Table 1 we recall the value of the turnpikes computed under the POL information structure and the corresponding turnpike values obtained with the S-adapted information structure. We can observe that these values are very close.

As in [11], we consider the following realization of the random process: the sequence of the modal change is 2, 3, 1, 2, 3, 1, 2, 1, 2, 3, and the jump times are 10, 13.5, 23.5, 52.6, 63.8, 65.2, 68.2, 71.6, 81.5. For this realization of the random process, the equilibrium trajectories are displayed in Figure 2. One sees that, after

Table 1: Turnpikes

Information	$\tilde{x}_1^1$	$\tilde{x}_2^1$	$\tilde{x}_1^2$	$\tilde{x}_2^2$	$\tilde{x}_1^3$	$\tilde{x}_2^3$
POL	1.04644	1.05067	0.92472	0.92854	0.80014	0.80353
S-adapted	1.056	1.060	0.927	0.931	0.788	0.791

[5]See [11] for the definition.

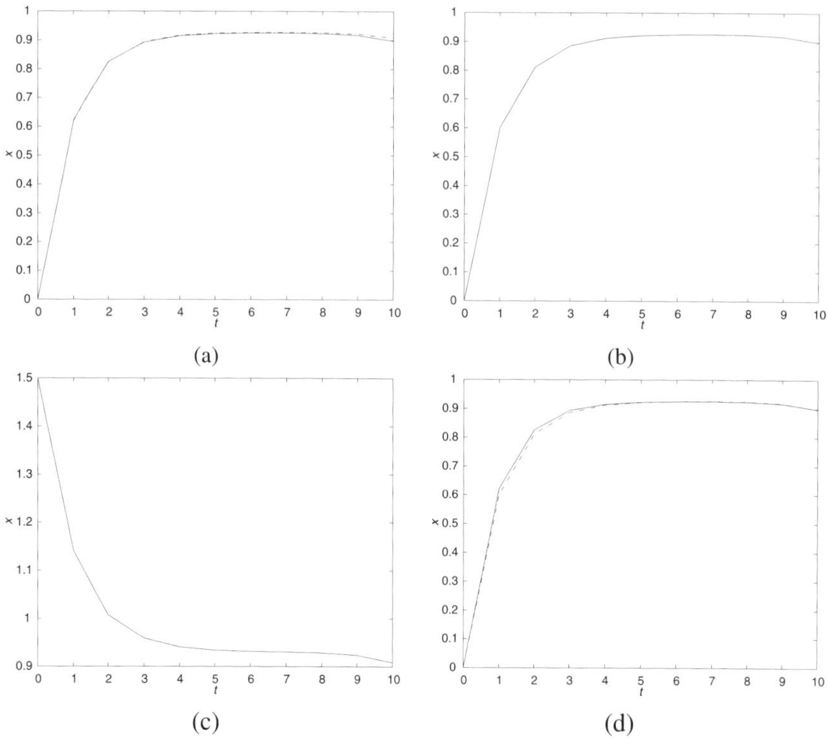

Figure 1: Equilibrium trajectories in state i=2. (a) $x_1(t)$ and $x_2(t)$ (dotted) for $x_1(0) = x_2(0) = 0$. (b) $x_1(t)$ for $x_1(0) = 0$, $x_2(0) = 1.5$. (c) $x_2(t)$ for $x_1(0) = 0$, $x_2(0) = 1.5$. (d) $x_1(t)$ for $x_1(0) = 0$, $x_2(0) = 0$ and for $x_1(0) = 0$, $x_2(0) = 1.5$ (dotted).

each jump, the equilibrium trajectories are attracted by the turnpikes associated with the current state i and remain close to them until the next jump occurs.

The value functions in state $i = 2$ for different initial states $x_i(0)$ are pictured in Figure 3.

We can see that the results for the S-adapted information structure are really close to the ones for the POL information structure.

5 Conclusion

In this paper we have proposed using a numerical technique based on the solution of an approximating variational inequality to compute a continuous-time Nash–Cournot ϵ-equilibrium. We have first considered the open-loop information structure for a deterministic differential game model and then the S-adapted information structure when the dynamic system is subject to jump Markov random disturbances. We have compared the S-adapted information structure with the POL information structure introduced in [9]–[11] and conjectured that, when the jump

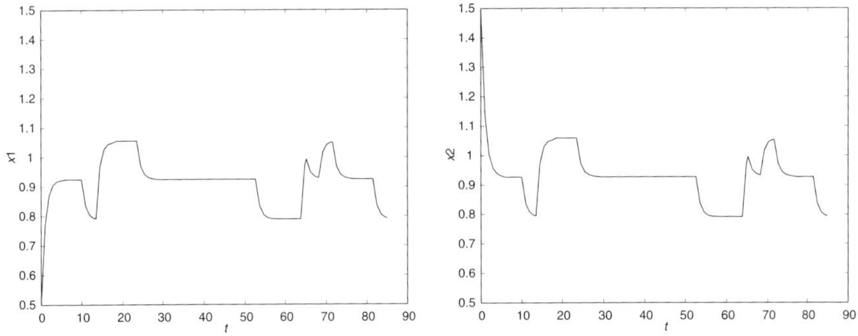

(a) Player 1 (b) Player 2

Figure 2: Equilibrium trajectories for a given realization of the random process.

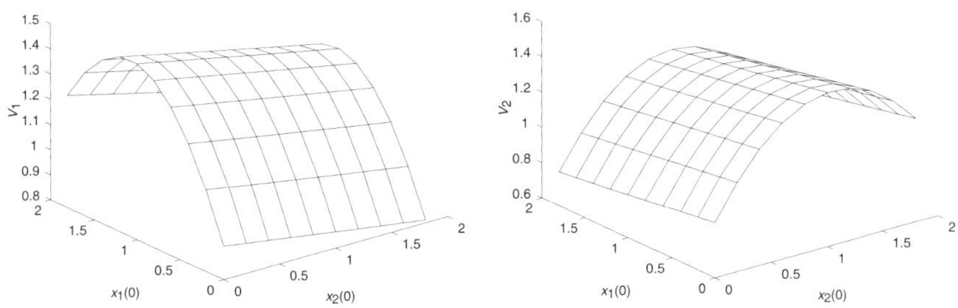

(a) Player 1 (b) Player 2

Figure 3: Value functions in state $i = 2$ for different initial states $x_i(0)$.

Markov process is uncontrolled, the two types of equilibria are likely to yield similar outcomes. We have verified this conjecture on the model already considered in [11], with the POL information structure. Finally, we have proved the proximity existing between the equilibrium strategies of the approximating game and those of the continuous-time game.

In all these developments we have used the dynamic oligopoly formalism of [4], [5], and [11]. Future extensions of this work should deal with a more general class of differential games and explore in greater detail the proximity of equilibria under POL and S-adapted information structures.

Appendix A Convergence Results for the Stochastic Case

A.1 Proof Outline

To simplify the notations, the proofs are done for a duopoly; they can be easily extended to the case of an oligopoly. We shall state a list of propositions to be proved in Section A.2.

Proposition A.1. *For any $\epsilon > 0$, there exists K_ϵ such that, for any approximating game of order $K > K_\epsilon$, the following holds*

$$\forall(\mathbf{u}_1^K, \mathbf{u}_2^K) \in U_1^K \times U_2^K, \quad |V_j^K(\mathbf{u}_1^K, \mathbf{u}_2^K) - V_j(\phi_1^K(\mathbf{u}_1^K), \phi_2^K(\mathbf{u}_2^K))| < \epsilon.$$

Proposition A.2. *Given γ_1 and γ_2, for any $\epsilon > 0$, there exists K_ϵ such that, for any approximating game of order $K > K_\epsilon$ the following holds:*

$$|V_j^K(\sigma_1^K(\gamma_1), \sigma_2^K(\gamma_2)) - V_j(\gamma_1, \gamma_2)| < \epsilon.$$

Proposition A.3. *Given γ_2, for any $\epsilon > 0$, there exists K_ϵ such that, for any approximating game of order $K > K_\epsilon$, the following holds:*

$$\forall \mathbf{u}_1^K \in U_1^K, \quad |V_1^K(\mathbf{u}_1^K, \sigma_2^K(\gamma_2)) - V_1(\phi_1^K(\mathbf{u}_1^K), \gamma_2)| < \epsilon.$$

Proposition A.4. *Given γ_1, for any $\epsilon > 0$, there exists K_ϵ such that, for any approximating game of order $K > K_\epsilon$, the following holds:*

$$\forall \mathbf{u}_2^K \in U_2^K, \quad |V_1^K(\sigma_1^K(\gamma_1), \mathbf{u}_2^K) - V_1(\gamma_1, \phi_2^K(\mathbf{u}_2^K))| < \epsilon.$$

Proof of Theorem 3.3. For a game of order K, consider $V_1^K(\mathbf{u}_1^K, \sigma_2^K(\gamma_2^*))$ for any admissible $\mathbf{u}_1^K$. If K is large enough ($K > K_\epsilon^1$), according to Proposition A.3 one has

$$V_1^K(\mathbf{u}_1^K, \sigma_2^K(\gamma_2^*)) \le V_1(\phi_1^K(\mathbf{u}_1^K), \gamma_2^*) + \frac{\epsilon}{2},$$

and by the equilibrium property of γ^*,

$$V_1(\phi_1^K(\mathbf{u}_1^K), \gamma_2^*) \le V_1(\gamma_1^*, \gamma_2^*).$$

Finally, according to Proposition A.2, we have

$$V_1(\gamma_1^*, \gamma_2^*) \le V_1^K(\sigma_1^K(\gamma_1^*), \sigma_2^K(\gamma_2^*)) + \frac{\epsilon}{2}$$

if K is large enough ($K > K_\epsilon^2$). Therefore, if $K > K_\epsilon = \sup\{K_\epsilon^1, K_\epsilon^2\}$, one has

$$V_1^K(\mathbf{u}_1^K, \sigma_2^K(\gamma_2^*)) \le V_1^K(\sigma_1^K(\gamma_1^*), \sigma_2^K(\gamma_2^*)) + \epsilon.$$

The same property holds for Player 2. The proof is complete. □

Proof of Theorem 3.4. Let $(\mathbf{u}_1^{K*}, \mathbf{u}_2^{K*})$ be an S-adapted equilibrium of the approximating game of order K. Consider, for any admissible policy γ_1 of the continuous-time game, the payoff $V_1(\gamma_1, \phi_2^K(\mathbf{u}_2^{K*}))$. According to Proposition A.4, if K is large enough ($K > K_\epsilon^1$), one has

$$V_1(\gamma_1, \phi_2^K(\mathbf{u}_2^{K*})) \le V_1^K(\sigma_1^K(\gamma_1), \mathbf{u}_2^{K*}) + \frac{\epsilon}{2}.$$

Since $(\mathbf{u}_1^{K*}, \mathbf{u}_2^{K*})$ is an S-adapted equilibrium pair for the approximating game of order K, one also has

$$V_1^K(\sigma_1^K(\gamma_1), \mathbf{u}_2^{K*}) \leq V_1^K(\mathbf{u}_1^{K*}, \mathbf{u}_2^{K*}).$$

Finally, according to Proposition A.1, one has

$$V_1^K(\mathbf{u}_1^{K*}, \mathbf{u}_2^{K*}) \leq V(\phi_1^K(\mathbf{u}_1^{K*}), \phi_2^K(\mathbf{u}_2^{K*})) + \frac{\epsilon}{2}$$

if K is large enough ($K > K_\epsilon^2$). Therefore, if $K > K_\epsilon = \sup\{K_\epsilon^1, K_\epsilon^2\}$, one has

$$V_1(\gamma_1, \phi_2^K(\mathbf{u}_2^{K*})) \leq V(\phi_1^K(\mathbf{u}_1^{K*}), \phi_2^K(\mathbf{u}_2^{K*})) + \epsilon.$$

The same property holds for the other player. □

A.2 Proof of the Propositions

We will first prove the following lemmas, which will be used later on.

Lemma A.1. *Suppose Assumption* 3.1 *holds. Then* $\forall t \in [0, T]$, $\forall x, y \in X_1 \times \cdots \times X_J$, $\forall u_j, v_j \in U_j$, *and* $\forall i, l \in I$, *we have*

$$|L_j^i(x(t), u_j(t)) - L_j^l(y(t), v_j(t))| \leq C.$$

Proof. This comes from the Lipschitz property L_j^i and the boundedness of the controls and state trajectories. □

Let $(\Omega, \mathcal{B}, P)$ be the probability space associated with the continuous-time stochastic process and $(\tilde{\Omega}^K, \tilde{\mathcal{B}}_K, \tilde{P}_K)$ the probability space associated with the discrete-time stochastic process of order K. Let $\pi_K^{-1}(\tilde{\xi}^K(\tilde{\omega}^K, \cdot)) = \{\xi(\omega, \cdot) : \pi_K(\xi(\omega, \cdot)) = (\tilde{\xi}^K(\tilde{\omega}^K, 1), \ldots, \tilde{\xi}^K(\tilde{\omega}^K, K))\}$, with π_K defined as in (15). Notice that the two Markov chains $\tilde{\xi}^K(\cdot)$ and $\pi_K(\xi(\cdot))$ associated with the projection π_K have the same sample path set. The convergence of the first Markov chains to the second one is stated in the following lemma.

Lemma A.2. $\tilde{P}_K(\tilde{\xi}^K(\tilde{\omega}^K, \cdot))$ *tends to* $P(\pi_K^{-1}(\tilde{\xi}^K(\tilde{\omega}^K, \cdot)))$, *uniformly in* $\tilde{\omega}^K$, *as K tends to infinity.*

Proof. As before, let $\delta = T/K$. Denote by S^K the transition probability matrix $[\text{Prob}(\xi(t + \delta) = h|\xi(t) = i)]$, $i, h \in \mathcal{I}$, and by $\tilde{S}^K$ the transition probability matrix $[\text{Prob}(\tilde{\xi}^K(k + 1) = h|\tilde{\xi}^K(k) = i)]$, $i, h \in \mathcal{I}$. The only thing to prove is that $\tilde{S}^K$, the generator of the Markov chain $\tilde{\xi}^K(\cdot)$, tends to S^K, the generator of the Markov chain $\pi_K(\xi(\cdot))$, as K tends to infinity.

Define $\|M\|$ the norm of an $I \times I$ matrix as follows:

$$\|M\| = \sup_i \sum_l |m_{il}|.$$

Applying the Taylor expansion, we have

$$\tilde{S}^K = \mathbf{1} + Q\delta + \tilde{R},$$

with $|\tilde{r}_{il}| \leq \delta^2 |q_{ii}| q_{il}/2$, which implies

$$\|\tilde{R}\| \leq \frac{\|Q\|^2\delta^2}{2}.$$

From [6], p. 149, we have the following:

$$S^K = e^{Q\delta} = \mathbf{1} + Q\delta + R,$$

with

$$\|R\| \leq \frac{\|Q\|^2\delta^2}{2}.$$

So we have

$$|\tilde{s}_{ij}^K(\delta) - s_{ij}^K(\delta)| \leq \|Q\|^2\delta^2. \qquad \square$$

To prove Proposition A.1, we will need the following lemma.

Lemma A.3. *For any $\tilde{\omega}^K \in \tilde{\Omega}^K$ and any $\omega \in \Omega$ such that $\pi_K(\xi(\omega,\cdot)) = \tilde{\xi}^K(\tilde{\omega}^K,\cdot)$, we have*

$$|\sum_{k=1}^{K} L_j^{\tilde{\xi}^K(\tilde{\omega}^K,k)}(x_1^K(\tilde{\omega}^K,k), x_2^K(\tilde{\omega}^K,k), u_j^K(\tilde{\omega}^K,k))e^{-\rho\delta k}\delta$$
$$- \sum_{k=1}^{K}\int_{t_{k-1}}^{t_k} L_j^{\tilde{\xi}^K(\tilde{\omega}^K,k)}(x_{1,\phi_1^K(\mathbf{u}_1^K)}(\omega,t), x_{2,\phi_2^K(\mathbf{u}_2^K)}(\omega,t), \phi_j^K(\mathbf{u}_j^K)(\omega,t))e^{-\rho t}\,dt|$$

tends uniformly in $\mathbf{u}_j^K$ to 0 as K tends to infinity. Recall that $x_{j,\phi_j^K(\mathbf{u}_j^K)}(\omega,\cdot)$ is the continuous-time trajectory generated by $\phi_j^K(\mathbf{u}_j^K)(\omega,\cdot)$, as introduced in Section 3.

Proof. Let $\tilde{\omega}^K \in \tilde{\Omega}^K$ and $\omega \in \Omega$ such that $\pi_K(\xi(\omega,\cdot) = \tilde{\xi}^K(\tilde{\omega}^K,\cdot)$. Let's first check that the term in $\mathcal{M}_j^i$ has a uniform convergence in $\mathbf{u}_j^K$:

$$|\sum_{k=1}^{K} \mathcal{M}_j^{\tilde{\xi}^K(\tilde{\omega}^K,k)}(u_j^K(\tilde{\omega}^K,k))e^{-\rho\delta k}\delta - \sum_{k=1}^{K}\int_{t_{k-1}}^{t_k} \mathcal{M}_j^{\tilde{\xi}^K(\tilde{\omega}^K,k)}(\phi_j^K(\mathbf{u}_j^K)(\omega,t))e^{-\rho t}\,dt|$$
$$= |\sum_{k=1}^{K} \mathcal{M}_j^{\tilde{\xi}^K(\tilde{\omega}^K,k)}(u_j^K(\tilde{\omega}^K,k))e^{-\rho\delta k}\delta[1 - \frac{e^{\rho\delta}-1}{\rho\delta}]|$$
$$\leq T\cdot C\cdot[1 - \frac{e^{\rho\delta}-1}{\rho\delta}] = T\cdot C\cdot[1 - \frac{e^{\rho\frac{T}{K}}-1}{\rho\frac{T}{K}}].$$

This tends to 0 (independently of $\mathbf{u}_j^K$) as K tends to infinity.

The convergence of the term in $\mathcal{L}_j^i$ comes from the properties of the x-trajectories. We defined in (6) the dynamics of the system in such a way that for $k = 0 \dots K$, $x^K_{j,\sigma^K_j(\gamma)}(\omega, t_k)$ equals $x^K_j(\tilde{\omega}^K, k)$. As the controls are bounded, the trajectories are Lipschitz and so the convergence is uniform in $\mathbf{u}^K_j$. □

Proof of Proposition A.1.

$$
\begin{aligned}
&|V^K_j(\mathbf{u}^K_1, \mathbf{u}^K_2) - V_j(\phi^K_1(\mathbf{u}^K_1), \phi^K_2(\mathbf{u}^K_2))| \\
&\quad = |\sum_{\tilde{\omega}^K} \tilde{P}_K(\tilde{\xi}^K(\tilde{\omega}^K, \cdot)) \sum_{k=1}^{K} L_j^{\tilde{\xi}^K(\tilde{\omega}^K,k)}(x^K_1(\tilde{\omega}^K, k), x^K_2(\tilde{\omega}^K, k), u^K_j(\tilde{\omega}^K, k)) e^{-\rho\delta k}\delta \\
&\qquad - E_\xi[\int_0^T L_j^{\xi(t)}(x_{1,\phi^K_1(\mathbf{u}^K_1)}(\omega, t), x_{2,\phi^K_2(\mathbf{u}^K_2)}(\omega, t), \phi^K_j(\mathbf{u}^K_j)(\omega, t))) e^{-\rho t}\, dt]|. \quad (16)
\end{aligned}
$$

Let F_l be the set of all realizations $\xi(\omega, \cdot)$ that have more than l jumps over $[0, T]$. We set $l = \lfloor\sqrt{K}\rfloor$, the largest integer smaller than the square root of K. The second term can be rewritten as

$$
\begin{aligned}
&E_{\xi|\xi(\omega,\cdot)\in F_l}[\sum_{k=1}^{K} \int_{t_{k-1}}^{t_k} \{L_j^{\xi(t_{k-1})}(x_{1,\phi^K_1(\mathbf{u}^K_1)}(\omega, t), x_{2,\phi^K_2(\mathbf{u}^K_2)}(\omega, t), \phi^K_j(\mathbf{u}^K_j)(\omega, t)) \\
&\qquad - L_j^{\xi(t)}(x_{1,\phi^K_1(\mathbf{u}^K_1)}(\omega, t), x_{2,\phi^K_2(\mathbf{u}^K_2)}(\omega, t), \phi^K_j(\mathbf{u}^K_j)(\omega, t)))\} e^{-\rho t}\, dt] \\
&+ E_{\xi|\xi(\omega,\cdot)\notin F_l}[\sum_{k=1}^{K} \int_{t_{k-1}}^{t_k} \{L_j^{\xi(t_{k-1})}(x_{1,\phi^K_1(\mathbf{u}^K_1)}(\omega, t), x_{2,\phi^K_2(\mathbf{u}^K_2)}(\omega, t), \phi^K_j(\mathbf{u}^K_j)(\omega, t)) \\
&\qquad - L_j^{\xi(t)}(x_{1,\phi^K_1(\mathbf{u}^K_1)}(\omega, t), x_{2,\phi^K_2(\mathbf{u}^K_2)}(\omega, t), \phi^K_j(\mathbf{u}^K_j)(\omega, t))\} e^{-\rho t}\, dt] \\
&+ E_\xi[\sum_{k=1}^{K} \int_{t_{k-1}}^{t_k} L_j^{\xi(t_{k-1})}(x_{1,\phi^K_1(\mathbf{u}^K_1)}(\omega, t), x_{2,\phi^K_2(\mathbf{u}^K_2)}(\omega, t), \phi^K_j(\mathbf{u}^K_j)(\omega, t)) e^{-\rho t}\, dt].
\end{aligned}
$$

According to Lemma A.1, and as $P(F_l)$ tend to 0 as l tends to infinity, the norm of the first term is smaller than ϵ for K big enough. For the second term we know that the realizations have less than l jumps. If in the interval $[t_{k-1}, t_k[$ we have no jumps, the two terms under the integral are identical. If there is one or more jump in the interval $[t_{k-1}, t_k[$, the integral over the interval is, according to Lemma A.1, less than C/K. As we have less than $l \leq \sqrt{K}$ jumps, the norm of the second term is smaller than $C/\sqrt{K}$ and so smaller than ϵ for K big enough. For the third term we can see that for two realizations ω_1 and ω_2 such that $\pi_K(\xi(\omega_1, \cdot)) = \pi_K(\xi(\omega_2, \cdot))$, the integral is the same. With any $\tilde{\omega}^K$, in Ω^K we associate $\varphi(\tilde{\omega}^K)$ in Ω such that $\xi(\omega, t) = \tilde{\xi}^K(\tilde{\omega}^K, k)$ and $k = \arg\min_s\{t_s | t_s \geq t\}$. The third term can thus be rewritten as

$$
\begin{aligned}
&\sum_{\tilde{\omega}^K} P(\pi_K^{-1}(\tilde{\xi}^K(\tilde{\omega}^K, \cdot)) \sum_{k=1}^{K} \int_{t_{k-1}}^{t_k} L_j^{\tilde{\xi}^K(\tilde{\omega}^K,k)}(x_{1,\phi^K_1(\mathbf{u}^K_1)}(\varphi(\tilde{\omega}^K), t), x_{2,\phi^K_2(\mathbf{u}^K_2)}(\varphi(\tilde{\omega}^K), t), \\
&\qquad\qquad \phi^K_j(\mathbf{u}^K_j)(\varphi(\tilde{\omega}^K), t)) e^{-\rho t}\, dt.
\end{aligned}
$$

To summarize, for K big enough, (16) becomes

$$|V^K_j(\mathbf{u}^K_1, \mathbf{u}^K_2) - V_j(\phi^K_1(\mathbf{u}^K_1), \phi^K_2(\mathbf{u}^K_2))| \leq \epsilon + \epsilon$$

$$+ \mid \sum_{\tilde{\omega}^K} \tilde{P}_K(\tilde{\xi}^K(\tilde{\omega}^K,\cdot)) \sum_{k=1}^{K} L_j^{\tilde{\xi}^K(\tilde{\omega}^K,k)}(x_1(\tilde{\xi}^K(\tilde{\omega}^K,\cdot),k), x_2(\tilde{\xi}^K(\tilde{\omega}^K,\cdot),k),$$

$$u_j(\tilde{\xi}^K(\tilde{\omega}^K,\cdot),k))e^{-\rho\delta k}\delta$$

$$+\sum_{\tilde{\omega}^K} P(\pi_K^{-1}(\tilde{\xi}^K(\tilde{\omega}^K,\cdot)) \sum_{k=1}^{K} \int_{t_{k-1}}^{t_k} L_j^{\tilde{\xi}^K(\tilde{\omega}^K,k)}(x_{1,\phi_1^K(\mathbf{u}_1^K)}(\varphi(\tilde{\omega}^K),t), x_{2,\phi_2^K(\mathbf{u}_2^K)}(\varphi(\tilde{\omega}^K),t),$$

$$\phi_j^K(\mathbf{u}_j^K)(\varphi(\tilde{\omega}^K),t))e^{-\rho t}\,dt\,|.$$

For K big enough, Lemmas A.1, A.2, and A.3 assert that the term in absolute value tends to 0 as K tends to infinity. Thus, we have proved that, for all $\epsilon > 0$, there exists a K_ϵ (uniform in $\mathbf{u}_1^K$ and $\mathbf{u}_2^K$) such that, for all K greater than K_ϵ, we have the following:

$$|V_j^K(\mathbf{u}_1^K, \mathbf{u}_2^K) - V_j(\phi_1^K(\mathbf{u}_1^K), \phi_2^K(\mathbf{u}_2^K))| \leq 3\epsilon. \qquad \square$$

To prove Proposition A.2, we will need the following lemma.

Lemma A.4. *For any admissible strategy γ_j, the norm $\|\phi_j^K \circ \sigma_j^K(\gamma_j) - \gamma_j\|$ tends almost surely to 0 as K tends to infinity.*

Proof. We have to prove that, for any ω in Ω and any admissible strategy γ_j, $\|\phi_j^K \circ \sigma_j^K(\gamma_j)(\omega,\cdot) - \gamma_j(\omega,\cdot)\|$ tends to 0 as K tends to infinity, except perhaps on a set of probability 0.

With any ω in Ω, we associate $\psi(\omega, K)$ in Ω as follows: $\xi(\psi(\omega,K),t) = \xi(\omega,t_k)$, where $t_k = \min_s\{t_s : t_s \geq t\}$. We thus obtain an approximation of the sample path $\xi(\omega,\cdot)$ through a step function with jumps at times t_k only. Suppose that $\xi(\omega,t)$ has finitely many jumps; then for all ϵ positive, there exists K_ϵ such that $K > K_\epsilon$ implies $d(\omega, \psi(\omega,K)) < \epsilon$. If Assumption 3.1 holds, this implies the following: for all ϵ positive, there exists K_ϵ such that $K > K_\epsilon$ implies $\|\gamma_j(\psi(\omega,K),\cdot) - \gamma_j(\omega,\cdot)\| < \epsilon$. That means that the control for any realization ω, except perhaps the ones with infinitely many jumps (which is a null set), can be approximated as closely as desired by the control for a realization where the jump times are multiples of T/K_ϵ. This means that for fixed ω the proof is complete if one proves that for all γ_j we have $\|\phi_j^K \circ \sigma_j^K(\gamma_j)(\psi(\omega,K),\cdot) - \gamma_j(\psi(\omega,K),\cdot)\|$ tends to 0 as K tends to infinity. To simplify the notation, we denote by $w(\cdot)$ the control $\gamma_j(\psi(\omega,K),\cdot)$ associated with the realization $\psi(\omega,K)$.

$$\|\phi_j^K \circ \sigma_j^K(w_j)(\cdot) - w_j(\cdot)\| = \sum_{k=1}^{K} \int_{t_{k-1}}^{t_k} \left|\sigma_j^K(w_j)(k) - w_j(t)\right| dt.$$

We know that for all ϵ there exists a step function g_j^ϵ with $u_j^{\min} \leq g_j^\epsilon \leq u_j^{\max}$ such that $\|g_j^\epsilon - w_j\|_1 < \epsilon$,

$$\begin{aligned}
&\sum_{k=1}^{K} \int_{t_{k-1}}^{t_k} \left|\sigma_j^K(w_j)(k) - w_j(t)\right| dt \\
&\leq \sum_{k=1}^{K} \int_{t_{k-1}}^{t_k} \left|\sigma_j^K(w_j)(k) - g_j^\epsilon(t)\right| dt + \sum_{k=1}^{K} \int_{t_{k-1}}^{t_k} \left|g_j^\epsilon(t) - w_j(t)\right| dt \\
&\leq \sum_{k=1}^{K} \int_{t_{k-1}}^{t_k} \left|\sigma_j^K(w_j)(k) - g_j^\epsilon(t)\right| dt + \epsilon.
\end{aligned}$$

Define $u^m = \max_{j\in\mathcal{J}}\{\max(u_j^{\min}, u_j^{\max})\}$. Let ζ be the number of jumps of g_j^ϵ. Then we choose K greater than $\zeta T u^m/\epsilon$. So the total length of the intervals where we have a jump is $\zeta\delta = \zeta T/K$, which is less than ϵ/u^m. For an interval where we have no jump, we have the following:

$$\begin{aligned}
\int_{t_{k-1}}^{t_k} \left|\sigma_j^K(w_j)(k) - g_j^\epsilon(t)\right| dt &= \left|\sigma_j^K(w_j)(k) - g_j^\epsilon(t_{k-1})\right| \delta \\
&= \left|\frac{\mu}{1-e^{-\mu\delta}} \int_{t_{k-1}}^{t_k} w(s)e^{-\mu(t_k-s)}\, ds - g_j^\epsilon(t_{k-1})\right| \delta \\
&= \left|\frac{\mu}{1-e^{-\mu\delta}} \int_{t_{k-1}}^{t_k} [w(s) - g_j^\epsilon(t_{k-1})]e^{-\mu(t_k-s)}\, ds\right| \delta \\
&= \frac{\mu\delta}{1-e^{-\mu\delta}} \left|\int_{t_{k-1}}^{t_k} [w(s) - g_j^\epsilon(s)]e^{-\mu(t_k-s)}\, ds\right| \\
&\leq \frac{\mu\delta}{1-e^{-\mu\delta}} \int_{t_{k-1}}^{t_k} \left|w(s) - g_j^\epsilon(s)\right| e^{-\mu(t_k-s)}\, ds \\
&\leq \frac{\mu\delta}{1-e^{-\mu\delta}} \int_{t_{k-1}}^{t_k} \left|w(s) - g_j^\epsilon(s)\right| ds.
\end{aligned}$$

Therefore, the following holds true:

$$\begin{aligned}
\sum_{k=1}^{K} \int_{t_{k-1}}^{t_k} \left|\sigma_j^K(w_j)(k) - g_j^\epsilon(t)\right| dt &\leq \frac{\mu\delta}{1-e^{-\mu\delta}} \int_0^T \left|w(t) - g_j^\epsilon(t)\right| dt + \epsilon \\
&\leq \frac{\mu\delta}{1-e^{-\mu\delta}}\epsilon + \epsilon \\
&\leq \frac{\mu T}{1-e^{-\mu T}}\epsilon + \epsilon.
\end{aligned}$$

For the first inequality, the term ϵ is the majoration for the intervals where we have a jump. The last inequality is valid, as the fraction is an increasing function for δ positive and $\delta \leq T$. So the proof is complete. □

Proof of Proposition A.2.

$$
\begin{aligned}
&|V_j^K(\sigma_1^K(u_1), \sigma_2^K(u_2)) - V_j(u_1, u_2)| &(17)\\
&\leq |V_j^K(\sigma_1^K(u_1), \sigma_2^K(u_2)) - V_j(\phi_1^K \circ \sigma_1^K(u_1), \phi_2^K \circ \sigma_2^K(u_2))| \\
&+ |V_j(\phi_1^K \circ \sigma_1^K(u_1), \phi_2^K \circ \sigma_2^K(u_2)) - V_j(u_1, u_2)|. &(18)
\end{aligned}
$$

The first term in (18) tends to 0 according to Proposition A.1. The profit functions can be rewritten as

$$
V_j(\gamma_1, \gamma_2) = E_\gamma \left[\int_0^T e^{-\rho_j t} L_j^{\xi(t)}(x_1(t), \ldots, x_J(t), u_j(t))\, dt \right] = E_\xi[W_j^{\gamma_1,\gamma_2}(\xi)].
$$

As L_j^i are Lipschitz-continuous functions, and as the strategies satisfy the stability requirement of Assumption 3.1, Lemma A.4 implies that the random variable $W_j^{\gamma_1,\gamma_2}(\cdot)$ converges almost surely to the random variable $W_j^{\phi_1^K \circ \sigma_1^K(\gamma_1), \phi_2^K \circ \sigma_2^K(\gamma_2)}(\cdot)$. This implies that the second term in (18) tends to 0. □

Proof of Proposition A.3.

$$
\begin{aligned}
&|V_1^K(\mathbf{u}_1^K, \sigma_2^K(u_2)) - V_1(\phi_1^K(\mathbf{u}_1^K), u_2)| &(19)\\
&\leq |V_1^K(\mathbf{u}_1^K, \sigma_2^K(u_2)) - V_1(\phi_1^K(\mathbf{u}_1^K), \phi_2^K \circ \sigma_2^K(u_2))| \\
&+ |V_1(\phi_1^K(\mathbf{u}_1^K), \phi_2^K \circ \sigma_2^K(u_2)) - V_1(\phi_1^K(\mathbf{u}_1^K), u_2)|. &(20)
\end{aligned}
$$

The first term in (20) tends to 0 according to Proposition A.1. The same argument as in the previous proof shows that the second term in (20) tends to 0. The uniformity in $\{\mathbf{u}_1^K\}_{K=1}^\infty$ is guaranteed by the Lipschitz property of the functions L_j^i. □

The proof of Proposition A.4 is similar to the last proof.

Appendix B Convergence Results for the Deterministic Case

As the deterministic oligopoly is a special case of the stochastic one, all the theorems and lemmas in Appendix A are valid for the deterministic oligopoly. In particular, Theorem 2.4 is a special case of Theorem 3.3, which has been proved in Appendix A. However, we have to prove Theorem 2.5 since it is stronger than Theorem 3.4 when it is applied to the deterministic oligopoly.

We will endow the control and state spaces with norms defined as follows: $\|u_j\| = \int_0^T |u_j(t)|\, dt$, a norm on U_j; $\|\mathbf{u}_j^K\| = \sum_k |u^K(k)|\delta$, a norm on U_j^K; $\|x_j\| = \max_{t\in[0,T]} |x_j^K(t)|$, a norm on X_j; and $\|\mathbf{x}_j^K\| = \max_k |x(k)|$, a norm on X_j^K. Before proving Theorem 2.5, we will state and prove the following lemma.

Lemma B.1. *Assume that, for all j, $L_j(x, u_j)$ is Lipschitz-continuous in x and u_j. Then $V_j^K(\mathbf{u}_1^K, \mathbf{u}_2^K)$ is uniformly continuous in $\mathbf{u}_j^K \in U_j^K$.*

Proof. First we show that two controls $\mathbf{u}_j^K$ and $\mathbf{v}_j^K$, which are close to each other, generate two trajectories $\mathbf{x}_j^K$ and $\mathbf{y}_j^K$ that are close to each other.

$$\begin{aligned}
|x_j^K(k) - y_j^K(k)| &= \left| \sum_{l=0}^{k} \sum_{l=1}^{k} [u_j^K(l) - v_j^K(l)] \cdot e^{-\mu_j(t_k - t_l)} \frac{1 - e^{-\mu_j \cdot \delta}}{\mu_j} \right| \\
&\leq \left| e^{-\mu_j(t_k - t_l)} \frac{1 - e^{-\mu_j \cdot \delta}}{\mu_j \delta} \right| \sum_{l=0}^{k} \left| u_j^K(l) - v_j^K(l) \right| \delta \\
&= \alpha_K \sum_{l=0}^{k} \left| u_j^K(l) - v_j^K(l) \right| \delta \\
&\leq \alpha_K \|\mathbf{u}_j^K - \mathbf{v}_j^K\|.
\end{aligned}$$

So we have

$$\|\mathbf{x}_j^K - \mathbf{y}_j^K\| = \max_k |x_j^K(k) - y_j^K(k)| \leq \alpha_K \|\mathbf{u}_j^K - \mathbf{v}_j^K\|.$$

We now prove the continuity of $V_1^K(\mathbf{u}_1^K, \mathbf{u}_2^K)$ in $\mathbf{u}_1^K$. The proof of the continuity in $\mathbf{u}_2^K$ is similar. We have

$$\begin{aligned}
&|V_1^K(\mathbf{u}_1^K, \mathbf{u}_2^K) - V_1^K(\mathbf{v}_1^K, \mathbf{u}_2^K)| \\
&\quad = \left| \sum_{k=1}^{K} e^{-\rho_j t_k} [L_1(x_1^K(k), x_2^K(k), u_1^K(k)) - L_1(y_1^K(k), x_2^K(k), v_1^K(k))] \,\delta \right| \\
&\quad \leq \sum_{k=1}^{K} |L_1(x_1^K(k), x_2^K(k), u_1^K(k)) - L_1(y_1^K(k), x_2^K(k), v_1^K(k))| \,\delta \\
&\quad \leq \sum_{k=1}^{K} |L_1(x_1^K(k), x_2^K(k), u_1^K(k)) - L_1(y_1^K(k), x_2^K(k), u_1^K(k))| \,\delta \\
&\qquad + \sum_{k=1}^{K} |L_1(y_1^K(k), x_2^K(k), u_1^K(k)) - L_1(y_1^K(k), x_2^K(k), v_1^K(k))| \,\delta \\
&\quad \leq \sum_{k=1}^{K} C|x_1^K(k) - y_1^K(k)|\delta + \sum_{k=1}^{K} C|u_1^K(k) - v_1^K(k)|\delta \\
&\quad = CT\|\mathbf{x}_1^K - \mathbf{y}_1^K\| + C\|\mathbf{u}_1^K - \mathbf{v}_1^K\| \\
&\quad \leq C(T\alpha_K + 1)\|\mathbf{u}_1^K - \mathbf{v}_1^K\|.
\end{aligned}$$

□

Proof of Theorem 2.5. The set X_j of the trajectories with the norm $\|x_j\| = \max |x_j(t)|$ is compact. Let $M_j(u_j) = \int_0^T \mathcal{M}_j(u_j(t))\, dt$ and $N_j = \{M_j(u_j) : u_j \in U_j\}$. As M_j is Lipschitz and U_j is compact, N_j is compact. Therefore, we can extract from the sequence $(\phi_1^K(\mathbf{x}_1^{K*}), \phi_2^K(\mathbf{x}_2^{K*}), M_1(\phi_1^K(\mathbf{u}_1^{K*})), M_2(\phi_2^K(\mathbf{u}_2^{K*})))$ a converging subsequence $(\phi_1^{K_p}(\mathbf{x}_1^{K_p*}), \phi_2^{K_p}(\mathbf{x}_2^{K_p*}), M_1(\phi_1^{K_p}(\mathbf{u}_1^{K_p*})), M_2(\phi_2^{K_p}(\mathbf{u}_2^{K_p*})))$. Denote by $(\tilde{x}_1, \tilde{x}_2, \tilde{m}_1, \tilde{m}_2)$ the limit of this subsequence. Denote by $(\tilde{u}_1, \tilde{u}_2)$ a control that generates the trajectory $(\tilde{x}_1, \tilde{x}_2)$. Here we must emphasize that we could have $M_j(\tilde{u}_i) \neq \tilde{m}_i$.

We first show that the case where $M_j(\tilde{u}_i) < \tilde{m}_i$ is impossible. Suppose $M_j(\tilde{u}_i) < \tilde{m}_i$. We rewrite the function $V_1(u_1, u_2)$ as $\tilde{V}_1(x_1, x_2, M_1(u_1))$ to emphasize that it is a sum of two terms, one depending on (x_1, x_2) and the other depending on $M_1(u_1)$. Thus, we have $\tilde{V}_1(\tilde{x}_1, x_2, \tilde{m}_1) < \tilde{V}_1(\tilde{x}_1, x_2, M_1(u_1))$. So there exist $\Delta > 0$ such that

$$\tilde{V}_1(\tilde{x}_1, x_2, \tilde{m}_1) = \tilde{V}_1(\tilde{x}_1, x_2, M_1(u_1)) - \Delta. \tag{21}$$

For K_p sufficiently big, we have the following:

$$\begin{aligned} V_1^{K_p}(\mathbf{u}_1^{K_p*}, \mathbf{u}_2^{K_p*}) &\leq V_1(\phi_1^{K_p}(\mathbf{u}_1^{K_p*}), \phi_2^{K_p}(\mathbf{u}_2^{K_p*})) + \epsilon \\ &\leq V_1(\tilde{u}_1, \phi_2^{K_p}(\mathbf{u}_2^{K_p*})) + 2\epsilon - \Delta \\ &\leq V_1^{K_p}(\sigma_1^{K_p}(\tilde{u}_1), \sigma_2^{K_p} \circ \phi_2^{K_p}(\mathbf{u}_2^{K_p*})) + 3\epsilon - \Delta \\ &= V_1^{K_p}(\sigma_1^{K_p}(\tilde{u}_1), \mathbf{u}_2^{K_p*}) + 3\epsilon - \Delta. \end{aligned}$$

The first inequality is implied by Proposition A.1. The second inequality is implied by the convergence of $\phi_1^{K_p}(\mathbf{x}_1^{K_p*})$ to $\tilde{x}_1$, the continuity of $\phi_j^{K_p}$, the continuity of $V_j^{K_p}$ (Lemma B.1), and (21). The third inequality comes from Proposition A.2. The last inequality is valid as $\sigma_j^K \circ \phi_j^K = \mathbf{1}$. As ϵ can be chosen as small as desired for K_p big enough, and as $\Delta > 0$, we conclude that

$$V_1^{K_p}(\mathbf{u}_1^{K_p*}, \mathbf{u}_2^{K_p*}) < V_1^{K_p}(\sigma_1^{K_p}\tilde{u}_1, \mathbf{u}_2^{K_p*}),$$

which contradicts the fact that $(\mathbf{u}_1^{K_p*}, \mathbf{u}_2^{K_p*})$ is an equilibrium for the approximating game.

So we can assume that $M_j(\tilde{u}_i) \geq \tilde{m}_i$. Then, for all u_1 in U_1 and K_p big enough, the following holds:

$$\begin{aligned} V_1(u_1, \tilde{u}_2) &\leq V_1^{K_p}(\sigma_1^{K_p}(u_1), \sigma_2^{K_p}(\tilde{u}_2)) + \epsilon \\ &\leq V_1^{K_p}(\sigma_1^{K_p}(u_1), \mathbf{u}_2^{K_p*}) + 2\epsilon \\ &\leq V_1^{K_p}(\mathbf{u}_1^{K_p*}, \mathbf{u}_2^{K_p*}) + 2\epsilon \\ &\leq V_1(\phi_1^{K_p}(\mathbf{u}_1^{K_p*}), \phi_2^{K_p}(\mathbf{u}_2^{K_p*})) + 3\epsilon \\ &\leq V_1(\tilde{u}_1, \tilde{u}_2) + 4\epsilon. \end{aligned}$$

The first inequality is implied by Proposition A.2. The second inequality is implied by the convergence[6] of $\phi_j^{K_p}(\mathbf{x}_j^{K_p*})$ to $\tilde{x}_j$, the continuity of $\phi_j^{K_p}$, the continuity of $V_j^{K_p}$ (Lemma B.1), and the property $\sigma_j^K \circ \phi_j^K = \mathbf{1}$. The third inequality is valid as $(\mathbf{u}_1^{K_p*}, \mathbf{u}_2^{K_p*})$ is the equilibrium for the discrete-time oligopoly of order K_p. The

[6] Recall that $V_1^{K_p}$ does not depend on $M_j(\phi_j^{K_p}(\mathbf{u}_j^{K_p*}))$, $j \neq 1$.

fourth inequality is obtained applying Proposition A.1. The last inequality is valid as $(\phi_1^{K_p}(\mathbf{x}_1^{K_p*}), \phi_2^{K_p}(\mathbf{x}_2^{K_p*}))$ converges to $(\tilde{x}_1, \tilde{x}_2)$ when $M_1(\phi_1^{K_p}(\mathbf{u}_1^{K_p*}))$ converges to something smaller than or equal to $M_1(\tilde{u}_1)$ and since V_j is a continuous function in all arguments. A similar property can be obtained for any other player j. As ϵ can be taken as small as desired, $(\tilde{x}_1, \tilde{x}_2)$ is the equilibrium trajectory of the continuous-time oligopoly. As the equilibrium is unique, the whole sequence $(\phi_1^K(\mathbf{x}_1^{K*}), \phi_2^K(\mathbf{x}_2^{K*}))$ converges to $(\tilde{x}_1, \tilde{x}_2) = (x_1^*, x_2^*)$. We will now show that we can exclude the case $M_j(u_i^*) = M_j(\tilde{u}_i) > \tilde{m}_i$. By the same argument as before, we can show that there exist $\Delta > 0$ such that

$$\tilde{V}_1(x_1^*, x_2^*, M_j(u_1^*)) = \tilde{V}_1(x_1^*, x_2^*, \tilde{m}_1) - \Delta.$$

As $(\phi_1^{K_p}(\mathbf{x}_1^{K_p*}), M_1(\phi_1^{K_p}(\mathbf{u}_1^{K_p*})))$ converges to $(x_1^*, \tilde{m}_1)$, as $\phi_j^{K_p}$ is a continuous function, and as $\tilde{V}_1$ is a continuous function in all arguments, for all ϵ there exists K_p big enough such that

$$\tilde{V}_1(x_1^*, x_2^*, \tilde{m}_1) \leq \tilde{V}_1(\phi_1^{K_p}(\mathbf{x}_1^{K_p*}), x_2^*, M_1(\phi_1^{K_p}(\mathbf{u}_1^{K_p*}))) + \epsilon.$$

Thus, from the last two statements, we obtain

$$V_1(u_1^*, u_2^*) < V_1(\phi_1^{K_p}(\mathbf{u}_1^{K_p*}), u_2^*),$$

which is a contradiction; hence, $M_j(u_i^*) = M_j(\tilde{u}_i) = \tilde{m}_i$. □

REFERENCES

[1] Altman, E., Jamenez, T. and Koole, G. On optimal call admission control, in *IEEE Conference on Decision Control*, December 1998.

[2] Altman, E., Pourtallier, O., and Tidball, M. Approximations in dynamic zero-sum games, II, *SIAM J. Cont. Opt.*, **35**, pp. 2101–2117, 1997.

[3] Billingsley, P. *Convergence of Probability Measures*, John Wiley and Sons, New York, 1968.

[4] Brock, W. Differential games with active and passive variables, pp. 34–52 in *Mathematical Economics and Game Theory: Essays in Honor of Oskar Morgenstern*, Henn and Moeschlin, eds., Springer-Verlag, Berlin, 1977.

[5] Carlson, D. and Haurie, A. A turnpike theory for infinite-horizon open-loop competitive processes, *SIAM J. Cont. Opt.*, **34**, 1996.

[6] Freedman, D. *Markov Chains*, Springer-Verlag, New York, 1983.

[7] Gürkan, G., Özge, Y., and Robinson, S. M. Sample-path solution for simulation optimization problems and stochastic variational inequalities. To appear in *Advances in Stochastic Optimization, Logic Programming, and Heuristic Search: Interfaces in Computer Science and Operations Research*, D. W. Woodruff, ed., Kluwer Academic Publishers (1998), 1997.

[8] Gürkan, G., Özge, Y., and Robinson, S. M. Sample-path solution of stochastic variational inequalities. To appear in *Mathematical Programming*, 1998.

[9] Haurie, A. Stochastic differential games in economic modeling, pp. 90–108, in *System Modelling and Optimization*, J. Henry and J. -P. Yvon, eds., Vol. 197 of Lecture Notes in Control and Information, Springer-Verlag, 1994.

[10] Haurie, A. and Leitmann, G. On the global stability of equilibrium solutions for open-loop differential games, *Large Scale Systems*, **6**, pp. 107–122, 1984.

[11] Haurie, A. and Roche, M. Turnpikes and computation of piecewise open-loop equilibria in stochastic differential games, *J. Econ. Dyn. Cont.*, **18**, pp. 317–344, 1994.

[12] Haurie, A., Zaccour, G., Legrand, J., and Smeers, Y. *A Stochastic Dynamic Nash-Cournot Model for European Gas Market*, tech. rep., HEC Montréal, 1987.

[13] Haurie, A., Zaccour, G., and Smeers, Y. Stochastic equilibrium programming for dynamic oligopolistic markets, *J. Opt. Th. Appl.*, **66**, pp. 243–253, 1990.

[14] Konnov, I. Combined relation methods for finding equilibrium points and solving related problems, *Izvestiya VUZ, Mathematika*, **37**, pp. 44–51, 1993.

[15] Pourtallier, O. *Approximations in Static Non-zero-sum Games*, Working paper INRIA, 1998.

[16] Rosen, J. B. Existence and uniqueness of equilbrium points for concave n-games, *Econometrica*, **33**, pp. 520–534, 1965.

Comparison of Two Numerical Approaches for the Barrier and Value of a Simple Pursuit–Evasion Game

Pierre Bernhard
University of Nice
and
I3S CNRS-UNSA
Sophia Antipolis, France

Stephane Crepey
INRIA
Sophia Antipolis, France

Alain Rapaport
INRIA
Montpelier, France

Abstract

We investigate the barrier of a simple pursuit–evasion game for which we are able to compare two theoretical and numerical approaches. One is based directly on the capture time, and the second one, introduced by one of the authors, transforms the game in one of approach (or L_∞-criterion). This second approach gives both a new characterization of barriers and a new, potentially more robust, numerical method for the determination of barriers. We provide a detailed analytical solution of the various problems thus raised and use it as a benchmark for the numerical method.

1 Introduction

We revisit a well-known, one-dimensional, second-order servomechanism problem, proposed by Bernhard in [7], with a new approach that transforms the game into one of approach (or L_∞-criterion). This simple pursuit–evasion game allows us to compare the traditional approach with this new one, both on theoretical and numerical points of view.

We present numerical methods for the computation of the value functions of the two versions of the game (the *game in time* and the *game in distance*), with a particular emphasis on the determination of the *barrier* of the pursuit–evasion game. Our methods use the theory of *viscosity solutions* for the Isaacs equation (see [4] or [12] for the state of the art), which is an alternative to the *viability* approach proposed in [9] and [10] or the *minimax* solutions of [20].

The first method is based on a finite-difference approximation of the discounted capture-time function, involving viscosity lower-envelope solutions of the Isaacs equation (cf. the work of Bardi et al., [1]). The associated numerical scheme computes an approximation by discrete stochastic games, introduced by Pourtallier and Tidball [18] following the work of Kushner [16].

Nevertheless, when a barrier occurs in the capture–evasion game, splitting the state space into capture and evasion areas, a detection of an infinite value of the capture-time function is required in order to characterize this manifold. (See [8] for a state-of-the-art description of barriers of differential games.) From a numerical point of view, this previous method does not seem to be well suited for accurate detection since the barrier sought appears as the boundary of the set where the discounted value function is strictly less than 1, a level it reaches with zero slope.

The second approach considers an approximation of the minimum oriented distance from the target, involving viscosity upper-envelope solutions of a variational inequality (see [19]). The oriented distance from the target needs to be known, which may require a numerical computation for an arbitrary target. Nevertheless, for many games (such as the ones studied by Isaacs [15]), the target is given by a simple analytic expression, and then its oriented distance is also an analytic function, easy to compute. The numerical scheme computes a monotone sequence of continuous solutions for a sequence of perturbated Hamiltonians, again using approximation by discrete stochastic games (see [13]). The barrier for the game in time is then determined by the zero-level set of the value function for the game in distance, an intrinsically robust determination, as the gradient is not 0 there. Moreover, this gradient also measures sensitivity of the barrier location with respect to the target.

Finally, we illustrate these methods with numerical experimentations, using the analytical solutions to benchmark the numerical results we obtain.

2 Presentation of the Game

Consider a one-dimensional, second-order plant:

$$\ddot{y} = \beta\, v, \qquad |v| \leq 1,$$

where the objective is to keep y as close as possible to a set point z subject to an unknown drift:

$$\dot{z} = \alpha\, u, \qquad |u| \leq 1.$$

More precisely, for a given positive number γ, we are looking for a (state feedback) control law $v^*(\ )$ that guarantees $|y(t) - z(t)| \leq \gamma$ for all $t \geq 0$ whatever the disturbance $u(.)$.

Considering the state vector

$$x := \begin{pmatrix} x_1 \\ x_2 \end{pmatrix} = \begin{pmatrix} y - z \\ \dot{y} \end{pmatrix},$$

this problem can be formulated as a pursuit–evasion game, whose dynamics are

$$\begin{cases} x(0) = x_0, \\ \dot{x}(t) = f(x(t), u(t), v(t)) := \begin{pmatrix} x_2(t) - \alpha u(t) \\ \beta v(t) \end{pmatrix}, \qquad |u(t)| \leq 1, \quad |v(t)| \leq 1; \end{cases}$$

with the target set

$$\mathcal{T} := \{x \in \mathbb{R}^2 \mid |x_1| \geq \gamma\}$$

(the player u is the "pursuer" and the player v the "evader"). The usual way to study the existence of such a control law v^* is to study the *game in time* (cf. [15]):

$$V(x_0) = \sup_{\psi[\,]} \inf_{u(\cdot)} t^c(x_0, u, v),$$

where $t^c(x_0, u, v) = \inf\{t \geq 0 \mid x(t) \in \mathcal{T}\}$ is the capture time, $(v(\cdot) = \psi[u(\cdot)], u(\cdot))$ are admissible controls, and $\psi[\,]$ belongs to a set of strategies defined below. This game has been investigated in detail by Masle [17] and Bernhard in [7]. A particular emphasis is made on the existence and the characterization of the barrier that splits the state space between initial positions for which there exists a strategy for player v avoiding a termination in finite time from its complementary.

Alternatively, we study another criterion related to the *game in distance*:

$$W(x_0) = \sup_{\psi[\,]} \inf_{u(\cdot)} \left[\inf_t d^o(x(t), \mathcal{T})\right],$$

where (u, v) belong to the same sets of strategies as for the previous game and d^o is the oriented distance function:

$$d^o(x, \mathcal{T}) = \begin{cases} d(x, \mathcal{T}) & \text{if } x \notin \mathcal{T}; \\ -d(x, \partial\mathcal{T}) & \text{otherwise.} \end{cases}$$

Here, $d^o(x, \mathcal{T}) = \gamma - |x_1|$, and we shall propose a new analytical resolution of this game. The barrier of the game in time is then determined by the set of points $\mathcal{B} = \{x \mid W(x) = 0\}$. Although this criterion does not provide any information on the capture time, it characterizes the sensitivity with respect to the target, which is of complementary interest compared with the traditional approach.

3 Analytical Solutions

3.1 Preliminaries

We shall define more precisely for which class of strategies the value functions V and W defined above should be considered.

Definition 3.1 (VREK Strategies). Let $\mathcal{U}, \mathcal{V}$ be the sets of measurable functions from $\mathbb{R}^+$ to $[-1, 1]$ or *open-loop* controls. $u(\cdot)$ is sought among $\mathcal{U}$ and $\psi[\,]$ among the *nonanticipative VREK strategies*

$$\{\psi[\,] : u \in \mathcal{U} \longmapsto \psi[u] \in \mathcal{V}\}$$

such that $\forall u \in \mathcal{U}$, $[\forall t \leq t', u(t) = u'(t)] \Longrightarrow [\forall t < t', \psi[u](t) = \psi[u'](t)]$.

Similarly, we can consider strategies for a reversed order of the players: $v(\cdot)$ is then sought among $\mathcal{V}$ and $\phi[\,]$ among the *nonanticipative VREK strategies*

$$\{\phi[\,] : v \in \mathcal{V} \longmapsto \phi[v] \in \mathcal{U}\}$$

such that $\forall v \in \mathcal{V}$, $[\forall t \leq t', v(t) = v'(t)] \Longrightarrow [\forall t < t', \phi[v](t) = \phi[v'](t)]$.

These classes of strategies are well suited to characterize the value functions in terms of *viscosity solutions* (see [12] and [4]), for which we recall the following definition.

Definition 3.2 (Viscosity Solutions). Consider a first-order partial differential equation on an open domain Ω:

$$\begin{gathered} H(x, V(x), \nabla V(x)) = 0, \quad x \in \Omega \\ \text{(possibly with a boundary condition } V(x) = K, x \in \partial\Omega). \end{gathered} \tag{1}$$

Let $D^+V(x)$ (*resp.*, $D^-V(x)$) denote the Fréchet super- (*resp.*, sub-) differential of the locally bounded function V at x, that is, the set of formal gradients p, such that

$$V(y) \leq V(x) + < p, y - x > + \theta(y - x)||y - x||$$

$$(\text{resp.}, V(y) \geq V(x) + < p, y - x > - \theta(y - x)||y - x||),$$

for some continuous function θ, null at 0.

(i) A *subsolution* (resp., *supersolution*) of H on Ω is a *u.s.c.* (resp., *l.s.c.*) locally bounded function V *s.t.* $H(x, V(x), D^+V(x)) \geq 0$ (resp., $H(x, V(x), D^-V(x)) \leq 0$) on Ω.
(ii) A Dirichlet subsolution (resp., *Dirichlet supersolution*) of (1) on $\overline{\Omega}$ must also satisfy $V \leq K$ (*resp.*, $V \geq K$) on $\partial\Omega$.
(iii) If a *subsolution* (resp., *supersolution*) of H on Ω satisfies at least $H(x, V(x), D^+V(x)) \geq 0$ (resp., $H(x, V(x), D^-V(x)) \leq 0$) wherever it fails to satisfy $V \leq K$ (resp., $V \geq K$) on $\partial\Omega$, we shall call it a *subsolution* (resp., *supersolution*) of (1) on $\overline{\Omega}$.
(iv) A (resp., Dirichlet) *viscosity solution* means a function that is both a (resp., Dirichlet) sub- and a supersolution.
(v) The viscosity upper- (resp., lower-) envelope solution on Ω (resp., on $\overline{\Omega}$) means the largest viscosity sub-solution on Ω (resp., the smallest Dirichlet viscosity super solution on $\overline{\Omega}$).

Alternatively, we shall also consider classes of feedback strategies.

Definition 3.3 (Feedback Strategies). $\Phi \subset \{\phi : (t,x) \mapsto \phi(t,x) \in [-1,1]\}$ and $\Psi \subset \{\psi : (t,x) \mapsto \psi(t,x) \in [-1,1]\}$ are *admissible classes of feedback strategies* if

(i) Open-loops are admissible: $\mathcal{U} \subset \Phi$ and $\mathcal{V} \subset \Psi$.
(ii) Φ and Ψ are closed by concatenation (i.e., switching from one strategy in the set to another one, at an intermediate instant of time, is allowed).
(iii) $\forall(\phi,\psi) \in \Phi \times \Psi$, $\forall x_0$, there exists a unique solution of $\dot{x} = f(x, \phi(.,x), \psi(.,x))$ over $\mathbb{R}^+$, leading to measurable controls: $u(.) = \phi(., x(.)) \in \mathcal{U}$ and $v(.) = \psi(., x(.)) \in \mathcal{V}$.

These properties do not uniquely define the pair (Φ, Ψ), but it is clear that such classes exist and are subclasses of VREK nonanticipative strategies.

3.2 Game in Time

We sketch here the analysis of [7] and [17], according to the classic Isaacs–Breakwell theory. From dimensional analysis, it is easy to see that the only meaningful parameter in that game is the ratio

$$p = \frac{\beta\gamma}{\alpha^2}.$$

First we find the usable part of the capture set (i.e., the subset of the boundary of the target such that $\sup_v \inf_u < \nu(x), f(x,u,v) > \leq 0$, where $\nu(x)$ is the outer normal to $\mathcal{T}$ at x) that here is made up of two symmetric pieces: $\{x_1 = \gamma, x_2 > -\alpha\}$ and $\{x_1 = -\gamma, x_2 < \alpha\}$. The boundary of the usable part (BUP) is thus made up of the two points $(x_1 = \varepsilon\gamma, x_2 = -\varepsilon\alpha)$ for $\varepsilon = \pm 1$. From the BUP, we attempt to construct a natural barrier. The semipermeable normal is $(\nu_1 = -\varepsilon, \nu_2 = 0)$. Given the Hamiltonian of the game of kind,

$$H = \nu_1(x_2 - \alpha u) + \nu_2 \beta v,$$

we see on the one hand that the semipermeable controls are $u = \operatorname{sign} \nu_1$ and $v = \operatorname{sign} \nu_2$, and on the other hand that the adjoint equations give

$$\dot{\nu}_1 = 0;$$
$$\dot{\nu}_2 = -\nu_1.$$

Initialized with the proposed semipermeable ν's on the BUP, this yields two parabolas with the controls $u = v = -\varepsilon$ (we call t_1 the *final time*):

$$x_1(t) = \varepsilon[\gamma - \frac{\beta}{2}(t_1 - t)^2],$$
$$x_2(t) = \varepsilon[-\alpha + \beta(t_1 - t)].$$

These intersect the "other edge" of the game space, that is, the straight line $x_1 = -\varepsilon\gamma$, at $x_2 = \varepsilon(-\alpha + 2\sqrt{\beta\gamma})$. We must now distinguish two cases depending on whether these points are in the usable or nonusable part.

The simple case is when this intersection happens in the nonusable part, which is the case if $p > 1$. In that case, the two parabola together with the pieces of (nonusable) capture-set boundary that join them (the thick lines in Figure 2) indeed form a barrier, separating an escape zone "inside" from the capture zone outside.

That composite curve is indeed a barrier. At all the points where it is smooth, the semipermeability condition holds (or, on the capture-set boundary, a stronger inequality for the evader). At its points of nondifferentiability—the two intersections of the parabola with the opposite capture sets—the evader may play according to the parabola's dictum, that is, $v = \varepsilon$. This ensures that the state remains inside the escape zone since $\dot{x}_1$ has the desired sign whatever the controls are.

Outside that region, we can construct a complete field of trajectories that happen to be parabolas translated from the previous ones parallel to the x_1 axis. It is a simple matter to check that they define a value function

$$V(x) = \frac{1}{\beta}\left[\alpha + \varepsilon x_2 - \sqrt{(\alpha + \varepsilon x_2)^2 - 2\beta(\gamma - \varepsilon x_1)}\right],$$

with $\varepsilon = 1$ in the upper region and $\varepsilon = -1$ in the lower region. Inside the escape zone, of course, $V = +\infty$. (We should emphasize that the value function computed here is Isaacs', not the function V of the next paragraphs, which is its Kruskov transform.)

In the case $p < 1$, the two parabolas intersect each other inside the game space, delineating what we shall call the *lens*. This lens is *not* an escape zone, however: the corners "leak." Following the classical analysis of the intersection of barriers, we have an intersection with incoming trajectories that cross it. Therefore, the composite surface is *not* a barrier.

As a matter of fact, the lens is the *intersection* of the proposed safety zones defined by each parabola. Therefore, to stay in it, the state should cross none of the parabolas, which the pursuer cannot enforce since the required controls are $+1$ for one of the parabola and -1 for the other one. Upon reaching such a corner, the pursuer can keep its optimal control according to the incoming parabola, and the state necessarily leaves the "lens."

In that case, there is no escape zone. But the complete solution in terms of singularities of Isaacs' equation is extremely involved. A private communication of John Breakwell suggested that the number of commutations of the optimal controls from $+1$ to -1, and conversely, can be arbitrarily large, depending on the initial state and the value of p.

3.3 Game in Distance

In [19], it is proved that the value function W for the game in distance is the viscosity upper-envelope solution of the following variational inequality (under technical assumptions that guarantee W to be u.s.c.):

$$H(x, W(x), \nabla W(x)) = \min\left[d^o(x, \mathcal{T}) - W(x), \min_u \max_v \nabla W(x).f(x, u, v)\right] = 0. \tag{2}$$

Unfortunately, the technical assumptions proposed in [19] in a general framework are not fulfilled in this game. Nevertheless, we show here, thanks to analytical considerations, that the value function W is a continuous viscosity solution of (2).

When $W(x) < d^o(x, \mathcal{T})$, the characteristic fields of the considered game are obtained for $u^*(x) = \operatorname{sign} \partial_1 W(x)$ and $v^*(x) = \operatorname{sign} \partial_2 W(x)$:

$$\begin{cases} x_1(t) &= \epsilon\beta t^2/2 + (x_1(0) - \epsilon\alpha)t + x_1(0) \\ x_2(t) &= \epsilon\beta t + x_2(0) \end{cases} \quad \text{for } \epsilon = \pm 1. \tag{3}$$

A necessary condition for t_1 to minimize $t \mapsto \gamma - |x_1(t)|$ is to have $\dot{x}_1(t_1) = 0$, which gives

$$\gamma - x_1(t_1) = \gamma + \epsilon x_1(0) - \frac{(x_2(0) - \epsilon\alpha)^2}{2\beta}.$$

This leads us to consider the following candidate Z solution of the variational inequality.

Definition 3.4.

$$Z(x) = \begin{cases} \min(\gamma + x_1,\ P^+(x)) & \text{when } x_2 \geq \alpha, \\ \min(\gamma - x_1,\ P^-(x)) & \text{when } x_2 \leq -\alpha, \\ \min(\gamma - \alpha^2/\beta,\ P^+(x),\ P^-(x)) & \text{when } |x_2| \leq \alpha, \end{cases}$$

with

$$\begin{cases} P^+(x) &= \gamma - x_1 - \dfrac{(x_2 + \alpha)^2}{2\beta}, \\ P^-(x) &= \gamma + x_1 - \dfrac{(x_2 - \alpha)^2}{2\beta}. \end{cases}$$

Remark 3.1. 1. Z is maximal and constant equal to $\gamma - \alpha^2/\beta$ inside the "lens" delineated by two arcs of parabola:

$$\mathcal{L} := \{x \mid P^-(x),\ P^+(x) \geq \gamma - \alpha^2/\beta\} \cap \{|x_2| \leq \alpha\}.$$

The constant value inside lens $\mathcal{L}$ is equal to the common value kept by the three functions $\gamma - |x_1|$, $P^-(x)$, and $P^+(x)$ at points x such that $|x_2| \leq \alpha$ and where they are equal, which are exactly the two points $A = (-\alpha^2/\beta, \alpha)$ and $B = (\alpha^2/\beta, -\alpha)$ (see Figure 1).

2. The set of points where the function Z is null is

(i) void if $\gamma - \alpha^2/\beta < 0$,

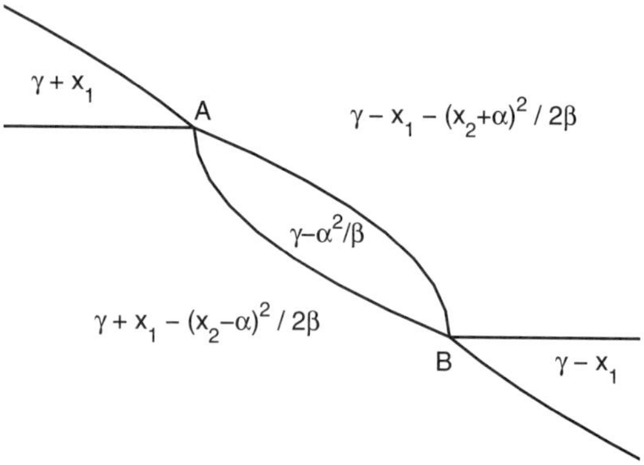

Figure 1: Different areas defining the function Z.

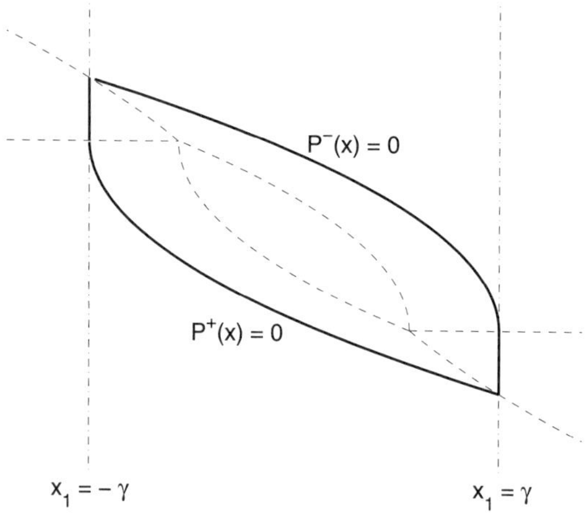

Figure 2: The set of points x where $Z(x) = 0$ (when $\gamma - \alpha^2/\beta > 0$).

(ii) otherwise equal to

$$\{P^+(x) = 0, x_1 \geq -\gamma, x_2 \geq -\alpha\} \cup \{P^-(x) = 0, x_1 \leq \gamma, x_2 \leq \alpha\}$$
$$\cup\{-\gamma\} \times [\alpha, 2\sqrt{\beta\gamma} - \alpha] \cup \{\gamma\} \times [-\alpha, \alpha - 2\sqrt{\beta\gamma}]$$

(see Figure 2).

We recognize in this last expression exactly the barrier found by Bernhard [7] for the game in time.

Proposition 3.1. *Z is a continuous viscosity solution of* (2).

Proof. Z is clearly continuous, nowhere above $d^o(.,\mathcal{T})$. Notice that requiring Z to be a continuous viscosity solution of (2) is then equivalent to

$$\begin{cases} \min_u \max_v p.f(x,u,v) \geq 0, \quad \forall p \in D^+Z(x), \\ Z(x) = d^o(x,\mathcal{T}) \text{or } \min_u \max_v p.f(x,u,v) \leq 0, \quad \forall p \in D^-Z(x). \end{cases}$$

Direct computation shows that Z satisfies the variational inequality (2) at its differentiable points. At nondifferentiable points, using nonsmooth calculus rules (see, for instance, [11]), we have the following conditions.

(i) For x such that $P^+(x) = \gamma + x_1$ and $x_2 \geq \alpha$,

$$D^+Z(x) = \left\{ -\lambda \begin{pmatrix} 1 \\ (x_2+\alpha)/\beta \end{pmatrix} + (1-\lambda)\begin{pmatrix} 1 \\ 0 \end{pmatrix} \right\}_{\lambda \in [0,1]}, \qquad D^-Z(x) = \emptyset.$$

So $\min_u \max_v p.f(x,u,v) = (1-2\lambda)x_2 - |1-2\lambda|\alpha + \lambda(x_2+\alpha) \geq 0$, $\forall p \in D^+Z(x)$.
(By symmetry, we have the same inequalities at points x such that $P^-(x) = \gamma - x_1$ and $x_2 \leq -\alpha$)

(ii) For x such that $P^+(x) = \gamma - \alpha^2/\beta$ and $|x_2| \leq \alpha$,

$$D^+Z(x) = \left\{ -\lambda \begin{pmatrix} 1 \\ (x_2+\alpha)/\beta \end{pmatrix} + (1-\lambda)\begin{pmatrix} 0 \\ 0 \end{pmatrix} \right\}_{\lambda \in [0,1]}, \qquad D^-Z(x) = \emptyset.$$

So $\min_u \max_v p.f(x,u,v) = -\lambda(x_2+\alpha) + \lambda(x_2+\alpha) = 0, \forall p \in D^+Z(x)$.
(By symmetry, we have the same inequalities at points x such that $P^-(x) = \gamma - \alpha^2/\beta$ and $|x_2| \leq \alpha$.)

(iii) For A,

$$D^+Z(A) = \left\{ -\lambda_1 \begin{pmatrix} 1 \\ 2\alpha/\beta \end{pmatrix} + \lambda_2 \begin{pmatrix} 1 \\ 0 \end{pmatrix} + (1-\lambda_1-\lambda_2)\begin{pmatrix} 0 \\ 0 \end{pmatrix} \right\}_{\substack{\lambda_1,\lambda_2 \geq 0, \\ \lambda_1+\lambda_2 \leq 1}},$$

$$D^-Z(A) = \emptyset.$$

So $\min_u \max_v p.f(A,u,v) = (\lambda_2 - \lambda_1)\alpha - |\lambda_2 - \lambda_1|\alpha + 2\lambda_1\alpha \geq 0$, $\forall p \in D^+Z(A)$. (By symmetry, we have the same inequalities at point B.)
Z is then a viscosity solution of the variational inequality (2). □

Proposition 3.2. *Z is the value function with feedback strategies (for the game in distance).*

Proof. Take a number c less than or equal to $\gamma - \alpha^2/\beta$, then $(\gamma - c) \geq \alpha^2/\beta$ and so, according to Section 3.2, there exists a barrier $\mathcal{B}_c$ for the game in time with the target $\mathcal{T}_c := \{x \in \mathbb{R}^2 \mid |x_1| \geq \gamma - c\}$. We notice also that the candidate Z is such that the manifold $\{x \mid Z(x) = c\}$ coincides exactly with the barrier $\mathcal{B}_c$ determined in the previous section. As the exterior of $\mathcal{B}_c$ is a guaranteed capture zone, the value function $W_{\gamma-c}(x_0)$ for the game in distance (with the target $\mathcal{T}_c$) is necessarily nonpositive as soon as x_0 is such that $Z(x_0) > c$. Similarly, the interior of $\mathcal{B}_c$ is a guaranteed evasion zone, so $W_{\gamma-c}(x_0)$ is nonnegative as soon as x_0 is such that $Z(x) < c$. Note also that the $W(x_0) = W_{\gamma-c}(x_0) + c$, $\forall x_0 \in \mathbb{R}^2$.

Now consider polar coordinates (r, θ) in the plane. For any θ, there exists $r > 0$ such that $re^{i\theta} \in \mathcal{B}_c$. A point $se^{i\theta}$ belongs to the exterior (resp., the interior) of $\mathcal{B}_c$ as soon as $s > r$ (resp., $s < r$), and so $W(se^{i\theta}) \leq c = Z(re^{i\theta})$ (resp., $W(se^{i\theta}) \geq c = Z(re^{i\theta})$). This can be achieved for any c such that $c \leq \gamma - \alpha^2/\beta$, namely, outside the lens,

$$\mathcal{L} := \{x \mid P^-(x),\ P^+(x) \geq \gamma - \alpha^2/\beta\} \cap \{|x_2| \leq \alpha\}.$$

Thus, we have

$$x_0 = re^{i\theta} \notin \mathcal{L} \Rightarrow \begin{cases} s > r \Rightarrow W(se^{i\theta}) \leq Z(x_0), \\ s < r \Rightarrow W(re^{i\theta}) \geq Z(x_0). \end{cases}$$

Because Z is a continuous function, we conclude that $W(x_0) = Z(x_0), \forall x_0 \notin \mathcal{L}$. We also deduce that $W(x_0) \geq \gamma - \alpha^2/\beta, \forall x_0 \in \mathcal{L}$. Note also that the above reasoning could be done with the "lower" value W^- (instead of the "upper" value W):

$$W^-(x_0) = \inf_{\phi[]} \sup_{v \in \mathcal{V}} \left[\inf_{t \geq 0} d^o(x(t), \mathcal{T}) \right] \leq W(x_0),$$

that is, we have $W = W^- = Z$ on $\mathbb{R}^2 \setminus \mathcal{L}$ and $W^- \geq \gamma - \alpha^2/\beta$ on $\mathcal{L}$.

Consider the state space divided into the three domains

$$\begin{aligned} \mathcal{S} &:= \left\{x | Z(x) = P^+(x) \text{ or } Z(x) = \gamma - x_1\right\}, \\ \mathcal{I} &:= \left\{x | Z(x) = P^-(x) \text{ or } Z(x) = \gamma + x_1\right\}, \\ \overset{\circ}{\mathcal{L}} &:= \mathbb{R}^2 \setminus (\mathcal{S} \cup \mathcal{I}). \end{aligned}$$

and the following feedback strategies:

$$\tilde{u}(x) = \left| \begin{array}{ll} 1 & \text{if } x \in \mathcal{I}, \\ -1 & \text{otherwise,} \end{array} \right.$$

and $\tilde{v}(x) = \tilde{u}(x)$. From Section 3.2 and the result above, it is clear that the pair $(\tilde{u}, \tilde{v})$ realizes $W(x_0)$ for any initial condition x_0 outside the lens and give a minimal oriented distance larger than or equal to $\gamma - \alpha^2/\beta$ for $x_0 \in \mathcal{L}$.

Consider now $x_0 \in \mathcal{L}$ and the pair $(\tilde{u}, v)$ for an arbitrary open-loop control $v \in \mathcal{V}$. Remember that inside the lens $x_2 \geq -\alpha$, so let $l = \inf_{t \geq 0}\{x_2(t) + \alpha\}$. Inside the lens, the dynamics in x_1 are $\dot{x}_1 = x_2 + \alpha > l$.

If $l > 0$, the trajectory leaves $\mathcal{L}$ in finite time, say at t_e. But then we have

$$\forall \epsilon > 0, \quad \exists \eta > 0 \text{ s.t. } |x_1(t)| \leq \alpha^2/\beta - \epsilon \Longrightarrow \dot{x}_1(t) \geq \eta, \quad \forall t \geq t_e$$

(if the trajectory reaches $\mathcal{S}$), and we conclude that $\inf_{t \geq t_e}\{\gamma - |x_1(t)|\} \leq \gamma - \alpha^2/\beta$ (by symmetry, we have the same inequality if the trajectory reaches $\mathcal{I}$).

If $l \leq 0$ and the trajectory does not leave the lens in finite time, it converges asymptotically inside $\mathcal{L}$ toward the corner point B and we have:

$$\inf_{t\geq 0}\{\gamma - |x_1(t)|\} \leq \max_{\xi\in\partial\mathcal{L}} Z(\xi) = Z(x_0), \quad \forall v \in \mathcal{V}.$$

So, in any case, we obtain

$$Z(x_0) = \gamma - \alpha^2/\beta \geq \sup_{v\in\mathcal{V}} \left[\inf_{t\geq 0} d^o(x(t), \mathcal{T})\right] \geq W^-(x_0), \quad \forall x_0 \in \mathcal{L}.$$

We then conclude that $Z(x_0) = W(x_0) = W^-(x_0), \forall x_0 \in \mathbb{R}^2$ (and that the game admits a saddle point for VREK strategies). We also conclude that the feedback strategies $(\tilde{u}, \tilde{v}) \in \Phi \times \Psi$ are optimal. □

Remark 3.2. W is necessarily nonpositive (resp., nonnegative) at capture points (resp., evasion points). So a barrier separates points x_c where $W(x_c) \leq 0$ from points x_e where $W(x_e) \geq 0$. Conversely, a point x belongs to a capture area (resp., evasion area) if $W(x) < 0$ (resp., $W(x) > 0$). So a hypersurface separating points where W is strictly negative from points where W is strictly positive is a barrier.

In this game, W is continuous and equal to Z (the value with feedback strategies). So the barrier of the capture–evasion game with feedback strategies is nothing more than the zero-level set of the function Z. The existence condition and determination of the barrier are both derived explicitly from Z, and the analysis is independent of the ratio p (introduced in the previous section). In addition, we have proved that a barrier does not exist when $p < 1$.

4 Numerical Methods

We shall study numerical approximations of the value functions V and W on a given subset of the state space $\mathcal{E} = \mathbb{R}^2$. For the game in time, the domain of definition of the value function V is then $\Omega = \mathcal{E} \setminus \mathcal{T}$ (for the game in distance, we shall simply say that $\Omega = \mathcal{E}$).

4.1 Preliminaries

Definition 4.1 (Kruskov transformation). Let U denote the Kruskov transform of the value function of the game, where, by definition, the Kruskov transform is $\phi(\xi) = 1 - \exp(-\xi)$.

We recall how U is related to the discounted version of the game.

Proposition 4.1. *U is the value function (in the same meaning) of the discounted differential game, where by "discounted game" we mean the game with the same dynamics as before and the criterion t^c or $\inf_t d^o$ replaced by $\phi(t^c)$ or $\inf_t \phi(d^o)$, by monotonicity of ϕ.*

From now on, V and W will refer to the value functions of the discounted games, which have the numerical advantage to be bounded from above by 1.

Definition 4.2 (Discounted Hamiltonians). To the discounted games, we associate the following Hamiltonians:

(i) for the game in time:

$$H(x, s, p) = \min_u \max_v \langle p, f(x, u, v)\rangle + 1 - s, \tag{4}$$

(ii) for the game in distance:

$$H(x, s, p) = \min \left[\phi(d^o(x, \mathcal{T})) - s, \min_u \max_v \langle p, f(x, u, v)\rangle\right]. \tag{5}$$

We introduce a finite-difference scheme to approximate values of both differential games (game in time or game in distance). This scheme is merely a classical *upwind* finite-difference scheme merely for first-order p.d.e., adapted by Kushner to optimal control problems [16] and later to differential games by Pourtallier and Tidball [18]. This scheme can also be interpreted as approximation by discrete stochastic games: this leads to proofs of the convergence results, which are alternative to the ones used for standard p.d.e. (see [16] for the one-player case and [13] for two-player games).

Definition 4.3 (Starred Mesh). A *starred mesh* of step h on $\mathcal{E}$ is given by

1. a discrete set of nodes $\mathcal{E}^h \subseteq \mathcal{E}$,
2. a local triangulation of the space around each node.

The last point means that about each node $x \in \mathcal{E}^h$ we choose a finite set of r-simplices, or *cells*, with edges linking the nodes of $\mathcal{E}^h$ (a typical instance is the square mesh we effectively use in the algorithms). These simplices must meet at x and fit together to cover the space about x just once. Roughly said, about each node x a finite sequence of boxes is constructed on the nodes of $\mathcal{E}^h$ so that these boxes meet at x and partition the space around x. The set of vertices of cells at x (included) will be noted $\mathcal{W}^h(x)$.

More precisely, we shall consider families $(\mathcal{E}^h)_{h>0}$ of starred meshes that fill $\mathcal{E}$, in the sense that the union of the nodes of all $\mathcal{E}^h$ ($h > 0$) is dense in $\mathcal{E}$. Moreover, we shall assume the nondegeneracy of these families

Assumption 4.1 (Nondegeneracy). There exist $\underline{\pi} \in (\pi/2, \pi)$, positive functions $\underline{\delta}(h)$ and $\overline{\delta}(h)$ going to 0 with h, such that for every $h > 0$, $x \in \mathcal{E}^h$, and $y \in \mathcal{W}^h(x)$,

(i) $\underline{\delta}(h) \le ||y - x|| \le \overline{\delta}(h)$,
(ii) The angle between an edge $\overrightarrow{xy}$ of a cell at x and the opposite face is less than or equal to $\underline{\pi}$.

Remark 4.1. The classical square mesh satisfies all the above requirements.

Proposition 4.2. *For every $(x, u, v) \in (\mathcal{E}^h \times \mathcal{U} \times \mathcal{V})$, there is a unique family of nonnegative $f^y(x, u, v)$ $(y \in \mathcal{W}^h\{x\})$ such that*

$$f(x, u, v) = \sum_{y \in \mathcal{W}^h(x)} f^y(x, u, v)(y - x)$$

and $f^y(x, u, v) = 0$ if $y \in \mathcal{W}^h(x)$ does not belong to the intersection of the cells at x that meet $x + \mathbb{R}^+_\star f(x, u, v)$.

Proof. It is a decomposition of a vector on a basis. □

Now let $\Delta t(x, u, v)$ be a notation for $\left(\sum_{z \in \mathcal{W}^h(x)} f^z(x, u, v)\right)^{-1}$ when $f(x, u, v) \neq 0$, $+\infty$ otherwise. We shall need the concept of *weak limit* introduced in [5]:

Definition 4.4 (Weak Limits). For any family of functions V_h on $\mathcal{E}^h$ $(h > 0)$, we introduce lower and upper weak limits $\underline{V}, \overline{V} : \mathcal{E} \to \overline{\mathbb{R}}$ when h tends to 0:

$$\underline{V}(x) = \liminf_{\substack{x^h \in \mathcal{E}^h, x^h \to x \\ h \to 0}} V_h(x^h) \leq \limsup_{\substack{x^h \in \mathcal{E}^h, x^h \to x \\ h \to 0}} V_h(x^h) = \overline{V}(x). \tag{6}$$

In order to deal with discontinuities, we also need the following definition, which is Definition 2.2 in [1]:

Definition 4.5 (Double Convergence). We call the doubly indexed family $(V_{\epsilon,h})_{\epsilon,h>0}$ *doubly convergent* toward V at $x \in \mathcal{E}$, where $V_{\epsilon,h}$ and V are real functions on $\mathcal{E}^h$, and we write

$$V(x) = \lim_{\substack{x^h \in \mathcal{E}^h, x^h \to x \\ h(\epsilon) \to 0}} V_{\epsilon,h}(x^h)$$

if, for any $\gamma > 0$, there exists a function $\overline{h} : (0, +\infty) \to (0, +\infty)$, and $\overline{\epsilon} > 0$, such that

$$|V_{\epsilon,h}(x^h) - V(x)| \leq \gamma,$$

for all $\epsilon \leq \overline{\epsilon}$, $h \leq \overline{h}(\epsilon)$, and $x^h \in \mathcal{E}^h$ such that $||x - x^h|| \leq \overline{h}(\epsilon)$.

For the next sections, we need the following assumption.

Assumption 4.2.

(a) f is continuous, Lipschitz-continuous w.r.t. x, uniformly in (u, v) and has linear growth.
(b) There exist constants $\underline{f}, \overline{f}$ such that

$$0 < \underline{f} \leq f(x, u, v) \leq \overline{f} < +\infty, \quad \forall (x, u, v).$$

(c) There exists a constant $\underline{d^o}$ such that

$$-\infty < \underline{d^o} \leq d^0(x, \mathcal{T}), \quad \forall x.$$

(Note that the function $d^0(., \mathcal{T})$ is Lipschitz-continuous for all the target sets $\mathcal{T}$).

4.2 Game in Time

Definition 4.6 (Discrete Stochastic Game). On the discrete space $\mathcal{E}^h$, we define a stochastic game (cf. [14]) composed of the following elements:

(i) a discrete target $\mathcal{T}^h = \mathcal{E}^h \cap \mathcal{T}$ and domain $\Omega^h = \mathcal{E}^h \setminus \mathcal{T}^h$,
(ii) transition probabilities

$$p(x, y \mid u, v) = \begin{cases} f^y(x, u, v)\Delta t(x, u, v) & \text{if } x \in \Omega^h,\ y \in \mathcal{W}^h(x), \\ 1 & \text{if } y = x \in \mathcal{T}^h, \\ 0 & \text{otherwise.} \end{cases}$$

For the particular case when $f(x, u, v) = 0$, we take

$$p(x, y \mid u, v) = \begin{cases} 1 & \text{if } y = x, \\ 0 & \text{otherwise.} \end{cases}$$

(iii) Instantaneous reward and discount factor:

$$k(x, u, v) = \begin{cases} \phi[\Delta t(x, u, v)] & \text{when } x \in \Omega^h, \\ 0 & \text{when } x \in \mathcal{T}^h, \end{cases}$$

$$\beta(x, u, v) = \begin{cases} \exp[-\Delta t(x, u, v)] & \text{when } x \in \Omega^h, \\ 0 & \text{when } x \in \mathcal{T}^h. \end{cases}$$

Classically, the value V_h of the discrete stochastic game so defined satisfies the following discrete averaged dynamic programming equation, known as the Shapley equation:

$$V_h = T_h V_h, \tag{7}$$

where, by definition, T_h is the following nonlinear operator from the metric complete space of all bounded, real sequences $\mathbb{R}_b^{\mathcal{E}^h}$ into itself:

$$[T_h V_h](x) = \min_u \max_v \left\{ k(x, u, v) + \beta(x, u, v) E_x^{u,v} V_h \right\}. \tag{8}$$

Here $E_x^{u,v} V_h$ means the expected value of V_h viewed as a functional on the Markov random field (7). $E_x^{u,v} V^h(x) = \sum_{y \in \mathcal{W}^h(x)} p(x, y \mid u, v) V_h(y)$. In particular, for $x \in \mathcal{T}^h$, (7) gives

$$V_h(x) = 0.$$

The following proposition is drawn from Prop. 3.1 [18].

Proposition 4.3. *Under Assumptions* 1, 2(a), *and* (2)b, T_h *is contractive from* $\mathbb{R}_b^{\mathcal{E}^h}$ *to itself so that Shapley equation* (7) *admits a unique solution* V_h.

Now we relate this Shapley solution with the viscosity Dirichlet lower-envelope solution of the Isaacs equation on $\overline{\Omega}$:

$$\begin{cases} H(x, V(x), \nabla V(x)) = 0, & x \in \Omega, \\ V(x) = 0, & x \in \partial\Omega. \end{cases} \tag{9}$$

Denoting $\underline{V}$ and $\overline{V}$ the weak limits when $h \to 0$ of V_h, solutions of Shapley fixed-point equations (7), we have the following result fundamental result.

Proposition 4.4. *Under Assumptions* 1, 2(a), *and* 2(b), $\overline{V}$ *(resp.,* $\underline{V}$*) is a viscosity subsolution (resp., supersolution) of Isaacs equation* (9) *on* $\overline{\Omega}$.

Proof. See [18] or [13] for an application to differential games, following ideas of Barles, and Souganidis [6]. □

When the discounted VREK value function V is continuous, it can be inferred that this scheme converges toward V, that is, $\lim_{h\to 0, x^h \to x} V_h(x^h) = V(x)$, under regularity assumptions on the boundary target, fulfilled for the present game (see [3]). But when the value function turns out to be discontinuous, which is the case when a barrier occurs, we have to consider a double approximating scheme, adding a dilatation of the target according to the ideas introduced in [1].

Definition 4.7 (Double Approximating Scheme). For $\epsilon > 0$, define $\mathcal{T}_\epsilon = \{x \in \mathcal{E} \mid d(x, \mathcal{T}) \le \epsilon\}$, while Ω_ϵ is $\mathcal{E} \setminus \mathcal{T}_\epsilon$. Let $V_{\epsilon,h}$ be the value function of the stochastic game with $\Omega := \Omega_\epsilon$ and $\mathcal{T} := \mathcal{T}_\epsilon$.

Following Theorem 2.5 in [1], we have the following:

Proposition 4.5. *Under Assumptions* 1 *and* 2(a)-2(b), $V_{\epsilon,h}$ *converges doubly towards the viscosity lower Dirichlet envelope solution of* (9) *on* $\overline{\Omega}$.

Proof. See the work of Bardi, et al. [1] or Crepey [13]. □

Remark 4.2.

1. For pursuit–evasion problems, Bardi et al. have shown that the viscosity Dirichlet lower-envelope solution of the Isaacs equation on $\overline{\Omega}$ is the value function for Friedman-like strategies as well as the limit when ϵ tends towards 0 of the VREK values for the target $\mathcal{T}_\epsilon$ [1]. For capture-time problems, proving that it is also the VREK value function is still an open problem except in the case where the VREK value function is continuous.
2. The dependence $h(\epsilon)$ between the sequences $h \to 0$ and $\epsilon \to 0$ required to guarantee the practical convergence of the scheme is also an open problem.
3. In our pursuit–evasion game in time, the dynamics do not satisfy Assumption 2(b), but the numerical experiments (described in Section 5) confronted to the analytical study of Section 3.2 suggest that the scheme nevertheless converges toward the value of the game.

4.3 Game in Distance

Following [19], for a given positive number ϵ, we consider the ϵ-game

$$W^\epsilon(x_0) = \sup_{\psi} \inf_{u(\cdot),t} \left\{ d^o(x(t), \mathcal{T}) + \int_0^t \epsilon \, d\tau \right\}.$$

Proposition 4.6. *Under Assumptions* 2(a) *and* 2(c), W^ϵ *is a nonincreasing sequence of bounded, continuous functions, unique viscosity solutions of the variational inequalities*

$$\min \left[d^o(x, \mathcal{T}) - W^\epsilon(x), \min_u \max_v \nabla W^\epsilon(x).f(x,u,v) + \epsilon \right] = 0, \quad \forall x \in \mathcal{E}.$$

Proof. See [19]. □

The Hamiltonian associated to the discounted version of this ϵ-game is then

$$H^\epsilon(x,s,p) = \min \left[\phi(d^o(x,\mathcal{T})) - s, \min_u \max_v < p, f(x,u,v) > +\epsilon(1-s) \right]. \tag{10}$$

The scheme described in previous sections to numerically compute a continuous value can be adapted here to approximate W^ϵ. More precisely, the dynamic programming for an appropriate approximation W_h^ϵ of W^ϵ on a grid $\mathcal{E}_h$ yields the Shapley-like equation

$$W_h^\epsilon = T_h^\epsilon W_h^\epsilon, \tag{11}$$

with

$$[T_h^\epsilon W_h^\epsilon](x) = \min \left[\phi[d^o(x,\mathcal{T})] - W_h^\epsilon(x), \min_u \max_v k(x,u,v) + \beta(x,u,v) E_x^{u,v} W_h^\epsilon \right],$$
$$k(x,u,v) = \phi(\epsilon \Delta t(x,u,v)) \quad \text{and} \quad \beta(x,u,v) = \exp\left(-\epsilon \Delta t(x,u,v)\right) \tag{12}$$

(remember that there is no boundary condition for this game: $\mathcal{T}_h = \emptyset$ in the definition of transition probabilities for this version of the game).

T_h^ϵ is a contractive operator on $\mathbb{R}_b^{\mathcal{E}^h}$, as T_h defined through the Shapley equation used to be for the game in time (notice that for any three real numbers a, b, c, we have $|\min(a,b) - \min(a,c)| \leq |b-c|$). So the fixed-point equation (11) defines a unique W_h^ϵ.

Theorem 4.1. *Under Assumptions* 1 *and* 2(a)-2(b)-2(c), W_h^ϵ *converges doubly to the viscosity upper-envelope solution of* (2) *on* Ω.

Proof. See [13]. □

Remark 4.3.

1. It is still an open problem to know if the viscosity upper-envelope solution coincides with the VREK value function of this game. For general sufficient conditions ensuring such coincidence, see [19].

2. In the minimum-distance game, f and d^0 do not fulfill Assumptions 2(b) and 2(c), but numerical experiments confronted to the analytical solutions of Section 3.3 suggest that the scheme converges toward the value of the game.

5 Algorithms

In order to approximate the capture time or the minimum oriented distance, we are led to solve the Shapley equation (8) (with a target dilated by ϵ) or (11). But these equations are infinite algebraic systems, since an infinite number of nodes are needed to cover the whole state space $\mathcal{E}$. Thus, their numerical resolutions require us to localize a bounded window of interest. Classically, on the border of the discretized domain, the probabilities of transitions that would lead the state outside the domain have been chosen to be 0.

Moreover, we also need to discretize the control sets into U_f and V_f. We use a rough discretization as usual for such problems, without prejudice on the quality of the results (Indeed, most of the optimal controls are *bang bang* or median in this example).

In the following experiments, we have used the set of parameters $(\alpha, \beta, \gamma) = (3, 2, 5)$, a window of 20×20 centered at the origin, a grid $\mathcal{E}_b^h$ of about 10^5 nodes in this window, and sets of discretized controls of five values (experiments with more values have been made without any significant improvement on the precision of the results).

5.1 Game in Time

A first possible algorithm to solve the fixed-point equation (7) (with Ω replaced by Ω_ϵ for small $\epsilon > 0$) is the Shapley one, namely iterations on the values

$$V_h^{n+1}(x) = \min_{u \in U_f} \max_{v \in V_f} \{k(x, u, v) + \beta(x, u, v) E_x^{u,v} V_h^n\}, \, x \in \mathcal{E}_b^h.$$

This is a gradient method, as noted in [14]. Therefore, its convergence is quite slow and, consequently, it is not the algorithm we shall use in practice.

Another possible algorithm is the Hoffman–Karp one, making iterations on the policies, which consists of solving the linear systems ($n \in \mathbb{N}$)

$$V_h^n(x) = k(x, u^{n-1}(x), v^{nl-1}(x)) + \beta(x, u^{n-1}(x), v^{n-1}(x)) E_x^{u^{n-1}(x), v^{n-1}(x)} V_h^n, \quad x \in \mathcal{E}_b^h, \tag{13}$$

where $(u^{n-1}(x), v^{n-1}(x)) \in U_f \times V_f$ mini-maximizes $\{k(x, u, v) + \beta(x, u, v) E_x^{u,v} V_h^{n-1}\}$, and V_h^0 is arbitrary in $\mathbb{R}_b^{\mathcal{E}^h}$.

It is of the Newton–Raphson type (see [14]), converging much faster than the Shapley one, although its convergence is not proved in general. It is the one we shall use in practice. [The linear systems (13) have been solved iteratively using a Picard method].

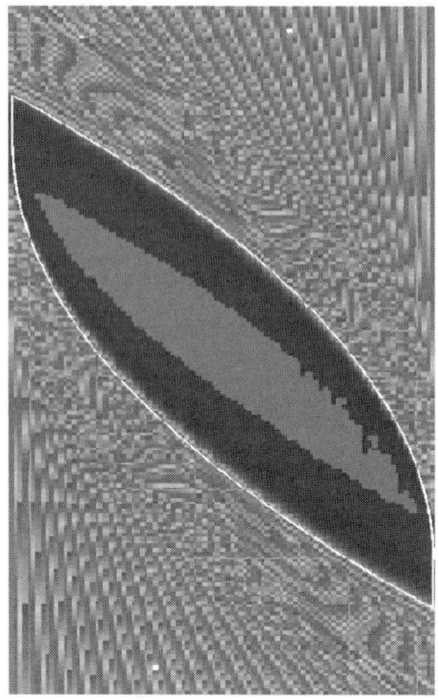

Figure 3: Iso-values for the game in time (the superposed white curve is the exact barrier).

Figure 3 shows the value $\widehat{V}$ obtained for small values of ϵ and h, after an hundred of Newton-Raphson iterations, which was the required amount of iterations to obtain the stabilization of the algorithm. On Figure 3, the results are presented in terms of level curves. Curves of level less than 0.9 are represented in light color, while those of greater level are darker. The lens that can be seen on this figure is the area $\{x \,|\, \widehat{V}(x) \geq 0.9\}$; therefore, it approximates the evasion zone. The existence and the general shape of this evasion zone are consistent with the analytical results obtained for this game by Bernhard ([7] and Section 3.2).

5.2 Game in Distance

Let $(W_h^n)_{n\in\mathbb{N}}$ be the sequence $W_h^n = T_h^\epsilon W_h^{n-1}$ $(n \in \mathbb{N}^*)$, where W_h^0 is arbitrary in $\mathbb{R}_b^{\mathcal{E}^h}$. By Picard's fixed-point theorem, W_h^ϵ is the uniform limit of W_h^n when $n \to \infty$. As before, however, we prefer to use a Newton–Raphson algorithm on the policies adapted from Hoffman–Karp, that is, we iteratively solve the linear systems

$$W_h^n(x) = \min[\phi[d(x, \mathcal{T})] - W_h^n(x), \tag{14}$$
$$k(x, u^{n-1}(x), v^{n-1}(x)) + \beta(x, u^{n-1}(x), v^{n-1}(x)) E_x^{u^{n-1}(x), v^{n-1}(x)} W_h^n], \quad x \in \mathcal{E}_b^h,$$

where $(u^{n-1}(x), v^{n-1}(x)) \in U_f \times V_f$ mini-maximizes $k(x, u, v) + \beta(x, u, v)$ $E_x^{u,v} W_h^{n-1}$ and $W_h^0 = \phi(d(\cdot, \mathcal{T}))$ on $\mathcal{E}^h$.

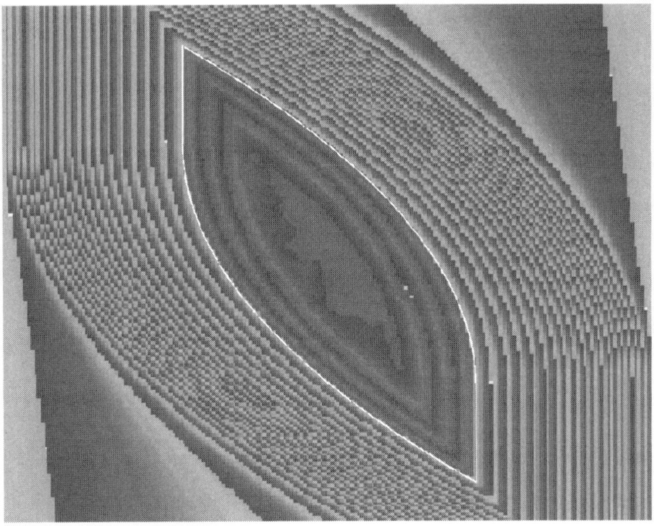

Figure 4: Iso-values for the game in distance (the superposed white curve is the exact barrier).

Figure 4 shows discounted value $\widehat{W}$ obtained for small positive ϵ and h: The curves of negative level are represented in light color, while those of positive level are darker. These numerical results are consistent with the theoretical study of Section 3.3, except in the anti-first diagonal corners of the window, where the edge effects are important. Looking at the figure, we recognize the optimal fields (3), and their separation along the abstract target $\mathcal{T}^\star := \{x \,|\, W(x) = d^o(x, \mathcal{T})\}$. The *oppidum* that should split the optimal fields is clearly visible (it corresponds to the inner lens whose upper border is well drawn). $\widehat{W}$ is roughly constant at its maximum value in this approximate *oppidum*, as expected.

5.3 Comparison

To compare the two methods, we have chosen the following criteria of convergence

1. for any fixed control, linear systems (13) and (14) have been solved using Picard iterations up to a relative error (between two iterations) less than or equal to 10^{-5}.
2. Then iterations on the controls have been achieved until a relative error of 10^{-4} on the fixed points V_h, W_h has been reached.

We have experimented approximatively the same amount of iterations for both methods.

Figures 3 and 4 allow one to compare the time and distance approaches as far as the determination of the barrier of the capture–evasion game is concerned. As mentioned, if we consider level sets of $\widehat{V}$ less than $1 - \mu$ (for small μ), these domains depend strongly on the arbitrary value of μ, so they are numerically very sensitive. Indeed, we can see in Figure 3 that the level curves are very sparse inside the dark lens.

On the opposite, consider once again Figure 4 illustrating the approach in distance. This time, the level curves are very close to each other about the border of the lens $\{x \mid \widehat{W}(x) < 0\}$. Indeed, there is no reason why W should be flat about the level curve 0. Therefore, it is not a surprising that the lens $\{x \mid W(x) < 0\}$ is less numerically sensitive than $\{x \mid V(x) < 1\}$.

6 Conclusion

Viscosity solutions of the Isaacs equation provide two complementary viewpoints on the solution of our capture–evasion game. The p.d.e. equation allows one to investigate the capture time, whereas the variational inequality is an efficient way to investigate the barrier.

Both approaches allow us to construct candidate solutions and lead to similar numerical approximation scheme, doubly indexed by two parameters (grid mesh h and dilatation ϵ) due to the possible discontinuities of the value function.

Numerical experiments performed on this analytical example suggest that the scheme should obtain, in general, good results for both time and distance approaches (although this requires being checked on further examples ...). The current major drawback is that we do not know how to choose ϵ when h goes toward 0, at least theoretically. (In this respect, the viability framework appears to be more satisfactory from a numerical point of view; see [9], and [10].) Mixing with other approaches could be a fruitful future task: for instance, using construction techniques from Isaacs–Breakwell theory and completing the results obtained by numerical investigations.

REFERENCES

[1] Bardi, M., Bottacin, S., and Falcone, M. Convergence of discrete schemes for discontinuous value functions of pursuit-evasion games, pp. 273–304 in *New Trends in Dynamic Games and Applications*, 5th Intl. Symp. in Differential Games, Birkhaüser, 1995.

[2] Bardi, M. and Capuzzo-Dolcetta, I. *Optimal Control and Viscosity Solutions of Hamilton–Jacobi–Bellman Equations,* Birkhaüser, 1997.

[3] Bardi, M. and Soravia, P. A comparison result for Hamilton–Jacobi equations and applications to some differential games lacking controllability, *Funkc. Ekvacioj*, Ser. Intl. **37**(1):19–43, 1994.

[4] Barles, G. *Solutions de Viscosité des Équations de Hamilton–Jacobi*, Springer-Verlag, 1994.

[5] Barles, G. and Perthame, B. Discontinuous solutions of deterministic optimal stopping time problems, *Mathematical Modeling and Numerical Analysis*, **21**, pp. 557–579, 1987.

[6] Barles, G. and Souganidis, P. E. Convergence of approximation schemes for fully nonlinear second order equations, *Asymptotic Anal.*, (4):271–283, 1991.

[7] Bernhard, P. Contribution à l'étude des jeux différentiels à deux joueurs, somme nulle et information parfaite, Thèse d'état, Université de Paris 6, August 1978.

[8] Bernhard, P. Differential games lecture notes on the Isaacs–Breakwell theory, *Summer school on game theory, Cagliari*, 1993.

[9] Cardaliaguet, P., Quincampoix, M. and Saint-Pierre, P. Some algorithms for a game with two players and one target. *Mathematical Modeling and Numerical Analysis*, **28**, pp. 441–461, 1994.

[10] Cardaliaguet, P., Quincampoix, M. and Saint-Pierre, P. *Numerical Methods for Optimal Control and Differential Games*, Technical Report 9510, Ceremade, 1995.

[11] Clarke, F. H. *Optimization and Nonsmooth Analysis*. Wiley, 1983.

[12] Crandall, M., Ishii, H., and Lions, P.-L. User's guide to viscosity solutions of second order partial differential equations, *Bull. Amer. Math. Soc.*, 1992.

[13] Crepey, S. Contribution à des méthodes numeriques appliquées à la finance et ausejeux différentiels, PhD thesis, École Polytechnique, France, 2001.

[14] Filar, J. and Raghavan, T. Algorithms for stochastic games—a survey, *Math. Methods Oper. Res.*, (35):437–472, 1991.

[15] Isaacs, R. *Differential Games*. Wiley, New York, 1965.

[16] Kushner, H. *Probability Methods for Approximations in Stochastic Control and for Elliptic equations*, Academic Press, 1977.

[17] Masle, J-F. Problèmes qualitatif et quantitatif liés en jeux différentiels, Thèse de docteur ingénieur, Université Paris 9, 1976.

[18] Pourtallier, O. and Tidball, M. Approximation of the value function for a class of differential games with target, *Res. Report INRIA 2942*, 1996.

[19] Rapaport, A. Characterization of barriers of differential games, *J. Opt., Th. Appl.*, **97**(1):151–180, April 1998.

[20] Subbotin, A. I. *Generalized Solutions of First-Order PDEs, the Dynamical Optimization Perspective*. Birkhäuser, 1995.

PART IV
Dynamic Games: Applications

On Optimal Missile Guidance Upgrades with Dynamic Stackelberg Game Linearizations

Michael H. Breitner
Technische Universität Clausthal
Fachbereich Mathematik und Informatik
Clausthal-Zellerfeld, Germany

Uwe Rettig
Technische Universität München
Zentrum Mathematik
München, Germany

Oskar von Stryk
Technische Universität München
Zentrum Mathematik
München, Germany

Abstract

We investigate the most critical final homing phase for one intercepting missile versus one maneuverable ballistic missile. In the worst case a future ballistic missile will know the intercepting missile's guidance scheme and could maximize the minimum (miss) distance. An intercepting missile with three-dimensional proportional navigation guidance is studied. An optimal small guidance upgrade for this missile is calculated as a closed-loop solution of a dynamic Stackelberg game. The intercepting missile acts as leader, whereas the ballistic missile advantageously acts as follower. An optimal small upgrade is approximated numerically along very many optimal trajectories that cover the relevant final homing scenarios. The trajectories are computed efficiently by a direct collocation method. Along the trajectories a linearized minimum–maximum principle using the adjoint variables' estimates is applied. An upgrade with nearest points on neighboring optimal trajectories is synthesized. Alternative synthesis approaches with local Taylor-series expansions with global smoothing or with artificial neural networks are discussed. Simulations for many different scenarios show the improved interception capability for the upgraded guidance of the intercepting missile.

1 Introduction and Problem Formulation

In the near future, advanced defense systems against ballistic missiles (see Figure 1), which have been extensively developed worldwide since the Gulf War in 1991, will have to face smart evasive maneuvers by the ballistic missiles. Seeker-

Figure 1: Mobile firing battery launcher for the new Patriot PAC-3 (Patriot antitactical missile capability level 3) missiles.

guided intercepting missiles and projectiles must achieve reliable target acquisition and highest-possible hit probabilities under all circumstances; see [17] and [30]. We investigate the most critical final homing phase of the last few seconds for one intercepting missile versus one maneuverable ballistic missile.

Very likely, a future ballistic missile's guidance can take advantage of the knowledge of an intercepting missile's guidance. A dynamic Stackelberg (hierarchical) game formulation is appropriate with the intercepting missile (pursuer, or P for short) as leader and the ballistic missile (evader, or E for short) as follower; see [2], [13], and [29]. Given P's guidance scheme, E can solve optimal control problems trying to avoid interception, whereas P wants to optimize the guidance scheme.

Proportional navigation (PNG), which appears in many different versions, is still an important and widely used intercepting missile guidance scheme; see [1] and [30]. Although the Germans apparently knew PNG during World War II at Peenemünde, no applications on German missiles were reported. The Lark missile, which had its first successful test in December 1950, was the first missile to use PNG (see [30] for the history of PNG). The popularity of this guidance scheme is based on its simplicity, effectiveness, and ease of implementation. Without loss of generality, our upgrade technique is tested and illustrated for *standard* three-dimensional PNG. Note that the emphasis here is on the *theoretical and algorithmic aspects* of the proposed guidance upgrade technique. Different well-known missile guidance schemes are not compared.

Today in endoatmospheric missiles, guidance commands are usually implemented by moving fins or other control surfaces to obtain the required lift. In reality, lift is built up gradually and the commanded acceleration can be generated only with significant delay. Point-mass dynamics without control delay, so-called ideal control dynamics, cannot be used for a realistic model. Exoatmospheric missiles already use powerful thrust vector control, lateral divert engines, and squibs to achieve the commanded acceleration. In order to get short delays and altitude-independent high maneuverability, *future* intercepting missiles against fast ballistic targets likely will use nonaerodynamic lift generation, too. Thus, ideal control dynamics for both the intercepting and the ballistic missile's model are used in the sequel.

Almost all research on missile guidance is based on collision-course linearization of the nonlinear intercepting missile-target geometry, which yields an accurate approximation to the actual geometry for many cases. With a separation of the linearized vertical and horizontal dynamics, two two-dimensional guidance schemes can be considered. Whether linearization and separation leads to a realistic model depends on the missile type (fins, control surfaces, thrust vector controls, etc.). Ballistic missiles can go through enormous decelerations as they reenter the atmosphere. The magnitude of the ballistic missiles' deceleration increases with increasing speed and increasing reentry angle. Additionally, future ballistic missiles will use smart evasive maneuvers. Note that any ballistic missile acceleration perpendicular to the line of sight will appear as an evasive maneuver to the intercepting missile. It is best for a pursuing missile to engage the evading missile on a trajectory where little of the evading missile's deceleration is perpendicular to the line of sight. If, for practical reasons, the evading missile must be engaged under stressing conditions, the intercepting missile must be sized to have adequate acceleration capability. Advanced guidance schemes, such as predictive guidance, can significantly relax the intercepting missile's acceleration requirements if ballistic missile position information is available permanently. The question of whether a collision-course linearization and separation leads to accurate enough guidance schemes for future antitactical missiles is very difficult to answer today; see [6], [8], [9], [10], [11], [14], [16], and [20]–[25]. Here we focus on three-dimensional guidance schemes and nonlinear missile dynamics.

It is reasonable to assume that future ballistic missiles and intercepting missiles will have full, accurate, and undelayed state information; see [17] and [30]. This information can be provided by ground-based radar, satellites, AWACS[1] aircraft, and on-board radar, infrared, or multimode seekers (active or semiactive). Most likely, combined state measurement methods will be used in the near future to increase reliability and accuracy.

[1]Stands for "airborne early warning and control system."

2 Missile Dynamics and Standard Three-Dimensional PNG

In spherical coordinates the missiles' motion is governed by the dynamic equations

$$\frac{d}{dt}x_{E/P} = v_{E/P}\cos\gamma_{E/P}\cos\chi_{E/P}, \tag{1}$$

$$\frac{d}{dt}y_{E/P} = v_{E/P}\cos\gamma_{E/P}\sin\chi_{E/P}, \tag{2}$$

$$\frac{d}{dt}h_{E/P} = v_{E/P}\sin\gamma_{E/P}, \tag{3}$$

$$\begin{aligned}\frac{d}{dt}v_{E/P} = &-\frac{D_{0,E/P}\left(h_{E/P}, v_{E/P}\right) + {n_{E/P}}^2 D_{I,E/P}\left(h_{E/P}, v_{E/P}\right)}{m_{E/P}}\\ &- g\left(h_{E/P}\right)\sin\gamma_{E/P},\end{aligned} \tag{4}$$

$$\frac{d}{dt}\gamma_{E/P} = \frac{g\left(h_{E/P}\right)}{v_{E/P}}\left(n_{E/P}\cos\mu_{E/P} - \cos\gamma_{E/P}\right), \tag{5}$$

$$\frac{d}{dt}\chi_{E/P} = \frac{g\left(h_{E/P}\right)}{v_{E/P}}\frac{n_{E/P}\sin\mu_{E/P}}{\cos\gamma_{E/P}}, \tag{6}$$

where the independent variable t denotes time, state variables x and y denote horizontal position, h denotes altitude, v denotes velocity, γ denotes flight-path angle, χ denotes azimuth angle, and control variables n and μ denote load factor and bank angle, respectively. The functions D_0, D_I, and g denote zero lift drag, induced drag, and Earth's gravitational acceleration, respectively, and the constant m denotes missile's mass. To save space and to focus on the theoretical background, the missiles' model functions and parameters (see [17], [19] and [30]), are omitted here. Readers can obtain the complete models from the authors. The subscripts E and P refer to E's and P's variables. The missiles are modelled as mass points and have no thrust as we consider the final phase of the encounter; see (1)–(6).

We define the control vectors

$$\boldsymbol{u}_E := \begin{pmatrix} n_E \\ \mu_E \end{pmatrix} \quad \text{and} \quad \boldsymbol{u}_P := \begin{pmatrix} n_P \\ \mu_P \end{pmatrix}.$$

Admissible controls must satisfy the box constraints

$$\boldsymbol{u}_E \in U_E \tag{7}$$

and

$$\boldsymbol{u}_P \in U_P, \tag{8}$$

with

$$U_{E/P} := \left\{ \left(n_{E/P}, \mu_{E/P}\right)^T \middle| 0 \le n_{E/P} \le n_{E/P,\max}, -\pi \le \mu_{E/P} \le \pi \right\},$$

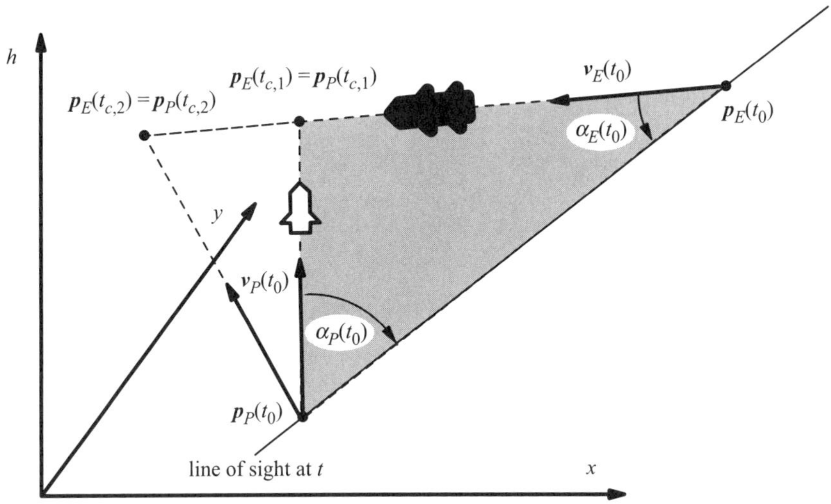

Figure 2: Collision triangles for three-dimensional PNG.

where $n_{E,\max} = 20$ and $n_{P,\max} = 50$ (defense against short- and medium-range ballistic missiles). We define the positions, the line-of-sight vector, the velocity vectors

$$\boldsymbol{p}_{E/P} := \begin{pmatrix} x_{E/P} \\ y_{E/P} \\ h_{E/P} \end{pmatrix}, \quad \boldsymbol{r} := \boldsymbol{p}_E - \boldsymbol{p}_P, \quad \boldsymbol{v}_{E/P} := \begin{pmatrix} v_{E/P} \cos \gamma_{E/P} \cos \chi_{E/P} \\ v_{E/P} \cos \gamma_{E/P} \sin \chi_{E/P} \\ v_{E/P} \sin \gamma_{E/P} \end{pmatrix}, \tag{9}$$

respectively, and initial positions $\boldsymbol{p}_E(t_0)$ and $\boldsymbol{p}_P(t_0)$.

Assuming *constant-velocity vectors*, a collision occurs if $t_c > t_0$ exists with

$$\boldsymbol{p}_E(t_c) = \boldsymbol{p}_E(t_0) + (t_c - t_0)\boldsymbol{v}_E(t_0) = \boldsymbol{p}_P(t_0) + (t_c - t_0)\boldsymbol{v}_P(t_0) = \boldsymbol{p}_P(t_c). \tag{10}$$

For a given final homing scenario $\boldsymbol{p}_E(t_0)$, $\boldsymbol{p}_P(t_0)$, $\boldsymbol{v}_E(t_0)$, and $v_P(t_0)$, (10) has two solutions, one solution, or no solution for $\gamma_P(t_0)$, $\chi_P(t_0)$, and t_c; see Figure 2. For *relevant* scenarios, two solutions $(\gamma_{P,1}(t_0), \chi_{P,1}(t_0), t_{c,1})$ and $(\gamma_{P,2}(t_0), \chi_{P,2}(t_0), t_{c,2})$ with $t_{c,1} < t_{c,2}$ always exist. To intercept as early as possible and as far away as possible from E's target $(\gamma_{P,1}(t_0), \chi_{P,1}(t_0), t_{c,1})$ is always taken. The line-of-sight angles α_E and α_P (see Figure 2), can be calculated with

$$\alpha_E = \arccos \frac{\boldsymbol{v}_E{}^T \boldsymbol{r}}{\|\boldsymbol{v}_E\|_2 \, \|\boldsymbol{r}\|_2}, \tag{11}$$

$$\alpha_P = \arcsin \left(\frac{v_E}{v_P} \sin \alpha_E \right). \tag{12}$$

With $\alpha_P(t_0)$ and $\alpha_E(t_0)$ from (11) and (12), the time to go until collision is

$$t_{c,1} - t_0 = \frac{\|\boldsymbol{r}(t_0)\|_2}{v_P(t_0)\cos\alpha_P(t_0) + v_E(t_0)\cos\alpha_E(t_0)}; \tag{13}$$

using (10), we have

$$\gamma_P(t_0) = \arcsin\left(\frac{(h_E - h_P)(v_P\cos\alpha_P + v_E\cos\alpha_E)}{v_P\|\boldsymbol{r}\|_2} + \frac{v_E}{v_P}\sin\gamma_E\right)\bigg|_{t_0}, \tag{14}$$

$\gamma_P \in \left]-\frac{\pi}{2}, \frac{\pi}{2}\right[$, and

$$\cos\chi_P(t_0) = \left(\frac{(x_E - x_P)(v_P\cos\alpha_P + v_E\cos\alpha_E)}{v_P\cos\gamma_P\|\boldsymbol{r}\|_2} + \frac{v_E\cos\gamma_E\cos\chi_E}{v_P\cos\gamma_P}\right)\bigg|_{t_0}, \tag{15}$$

$\chi_P \in [-\pi, \pi[$, such that

$$\sin\chi_P(t_0) = \left(\frac{(y_E - y_P)(v_P\cos\alpha_P + v_E\cos\alpha_E)}{v_P\cos\gamma_P\|\boldsymbol{r}\|_2} + \frac{v_E\cos\gamma_E\sin\chi_E}{v_P\cos\gamma_P}\right)\bigg|_{t_0}. \tag{16}$$

Singular $\gamma_P \approx \pm\pi/2$ are implemented numerically as the limit case. E and P are on collision course at time $t \geq t_0$ if (11)–(16) are fulfilled at t.

Three-dimensional PNG issues acceleration commands, which are proportional to the closing velocity $-(d/dt)\boldsymbol{r} = (d/dt)\boldsymbol{p}_P - (d/dt)\boldsymbol{p}_E = \boldsymbol{v}_P - \boldsymbol{v}_E$; see [1], [17], and [30]. P is on collision course at $t \geq t_0$ if and only if $-(d/dt)\boldsymbol{r}$ and $\boldsymbol{r}$ are parallel, that is, if and only if there holds

$$\frac{\boldsymbol{r}^T \frac{d}{dt}\boldsymbol{r}}{\|\boldsymbol{r}\|_2 \left\|\frac{d}{dt}\boldsymbol{r}\right\|_2} = -1. \tag{17}$$

Using three-dimensional PNG, P minimizes

$$\frac{\boldsymbol{r}^T \frac{d}{dt}\boldsymbol{r}}{\|\boldsymbol{r}\|_2 \left\|\frac{d}{dt}\boldsymbol{r}\right\|_2}, \tag{18}$$

with $0 \leq n_P < \infty$ and $-\pi \leq \mu_P \leq \pi$. With a heuristic guidance constant c_{pn}, $3 \leq c_{\text{pn}} \leq 5$, and distance adaption we have P's *commanded acceleration*

$$\boldsymbol{a}_{P,\text{pn,c}} = \begin{pmatrix} a_{P,\text{pn,c},1} \\ a_{P,\text{pn,c},2} \\ a_{P,\text{pn,c},3} \end{pmatrix} = \pm\frac{c_{\text{pn}}}{\|\boldsymbol{r}\|_2{}^2}\left(\boldsymbol{r} \times \frac{d}{dt}\boldsymbol{r}\right) \times \boldsymbol{v}_P. \tag{19}$$

Neglecting the minor influence of the term $(g/v_P)\cos\gamma_P$ on $(d/dt)\gamma_P$, and obeying constraints (8), (4)–(6) and their derivatives yield

$$n_{P,\text{pn}} = \min\left(\frac{\left\|\boldsymbol{a}_{P,\text{pn,c}}\right\|_2}{g}, n_{P,\max}\right), \tag{20}$$

$$\sin\mu_{P,\text{pn}} = \frac{a_{P,\text{pn,c},2}\cos\chi_P - a_{P,\text{pn,c},1}\sin\chi_P}{\left\|\boldsymbol{a}_{P,\text{pn,c}}\right\|_2}, \tag{21}$$

$$\cos \mu_{P,\text{pn}} = \frac{a_{P,\text{pn,c},3}}{\left\| \boldsymbol{a}_{P,\text{pn,c}} \right\|_2 \cos \gamma_P}, \tag{22}$$

and P's control vector

$$\boldsymbol{u}_{P,\text{pn}} = \begin{pmatrix} n_{P,\text{pn}} \\ \mu_{P,\text{pn}} \end{pmatrix}. \tag{23}$$

In (19) the sign + or − is taken, which minimizes the angle between $-(d/dt)\boldsymbol{r}$ and $\boldsymbol{r}$.

3 State-Space Reduction and Optimal Evasive Guidance of E

Without loss of generality, the state-space dimension can be reduced to 9 using the state variables $x := x_E - x_P$, $y := y_E - y_P$, and $\chi := \chi_E - \chi_P$ instead of x_E, x_P, y_E, y_P, χ_E, and χ_P. We define the state vector

$$z = (x, y, h_E, h_P, v_E, v_P, \gamma_E, \gamma_P, \chi)$$

and obtain, using (1)–(6), the pursuit–evasion dynamics

$$\frac{d}{dt} z = \boldsymbol{f}(z, \boldsymbol{u}_E, \boldsymbol{u}_P), \quad z(t_0) = z_0, \tag{24}$$

for $t \in [t_0, \infty[$ with feedback controls $\boldsymbol{u}_E(z)$ and $\boldsymbol{u}_P(z)$. Due to the $(y \equiv 0)$-plane symmetry of the scenarios, it is sufficient to consider $\chi(t_0) \in [0, \pi[$ only. Scenarios with $\chi(t_0) \in [-\pi, 0[$ are mapped to analogous scenarios with $\chi(t_0) \in [0, \pi[$.

Initial states z_0 for relevant final homing scenarios are (defense against short- and medium-range ballistic missiles)

$$\begin{aligned} &x_E(t_0) = 0, \quad y_E(t_0) \in [3.5\text{km}, 12\text{km}], \quad h_E(t_0) \in [6\text{km}, 20\text{km}], \\ &v_E(t_0) \in [1.4\text{km/s}, 2.8\text{km/s}], \quad \gamma_E(t_0) \in [-\frac{\pi}{9}, 0], \quad \chi_E(t_0) \in \left[-\frac{\pi}{2}, \frac{\pi}{2}\right], \\ &x_P(t_0) = 0, \quad y_P(t_0) = 0, \quad h_P(t_0) \in [6\text{km}, h_E(t_0)], \\ &v_P(t_0) \in [1.4\text{km/s}, 2.0\text{km/s}] \text{ and } \gamma_P(t_0) \text{ and } \chi_P(t_0) \text{ from (14) and (15).} \end{aligned} \tag{25}$$

Interception is impossible for some scenarios for which (14) and (15) have no solution. All other scenarios according to (25) define the *set of usable initial states* $S_{\text{uis}} \subset \mathbb{R}^9$. States $z(t)$ that may arise in the course of the final homing phase form the *set* $S_{\text{rs}} \subset \mathbb{R}^9$ *of relevant states*.

From P's point of view, the very likely worst case is that future ballistic missiles can take advantage of the knowledge of the intercepting missile's guidance. Given an admissible guidance scheme $\boldsymbol{u}_P(z)$ and $z_0 \in S_{\text{uis}}$, E can solve the optimal control problem

$$\underset{\boldsymbol{u}_E(z(t))}{\text{maximize}} \min_{t \geq t_0} \|\boldsymbol{r}(z(t))\|_2 \tag{26}$$

subject to

$$\frac{d}{dt}z = \boldsymbol{f}(z, \boldsymbol{u}_E, \boldsymbol{u}_P(z)), \quad z(t_0) = z_0, \quad \text{and} \quad \boldsymbol{u}_E(z(t)) \in U_E, \tag{27}$$

for all $t \geq t_0$. Consider optimal trajectories for all $z_0 \in S_{\text{uis}}$. An optimal evasive guidance scheme $\boldsymbol{u}_{E,\boldsymbol{u}_P}{}^*(z)$, the minimum (miss) distance

$$d_{\boldsymbol{u}_\mathrm{P}}{}^*(z) = \min_{t \geq t_i} \|\boldsymbol{r}(\tilde{z}(t))\|_2, \quad \tilde{z}(t_i) = z, \quad t_i \in [t_0, \infty[, \tag{28}$$

and its gradient grad $d_{\boldsymbol{u}_\mathrm{P}}{}^*(z)$ are known functions of z for all $z \in S_{\text{rs}}$. In accordance with the maximum principle, the Hamilton–Jacobi–Bellman equation

$$\left.\frac{d}{dt}d_{\boldsymbol{u}_\mathrm{P}}{}^*\right|_{\substack{z \\ \boldsymbol{u}_{\mathrm{E},\boldsymbol{u}_\mathrm{P}}{}^*(z) \\ \boldsymbol{u}_\mathrm{P}(z)}} = \operatorname{grad} d_{\boldsymbol{u}_\mathrm{P}}{}^*(z) \cdot \boldsymbol{f}\left(z, \boldsymbol{u}_{E,\boldsymbol{u}_P}{}^*\left(z, \operatorname{grad} d_{\boldsymbol{u}_\mathrm{P}}{}^*(z)\right), \boldsymbol{u}_P(z)\right) = 0, \tag{29}$$

with

$$\begin{aligned} \boldsymbol{u}_{E,\boldsymbol{u}_P}{}^*\left(z, \operatorname{grad} d_{\boldsymbol{u}_\mathrm{P}}{}^*(z)\right) &= \underset{\boldsymbol{u}_\mathrm{E} \in \mathrm{U}_\mathrm{E}}{\operatorname{argmax}} \left.\frac{d}{dt} d_{\boldsymbol{u}_\mathrm{P}}{}^*\right|_{\substack{z \\ \boldsymbol{u}_E \\ \boldsymbol{u}_P(z)}} \\ &= \underset{\boldsymbol{u}_\mathrm{E} \in \mathrm{U}_\mathrm{E}}{\operatorname{argmax}} \operatorname{grad} d_{\boldsymbol{u}_\mathrm{P}}{}^*(z) \cdot \boldsymbol{f}(z, \boldsymbol{u}_E, \boldsymbol{u}_P(z)) \end{aligned} \tag{30}$$

must be fulfilled by $d_{\boldsymbol{u}_\mathrm{P}}{}^*(z)$ for all $z \in S_{\text{rs}}$. Equation (29) is a partial differential equation of first-order for the state variables, $z_1, z_2, \ldots, z_9$ here, as independent variables. Note that (29) is *only a necessary, and not a sufficient*, condition. The method of characteristics provides a classical technique to solve any first-order partial differential equation; see, for instance, [3] and [13]. Since neither differentiability nor continuity of $d_{\boldsymbol{u}_\mathrm{P}}{}^*(z)$ in S_{rs} can be guaranteed, discontinuities of grad $d_{\boldsymbol{u}_\mathrm{P}}{}^*(z)$ in S_{rs} are possible. The method of characteristics leads to well-defined multipoint boundary-value problems with ordinary differential equations for z, grad $d_{\boldsymbol{u}_P}{}^*$, and $d_{\boldsymbol{u}_P}{}^*$. Solutions of these boundary-value problems are characteristics of (29), and thus grad $d_{\boldsymbol{u}_P}{}^*(z)$ is known along these characteristics. For the unprescribed final time t_f,

$$d_{\boldsymbol{u}_\mathrm{P}}{}^*\left(z(t_f)\right) = \left\|\boldsymbol{r}\left(z(t_f)\right)\right\|_2 \tag{31}$$

holds and t_f can be determined with the *necessary* condition

$$\frac{d}{dt}\left\|\boldsymbol{r}\left(z(t_f)\right)\right\|_2 = 0, \tag{32}$$

that is,

$$\boldsymbol{r}\left(z(t_f)\right)^T \frac{d}{dt}\boldsymbol{r}\left(z(t_f)\right) = \boldsymbol{r}\left(z(t_f)\right)^T \left(\boldsymbol{v}_E\left(z(t_f)\right) - \boldsymbol{v}_P\left(z(t_f)\right)\right) = 0. \tag{33}$$

The numerical solution of the multipoint boundary-value problems (e.g., with an indirect multiple shooting method or an indirect collocation method; see [26]) is a very difficult and tedious task; see [3], [5], [6], [14], [15], and [27]. Here we select the *direct* numerical method DIRCOL,[2] (see [27] and [28]), to approximate very many characteristics of (29), namely optimal trajectories for the optimal control problems (26) with $z_0 \in S_{\text{uis}}$. DIRCOL enables a comfortable, fast, and reliable numerical solution:

- Neither the adjoint differential equations nor the optimal control law (30) must be supplied.
- Even noncontinuous and nondifferentiable dynamics can be handled since no explicit numerical integration of the differential equations (24) is done.
- Poor initial guesses for control and state variable histories and t_f are sufficient.
- Large domains of convergence enable the automatic computation of many thousands of optimal trajectories for various $z_0 \in S_{\text{uis}}$.
- DIRCOL yields highly accurate estimates for $\operatorname{grad} d_{\boldsymbol{u}_P}{}^*(z)$ along optimal trajectories.

Briefly summarized, DIRCOL transcribes the optimal control problem (26) into a nonlinear optimization problem by full discretization of control and state variables. The resulting large, finite-dimensional, nonlinearly constrained optimization problem is solved by the sequential (successive) quadratic programming methods NPSOL or SNOPT; see [12]. Using $\boldsymbol{u}_{P,\text{pn}}$, for numerically robust DIRCOL computations, (33) is modified to

$$\left\|\boldsymbol{r}\left(\tilde{t_f}\right)\right\|_2 \boldsymbol{r}\left(z\left(\tilde{t_f}\right)\right)^T \left(\boldsymbol{v}_E\left(z\left(\tilde{t_f}\right)\right) - \boldsymbol{v}_P\left(z\left(\tilde{t_f}\right)\right)\right) = -0.1, \tag{34}$$

and the normalized time grid $t_0 = 0, t_1 = 0.2\tilde{t_f}, t_2 = 0.4\tilde{t_f}, t_3 = 0.6\tilde{t_f}, t_4 = 0.7\tilde{t_f}$, $t_5 = 0.8\tilde{t_f}$, $t_6 = 0.9\tilde{t_f}$, $t_7 = 0.98\tilde{t_f}$, and $t_8 = \tilde{t_f}$ for the discretization of both control and state variables can be used. Linear extrapolation of E's and P's motion with constant $\boldsymbol{v}_E(\tilde{t_f})$ and $\boldsymbol{v}_P(\tilde{t_f})$ yields

$$\begin{aligned} t_f - \tilde{t_f} &= \frac{\boldsymbol{r}\left(z\left(\tilde{t_f}\right)\right)^T \left(\boldsymbol{v}_P\left(z\left(\tilde{t_f}\right)\right) - \boldsymbol{v}_E\left(z\left(\tilde{t_f}\right)\right)\right)}{\left\|\boldsymbol{v}_E\left(z\left(\tilde{t_f}\right)\right) - \boldsymbol{v}_P\left(z\left(\tilde{t_f}\right)\right)\right\|_2} \\ &= \frac{0.1}{\left\|\boldsymbol{r}\left(z\left(\tilde{t_f}\right)\right)\right\|_2 \left\|\boldsymbol{v}_E\left(z\left(\tilde{t_f}\right)\right) - \boldsymbol{v}_P\left(z\left(\tilde{t_f}\right)\right)\right\|_2} \end{aligned} \tag{35}$$

and

$$\begin{aligned} d_{\boldsymbol{u}_{\text{P,pn}}}{}^{\tilde{}*}\left(z\left(t_f\right)\right) = \Bigg\| & \boldsymbol{p}_E\left(z\left(\tilde{t_f}\right)\right) - \boldsymbol{p}_P\left(z\left(\tilde{t_f}\right)\right) \\ & + \frac{0.1\left(\boldsymbol{v}_E\left(z\left(\tilde{t_f}\right)\right) - \boldsymbol{v}_P\left(z\left(\tilde{t_f}\right)\right)\right)}{\left\|\boldsymbol{r}\left(z\left(\tilde{t_f}\right)\right)\right\|_2 \left\|\boldsymbol{v}_E\left(z\left(\tilde{t_f}\right)\right) - \boldsymbol{v}_P\left(z\left(\tilde{t_f}\right)\right)\right\|_2} \Bigg\|_2, \end{aligned} \tag{36}$$

[2]This stands for the direct collocation method developed by the third author O. von Stryk since 1988.

with the slightly modified miss distance $d_{\boldsymbol{u}_{\tilde{P},\text{pn}}}{}^*$. Note that both modification error and linearization error is very small due to the very short $t_f - \tilde{t_f}$.

To fill S_{rs} as densely as necessary with optimal trajectories, 1830 optimal control problems with z_0 according to (25) have been computed. Almost 97% of the optimal trajectories have been obtained automatically using DIRCOL in homotopy sequences. Only a few times did the initial guesses for control and state variable histories and t_f have to be improved manually. The other 3% of the optimal trajectories have been computed interactively using a time-grid refinement technique. The computing time per problem is less than 10 minutes for almost all solutions on an IBM RISC 6000 with a single Power-PC 233-MHz processor. Using $z(t_0), z(t_1), z(t_2), \ldots, z(t_8)$ along the optimal trajectories, the set of approximation points

$$S_{\text{app}} = \left\{ \left(z_k, \operatorname{grad} d_{\boldsymbol{u}_{P,\text{pn}}}{}^* (z_k)^T \right), k = 1, 2, \ldots 16470 \right\} \tag{37}$$

has been defined. Note that, if necessary, beside these discretization points, z_k additional pairs along the optimal trajectories are easily computable .

4 Dynamic Stackelberg Game Linearizations

When P is close to its target, it detonates a warhead that extends its lethal radius by propagating blast pressure or fragments. In the best case the warhead detonates at the closest point of approach $z(t_f)$ with $d_{\boldsymbol{u}_P}{}^*(z(t_f)) = \|\boldsymbol{r}(z(t_f))\|_2$. The probability of E 's destruction is highly dependent on $d_{\boldsymbol{u}_P}{}^*$ and on the warhead's weight, see [17] and [30] and see Table 1. Some projects are even dedicated to future highly maneuverable anti-tactical missiles with very small or no warhead, that is a direct hit of the ballistic missile is necessary; see [22]. Given a designed intercepting missile, improved performance means improved miss-distance distribution, or smaller $d_{\boldsymbol{u}_P}{}^*(z)$ for many $z \in S_{\text{rs}}$.

Consider upgraded three-dimensional PNG schemes

$$\boldsymbol{u}_{P,\text{up}}(z) = \boldsymbol{u}_{P,\text{pn}}(z) + \Delta \boldsymbol{u}_{P,\text{up}}(z) = \begin{pmatrix} n_{P,\text{pn}}(z) + \Delta n_{P,\text{up}}(z) \\ \mu_{P,\text{pn}}(z) + \Delta \mu_{P,\text{up}}(z) \end{pmatrix}. \tag{38}$$

Then for all admissible guidance upgrades $\Delta \boldsymbol{u}_{P,\text{up}}(z) \in U_{P,\text{up}}(z)$, there holds

$$\max \left\{ \Delta n_{P,\text{up,min}}, -n_{P,\text{pn}}(z) \right\} \leq \Delta n_{P,\text{up}}(z) \leq \min \left\{ \Delta n_{P,\text{up,max}}, n_{P,\max} - n_{P,\text{pn}}(z) \right\} \tag{39}$$

Table 1: Miss distance versus estimated probability of E 's destruction for a medium-weight warhead.

$d_{\boldsymbol{u}_P}{}^*$	0–1 m	1–2 m	2–3 m	3–5 m	5–10 m	10–15 m	15–20 m	> 20 m
e.d.p.	99%	97%	90%	80%	70%	50%	20%	< 1%

and

$$\Delta\mu_{P,\text{up},\min} \leq \Delta\mu_{P,\text{up}}(z) \leq \Delta\mu_{P,\text{up},\max}. \tag{40}$$

Assume E also knows $\boldsymbol{u}_{P,\text{up}}(z)$ and evades optimally using $\boldsymbol{u}_{E,\boldsymbol{u}_{P,\text{up}}}{}^*(z)$ for $z \in S_{\text{rs}}$.

Then best $\boldsymbol{u}_{P,\text{up}}{}^*(z)$, $\boldsymbol{u}_{E,\boldsymbol{u}_{P,\text{up}}{}^*}{}^*(z)$ and grad $d_{\boldsymbol{u}_{P,\text{up}}{}^*}{}^*(z)$ are closed-loop solutions of the dynamic Stackelberg game stated in Section 1; see [2], [13], and [29]. Today this very complex dynamic Stackelberg game is solvable neither analytically nor numerically. However, with the following simplifying assumption, an approximate closed-loop solution can be calculated.

> *Small upgrade assumption.* All upgrades $\Delta\boldsymbol{u}_{P,\text{up}}(z) \in U_{P,\text{up}}(z)$ are *sufficiently small* such that grad $d_{\boldsymbol{u}_{P,\text{up}}}{}^*(z) \doteq \text{grad}\, d_{\boldsymbol{u}_{P,\text{pn}}}{}^*(z)$ holds for all $\Delta\boldsymbol{u}_{P,\text{up}}(z) \in U_{P,\text{up}}(z)$ and all $z \in S_{\text{rs}}$.

"$\doteq$" implies that in the sequel a numerically synthesized $d_{\boldsymbol{u}_{\text{P,pn}}}{}^*(z)$ is a sufficiently accurate approximation for grad $d_{\boldsymbol{u}_{P,\text{up}}}{}^*(z)$. Note that this assumption usually must be *verified a posteriori* along recomputed trajectories. Here "sufficiently small" means

$$\Delta n_{P,\text{up},\min} \in [-10, 0], \qquad \Delta n_{P,\text{up},\max} \in [0, 10], \tag{41}$$

$$\Delta\mu_{P,\text{up},\min} \in \left[-\frac{\pi}{18}, 0\right], \qquad \Delta\mu_{P,\text{up},\max} \in \left[0, \frac{\pi}{18}\right] \tag{42}$$

(defense against short- and medium-range ballistic missiles).

The approach presented here is quite different from the approach presented in [7]. On the one hand, our approach has the advantage that it is not limited to pursuit–evasion problems. Moreover, it is not limited to problems with separable dynamics and constraints or with performance index time to capture. On the other hand, there is the disadvantage that many dynamic games are not solvable because the fundamental small upgrade assumption cannot be fulfilled with *reasonable* upgrades. Here, with the game's separable dynamics [see (1)–(6)], the minimum-maximum principle yields

$$\begin{aligned}
\boldsymbol{u}_{E,\boldsymbol{u}_{P,\text{up}}{}^*}{}^*(z) &= \underset{\boldsymbol{u}_E \in U_E}{\text{argmax}} \left.\frac{d}{dt} d_{\boldsymbol{u}_{P,\text{pn}}}{}^*\right|_{\substack{z \\ \boldsymbol{u}_{\text{E}} \\ \boldsymbol{u}_{\text{P,up}}{}^*(z)}} \\
&= \underset{\boldsymbol{u}_E \in U_E}{\text{argmax}}\, \text{grad}\, d_{\boldsymbol{u}_{P,\text{pn}}}{}^*(z) \cdot \boldsymbol{f}\left(z, \boldsymbol{u}_E, \boldsymbol{u}_{P,\text{up}}{}^*(z)\right) \\
&= \underset{\boldsymbol{u}_E \in U_E}{\text{argmax}} \left(-\frac{\partial d_{\boldsymbol{u}_{P,\text{pn}}}{}^*}{\partial v_E}(z) \frac{D_{I,E}(z)}{m_E} n_E{}^2 \right. \\
&\quad \left. + \frac{g(z)}{v_E} n_E \left(\frac{\partial d_{\boldsymbol{u}_{P,\text{pn}}}{}^*}{\partial \gamma_E}(z) \cos\mu_E + \frac{\partial d_{\boldsymbol{u}_{P,\text{pn}}}{}^*}{\partial \chi_E}(z) \frac{\sin\mu_E}{\cos\gamma_E} \right) \right),
\end{aligned} \tag{43}$$

and, analogously,

$$\begin{aligned}\Delta \boldsymbol{u}_{P,\mathrm{up}}^{*}(z) = & \operatorname*{argmin}_{\Delta \boldsymbol{u}_{P,\mathrm{up}} \in U_{P,\mathrm{up}}(z)} \left. \frac{d}{dt} d_{\boldsymbol{u}_{P,\mathrm{pn}}}^{*} \right|_{\substack{z \\ u_{\mathrm{E},\boldsymbol{u}_{\mathrm{P,pn}}}^{*}(z) \\ u_{\mathrm{P,pn}}(z) + \Delta u_{\mathrm{P,up}}}} \\ = & \operatorname*{argmin}_{\Delta \boldsymbol{u}_{P,\mathrm{up}} \in U_{P,\mathrm{up}}(z)} \left(-\frac{\partial d_{\boldsymbol{u}_{P,\mathrm{pn}}}^{*}}{\partial v_P}(z) \frac{D_{I,P}(z)}{m_P} \left(n_{P,\mathrm{pn}}(z) + \Delta n_{P,\mathrm{up}} \right)^2 \right. \\ & + \frac{g(z)}{v_P} \left(n_{P,\mathrm{pn}}(z) + \Delta n_{P,\mathrm{up}} \right) \\ & \cdot \left(\frac{\partial d^{*}_{\boldsymbol{u}_{P,\mathrm{pn}}}}{\partial \gamma_P}(z) \cos\left(\mu_{P,\mathrm{pn}}(z) + \Delta \mu_{P,\mathrm{up}} \right) \right. \\ & \left. \left. + \frac{\partial d^{*}_{\boldsymbol{u}_{P,\mathrm{pn}}}}{\partial \chi_P}(z) \frac{\sin\left(\mu_{P,\mathrm{pn}}(z) + \Delta \mu_{P,\mathrm{up}} \right)}{\cos \gamma_P(z)} \right) \right). \qquad (44)\end{aligned}$$

With a numerically synthesized $d_{u_{\mathrm{P,pn}}}{}^{*}(z)$ using, for example, local Taylor-series expansions with global smoothing or artificial neural networks (see [3], [15], and [18]),

$$\Delta \boldsymbol{u}_{P,\mathrm{up}}^{*}\left(z, \operatorname{grad} d_{\boldsymbol{u}_{P,\mathrm{pn}}}^{*}(z)\right) = \begin{pmatrix} \Delta n_{P,\mathrm{up}}^{*}\left(z, \operatorname{grad} d_{\boldsymbol{u}_{P,\mathrm{pn}}}^{*}(z)\right) \\ \Delta \mu_{P,\mathrm{up}}^{*}\left(z, \operatorname{grad} d_{\boldsymbol{u}_{P,\mathrm{pn}}}^{*}(z)\right) \end{pmatrix} \qquad (45)$$

is computable as a solution of the constrained nonlinear minimization problem (44). Nevertheless, here the following direct synthesis approach for $\Delta \boldsymbol{u}_{P,\mathrm{up}}^{*}$, which is easier to implement, has turned out to be very successful. For $z \in S_{\mathrm{rs}}$, the closest points with respect to appropriate scaling $\mathbb{R}^9$

$$z_1(z) = \operatorname*{argmin}_{z_i \in S_{\mathrm{app}}} \|\operatorname{diag}(\boldsymbol{s})(z - z_i)\|_2, \ldots z_5(z) = \operatorname*{argmin}_{z_i \in S_{\mathrm{app}} \setminus \{z_1, \ldots z_4\}} \|\operatorname{diag}(\boldsymbol{s})(z - z_i)\|_2$$

are computed and used for the optimal upgrades

$$\Delta n_{P,\mathrm{up}}^{*}(z) = \begin{cases} \min\left\{ \Delta n_{P,\mathrm{up,max}}, \right. & \text{if } \|\boldsymbol{r}(z)\|_2 \geq r_{\mathrm{final}} \text{ and } \Delta n_{P,\mathrm{up}}^{*}(z_i) = \\ \quad \left. n_{P,\max} - n_{P,\mathrm{pn}}(z) \right\}, & \quad \min\left\{ \Delta n_{P,\mathrm{up,max}}, n_{P,\max} - n_{P,\mathrm{pn}}(z_i) \right\} \\ & \quad \text{for at least } n_{\mathrm{app},n} \text{ points } z_i(z),\ i = 1, \ldots, 5, \\ \max\left\{ \Delta n_{P,\mathrm{up,min}}, \right. & \text{if } \|\boldsymbol{r}(z)\|_2 \geq r_{\mathrm{final}} \text{ and } \Delta n_{P,\mathrm{up}}^{*}(z_i) = \\ \quad \left. -n_{P,\mathrm{pn}}(z) \right\}, & \quad \max\left\{ \Delta n_{P,\mathrm{up,min}}, -n_{P,\mathrm{pn}}(z_i) \right\} \\ & \quad \text{for at least } n_{\mathrm{app},n} \text{ points } z_i(z),\ i = 1, \ldots, 5, \\ 0, & \text{else,} \end{cases} \qquad (46)$$

and

$$
\Delta\mu_{P,\mathrm{up}}{}^{*}(z) = \begin{cases} \Delta\mu_{P,\mathrm{up,max}}, & \text{if } \|\boldsymbol{r}(z)\|_2 \geq r_{\mathrm{final}} \text{ and} \\ & \quad \Delta\mu_{P,\mathrm{up}}{}^{*}\left(z_i, \operatorname{grad} d_{\boldsymbol{u}_{P,\mathrm{pn}}}{}^{*}(z_i)\right) = \Delta\mu_{P,\mathrm{up,max}} \\ & \quad \text{for at least } n_{\mathrm{app},\mu} \text{ points } z_i(z),\ i = 1, \ldots, 5, \\ \Delta\mu_{P,\mathrm{up,min}}, & \text{if } \|\boldsymbol{r}(z)\|_2 \geq r_{\mathrm{final}} \text{ and} \\ & \quad \Delta\mu_{P,\mathrm{up}}{}^{*}\left(z_i, \operatorname{grad} d_{\boldsymbol{u}_{P,\mathrm{pn}}}{}^{*}(z_i)\right) = \Delta\mu_{P,\mathrm{up,min}} \\ & \quad \text{for at least } n_{\mathrm{app},\mu} \text{ points } z_i(z),\ i = 1, \ldots, 5, \\ 0, & \text{else.} \end{cases} \tag{47}
$$

5 Numerical Results

For the complex, real-life problem under investigation, an a priori verification of the small upgrade assumption is impossible. We have chosen 102 representative *additional test scenarios* with $z_0 \in S_{\mathrm{uis}}$ to confirm the theoretic results and highlight the improvement of the interception performance. Using three-dimensional PNG for P, we have computed E 's optimal evasive maneuvers with DIRCOL. For the DIRCOL computations, we have used a fine discretization (20 to 45 interior points) to get accurate optimal controls for E. Equations (1)–(6) have been integrated using E 's stored optimal evasive controls and three-dimensional standard PNG and optimally upgraded PNG, respectively. Stop condition (33) has been used to determine the miss distance exactly. These simulations (see Figure 4 for an illustrative example) have enabled a good choice of the upgrade parameters $\Delta n_{P,\mathrm{up,max}}$, $\Delta n_{P,\mathrm{up,min}}$, $\Delta\mu_{P,\mathrm{up,max}}$, $\Delta\mu_{P,\mathrm{up,min}}$, r_{final}, $n_{\mathrm{app},n}$, and $n_{\mathrm{app},\mu}$. The small upgrade assumption has been *verified numerically* a posteriori. For three-dimensional PNG the miss-distance distribution $d_{\boldsymbol{u}_{\mathrm{P,pn}}}{}^{*}(z_0)$ for the 1830 $z_0 \in S_{\mathrm{uis}}$ is depicted in Figure 3. For the 102 test scenarios, Table 2 shows the improvement of the interception performance dependent on $\Delta n_{P,\mathrm{up,max}}$, $\Delta n_{P,\mathrm{up,min}}$, $\Delta\mu_{P,\mathrm{up,max}}$, $\Delta\mu_{P,\mathrm{up,min}}$, $n_{\mathrm{app},n}$, and $n_{\mathrm{app},\mu}$. The best improvement is achieved applying the largest possible upgrade if and only if all five closest points indicate this largest upgrade. More than 94% of the test scenarios lead to an improved, (i.e., on the average, more than 60% smaller) miss distance. Table 3 shows the improvement of the interception performance dependent on r_{final}. The best improvement is achieved applying $r_{\mathrm{final}} = 1000$ m, that is, for $\|\boldsymbol{r}(z)\|_2 \leq 1000$ m, three-dimensional PNG is advantageous. A very important experience is sketched in Table 4. For the 102 test scenarios, Table 4 shows the improvement of the interception performance dependent only on $\Delta\mu_{P,\mathrm{up,max}}$ and $\Delta\mu_{P,\mathrm{up,min}}$. Note that the bank angle upgrade has *almost no* influence on the miss-distance distribution and can be omitted for the practical implementation of an upgraded three-dimensional PNG. In summary, the numerical simulations for the 102 test scenarios have validated success and real-life applicability of the upgrade technique outlined here.

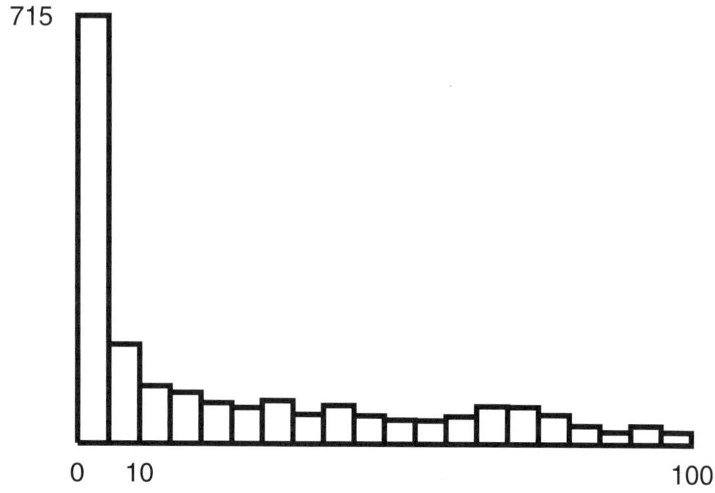

Figure 3: Miss-distance distribution $d^*_{\boldsymbol{u}_{P,\mathrm{pn}}}(z_0)$ for the 1830 optimal control problems computed (coarse discretization).

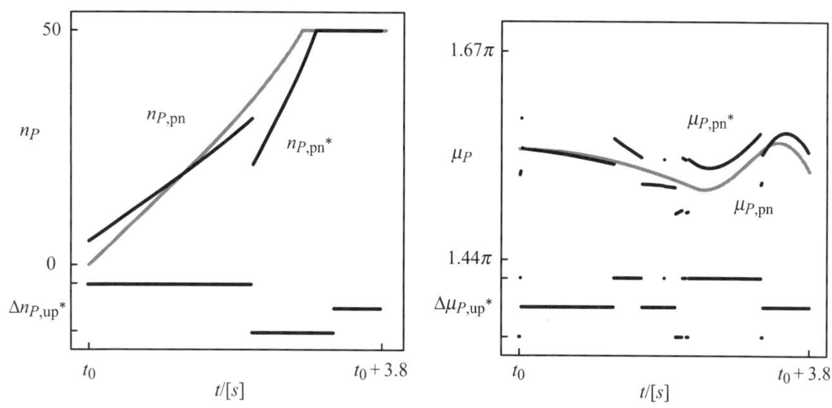

Figure 4: Comparison of illustrative histories of upgraded controls $n_{P,\mathrm{up}}{}^*$ and $\mu_{P,\mathrm{up}}{}^*$ (black) with three-dimensional PNG $n_{P,\mathrm{pn}}$ and $\mu_{P,\mathrm{pn}}$ (gray).

6 Conclusions and Outlook

The computer simulations encourage guidance scheme upgrades for intercepting missiles against future maneuvering short- and medium-range ballistic missiles. The outlined upgrade technique is *not limited to PNG* as a starting guidance scheme: any other starting guidance scheme can be used. Future research by the first author will be dedicated to the upgrade of more advanced guidance schemes. Various realistic final homing scenarios have been investigated. For most encounters (i.e., more than 94%), the original miss-distance $d_{u_{\mathrm{P,pn}}}{}^*(z)$ has been cut

Table 2: Miss-distance improvement dependent on $\Delta n_{P,\mathrm{up,max}}$, $\Delta n_{P,\mathrm{up,min}}$, $\Delta\mu_{P,\mathrm{up,max}}$, $\Delta\mu_{P,\mathrm{up,min}}$, $n_{\mathrm{app},n}$ and $n_{\mathrm{app},\mu}$.

$\Delta n_{P,\mathrm{up,max}} = -\Delta n_{P,\mathrm{up,min}}$	$\Delta\mu_{P,\mathrm{up,max}} = -\Delta\mu_{P,\mathrm{up,min}}$	r_{final} / [m]	$n_{\mathrm{app},n} = n_{\mathrm{app},\mu}$	tests scenarios with $d\boldsymbol{u}_{P,\mathrm{up}}{}^* < d\boldsymbol{u}_{P,\mathrm{pn}}{}^*$	average $\frac{(d\boldsymbol{u}_{P,\mathrm{pn}}{}^* - d\boldsymbol{u}_{P,\mathrm{up}}{}^*)}{d\boldsymbol{u}_{P,\mathrm{pn}}{}^*}$
1	π / 180	1000	4	90.20%	9.58%
1	π / 180	1000	5	94.12%	10.15%
5	π / 36	1000	4	91.18%	36.11%
5	π / 36	1000	5	94.12%	36.21%
10	π / 18	1000	4	91.18%	59.29%
10	π / 18	1000	5	94.12%	60.76%

Table 3: Miss-distance improvement dependent on r_{final}.

$\Delta n_{P,\mathrm{up,max}} = -\Delta n_{P,\mathrm{up,min}}$	$\Delta\mu_{P,\mathrm{up,max}} = -\Delta\mu_{P,\mathrm{up,min}}$	r_{final} / [m]	$n_{\mathrm{app},n} = n_{\mathrm{app},\mu}$	Test scenarios with $d\boldsymbol{u}_{P,\mathrm{up}}{}^* < d\boldsymbol{u}_{P,\mathrm{pn}}{}^*$	Average $\frac{(d\boldsymbol{u}_{P,\mathrm{pn}}{}^* - d\boldsymbol{u}_{P,\mathrm{up}}{}^*)}{d\boldsymbol{u}_{P,\mathrm{pn}}{}^*}$
1	π / 180	0	5	93.14%	10.23%
1	π / 180	1000	5	94.12%	10.12%
5	π / 36	0	5	94.12%	36.38%
5	π / 36	1000	5	94.12%	36.21%
5	π / 36	2000	5	95.10%	34.21%
10	π / 18	1000	5	94.12%	60.76%
10	π / 18	2000	5	94.12%	56.25%

Table 4: Miss-distance improvement dependent only on $\Delta\mu_{P,\mathrm{up,max}}$ and $\Delta\mu_{P,\mathrm{up,min}}$.

$\Delta n_{P,\mathrm{up,max}} = -\Delta n_{P,\mathrm{up,min}}$	$\Delta\mu_{P,\mathrm{up,max}} = -\Delta\mu_{P,\mathrm{up,min}}$	r_{final} / [m]	$n_{\mathrm{app},n} = n_{\mathrm{app},\mu}$	test scenarios with $d\boldsymbol{u}_{P,\mathrm{up}}{}^* < d\boldsymbol{u}_{P,\mathrm{pn}}{}^*$	Average $\frac{(d\boldsymbol{u}_{P,\mathrm{pn}}{}^* - d\boldsymbol{u}_{P,\mathrm{up}}{}^*)}{d\boldsymbol{u}_{P,\mathrm{pn}}{}^*}$
0	π / 360	1000	5	45.10%	0.31%
0	π / 180	1000	5	38.24%	0.77%
0	π / 90	1000	5	33.33%	1.87%
0	π / 36	1000	5	41.18%	0.60%

down significantly. Note that, if necessary, additional small upgrades $\Delta\boldsymbol{u}_{P,\mathrm{up},2}{}^*(z)$, $\Delta\boldsymbol{u}_{P,\mathrm{up},3}{}^*(z)$, ... can be calculated. Given P's guidance (i.e., $\boldsymbol{u}_{P,\mathrm{pn}}(z)$ plus former upgrades), all the optimal control problems (26) for E must be solved again. With many appropriate upgrades, P's guidance may converge to an optimal strategy for the Stackelberg game. Here the computation of more than one upgrade for the intercepting missile is not of practical importance. Alternative upgrade techniques based on the linearization of P's and/or E's guidance schemes are possible and worth being tried. The upgrade technique used here is not based on guidance scheme linearization. Roughly, P's control upgrades can be considered as instantaneously changeable parameters in the optimal control problem, which must be solved for the optimal evasive guidance of E.

The outlined approach is applicable to some other optimal control problems if unpredictable and unmeasurable influences are present; see [3], [4], [5], [6], [14], and [15]. Consider a space shuttle returning to Earth. The on-board guidance system is faced with unpredictable air-density fluctuations and wind turbulences. Or consider a car's on-board system for automatic or semiautomatic driving. On the basis of simplified mechatronics for a car, the on-board system has to avoid collision against various unpredictable maneuvers of other cars, including wrong-way drivers. For these problems, unpredictable and unmeasurable influences can be taken into account via worst-case analysis, and for some cases an antagonistic dynamic Stackelberg game can be formulated. Moreover, optimal strategies of some standard dynamic games coincide with optimal strategies of auxiliary dynamic Stackelberg games.

Acknowledgments

This research has been done in fruitful cooperation with the Bodenseewerk Gerätetechnik GmbH (BGT), Überlingen. The Daimler-Benz Aerospace Lenkflugkörpersysteme GmbH (DASA-LFK), Ottobrunn, has provided interesting and valuable information. This research has been supported by the Deutsche Forschungsgemeinschaft within the project "Echtzeitoptimierung großer Systeme" and FORTWIHR, the Bayerischer Forschungsverbund für Technisch-Wissenschaftliches Hochleistungsrechnen. The authors thank the associate editor and all reviewers for valuable comments and constructive criticism, which helped to *upgrade* this paper significantly. Last, but not least, the first author is greatly indebted to Josef Shinar, Technion Haifa (Israel Institute of Technology), for many fruitful discussions and helpful comments during the last decade.

REFERENCES

[1] Adler, F. P. Missile guidance by three-dimensional proportional navigation, *J. Appl. Phys.* **27**, 1954.

[2] Başar, T. and Olsder, G. J. *Dynamic Noncooperative Game Theory*, Academic Press, London, 1982 (2nd edition, 1995).

[3] Breitner, M. H. Robust optimale Rückkopplungssteuerungen gegen unvorhersehbare Einflüsse: Differentialspielansatz, numerische Berechnung und Echtzeitapproximation. Extended Ph.d. thesis, VDI Fortschritt-Bericht 596, Series 8 "Mess -, Steuerungs- und Regelungstechnik," VDI-Verlag, Düsseldorf, 1995.

[4] Breitner, M. H. and Heim, A. Robust optimal control of a reentering space shuttle, in *DGLR-Jahrbuch 1995,* Deutsche Gesellschaft für Luft- und Raumfahrt, Bonn, 1995.

[5] Breitner, M. H. and Pesch, H. J. Reentry trajectory optimization under atmospheric uncertainty as a differential game, *Ann. Inter. Soc. Dyn. Games*, **1**, 1994.

[6] Breitner, M. H., Pesch, H. J., and Grimm, W. Complex differential games of pursuit–evasion type with state constraints, Part 1: Necessary conditions for optimal open-loop strategies, Part 2: Numerical computation of optimal open-loop strategies, *J. Opt. Th. Appl.*, **78**, 1993.

[7] Ehtamo, H. and Raivio, T. A feasible direction method for saddlepoint problems. *Proc. 8th Intl. Symp. Dyn. Games and Appl.*, Vaals, 1998.

[8] Imado, F. Some aspects of a realistic three-dimensional pursuit-evasion game. *J. Guidance, Cont. Dyn.*, **16**, 1993.

[9] Imado, F. and Ishihara, T. Pursuit–evasion geometry analysis between two missiles and an aircraft, *Comp. Math. Appl.*. **26**, 1993.

[10] Imado, F., Kuroda, T., and Miwa, S. Optimal midcourse guidance for medium-range air-to-air missiles, *J. Guidance, Cont. Dyn.*, **13**, 1990.

[11] Imado, F., Kuroda, T. and Tahk, M. J. A trade-off study between conventional and augmented proportional navigation, submitted to *J. Guidance, Cont. Dyn.*, 1998.

[12] Gill, P. E., Murray, W., and Wright, M. H. *Practical optimization*, Academic Press, London 1997 (1st ed., 1981).

[13] Krasovskii, N. N. and Subbotin, A. I. *Game-Theoretical Control Problems*, Springer-Verlag, New York, 1988.

[14] Lachner, R., Breitner, M. H., and Pesch, H. J. Three-dimensional air combat analysis—An example for the numerical solution of complex differential games, *Ann. Intl. Soc. Dyn. Games*, **3**, 1996.

[15] Lachner, R., Breitner, M. H., and Pesch, H. J. Real-time collision avoidance against wrong drivers: Differential game approach, numerical solution and synthesis of strategies with neural networks, *Ann. Intl. Soc. Dyn. Games*, **5**, 1998.

[16] Lipman, Y. and Shinar, J. A linear pursuit–evasion game with a state constraint for a highly maneuverable evader, *Ann. Intl. Soc. Dyn. Games*, **3**, 1995.

[17] Macfadzean, R. H. M. *Surface-based Air Defense System Analysis*, Artech House, Boston, 1992.

[18] Pesch, H. J., Gabler, I., Breitner, M. H., and Miesbach, S. Synthesis of optimal strategies for differential games by neural networks, *Ann. Intl. Soc. Dyn. Games*. **3**, 1996.

[19] Rettig, U. Verbesserung von Regelalgorithmen mit direkten numerischen Optimierungsverfahren, Diploma thesis, Technische Universität Clausthal, Clausthal-Zellerfeld, 1998.

[20] Shinar, J. and Davidovitz, A. A two-target game analysis in line-of-sight coordinates, *Comp. Math. Appl.*, **13**, 1987.

[21] Shinar, J. and Gutman, S. Three-dimensional optimal pursuit and evasion with bounded controls, *IEEE Trans. Auto. Cont.*, **25**, 1980.

[22] Shinar, J. and Shima, T. A new application of Bernhard's certainty equivalence principle, *Proc. 8th Intl. Symp. Dyn. Games Appl.*, Vaals, 1998.

[23] Shinar, J. and Tabak, R. New results in optimal missile avoidance analysis, *J. Guidance, Cont. Dyn.*, **17**, 1994.

[24] Shinar, J. and Visser, H. G. Synthesis and validation of feedback guidance laws for air-to-air interceptions, *Cont. Dyn. Sys.*, **38**, 1990.

[25] Shinar, J., Well, K. H., and Järmark, B., *Near-Optimal Feedback Control for Three-dimensional Interceptions*, 15th ICAS Congress, London, 1986.

[26] Stoer, J. and Bulirsch, R. *Introduction to Numerical Analysis*, Springer-Verlag, New York 1993.

[27] von Stryk, O. Numerische Lösung optimaler Steuerungsprobleme: Diskretisierung, Parameteroptimierung und Berechnung der adjungierten Variablen. Ph.d. thesis, Fortschritt-Bericht 441, Series 8 "Meß-, Steuerungs- und Regelungstechnik," VDI-Verlag, Düsseldorf, 1994.

[28] von Stryk, O. Numerical solution of optimal control problems by direct collocation, in *Optimal Control, Calculus of Variations, Optimal Control Theory and Numerical Methods,* Birkhäuser, Intl. Series of Numerical Mathematics, **111**, Basel, 1993.

[29] Zakharov, V. V. Stackelberg differential games and problem of time consistency, *Intl. Yearbook Game Th. App.*, **1**, Nova Science Publishers, New York, 1993.

[30] Zarchan, P. *Tactical and Strategic Missile Guidance*, American Institute of Aeronautics and Astronautics, Progress in Astronautics and Aeronautics 157, Washington, DC. 1996.

Homicidal Chauffeur Game: Computation of Level Sets of the Value Function

Valery S.Patsko
Institute of Mathematics and Mechanics
Ekaterinburg, Russia

V.L.Turova
Center of Advanced European Studies and Research
Bonn, Germany

Abstract

Two differential games with the dynamics of the homicidal chauffeur are considered. The first problem is the Isaacs homicidal chauffeur differential game introduced in [6]. In this game, a chauffeur (the pursuer P) minimizes the capture time of a pedestrian (the evader E). The objective of the pedestrian is to avoid the capture or to maximize the capture time. The magnitude of the velocity is constant for the pursuer; his maneuverability is bounded through a minimum turn radius. The maneuverability of the evader is not bounded. The pursuer's control is the rate of turn; the evader governs the choice of directions of his velocity. The second problem is the conic surveillance–evasion game from [11]. In this game, the dynamics are the same as in the Isaacs problem, but the goals of the players differ from the classic formulation: an evader E minimizes the time of escaping from a detection set that is a two-dimensional semi-infinite cone. The detection set is attached to the velocity vector of a pursuer P whose objective is to keep the evader within the detection set for maximum time. The paper describes the computation of level sets of the value functions for these games. The algorithm proposed by the authors is used.

1 Introduction

The homicidal chauffeur game was formulated by Isaacs more than 30 years ago [6]. Since that time, many authors have studied this problem in various ways. The most complete qualitative solution was given in the Ph.D. dissertation of Merz [12].

Very often (see, for example, [10], [11], and [4]) the dynamics of the homicidal chauffeur game were used, but the statement of the problem differed from the one of Isaacs. In [11], a surveillance–evasion game with the pursuer's detection zone in the shape of a cone, and a qualitative solution to this problem, are given. Similarly to [12], the parameter space of the problem is divided into subregions. In each subregion, the type of solution (possible singular lines and the strategies of the players) is described.

Many papers on differential games are devoted to the development of algorithms for solving nonlinear differential games in the plane [1], [2], [5], [18], and [21].

In this paper, the problems with the dynamics of the homicidal chauffeur are investigated using the algorithm the authors propose for computing level sets of the value function. Our method is based on the general theory of differential games [8] and [9]. The algorithm is a natural extension of the algorithms from [19] and exploits ideas of the algorithms [15], [16], and [17] for linear time-optimal differential games in the plane. Some experience ([14], [15], [19], and [20]) in solving differential games of kind [6] in the plane helps to find very complicated types of solutions and to verify the solution's validity. The computation results are consistent with those obtained in [11] and [12].

2 Statement of the Problem

2.1 The Homicidal Chauffeur Game in the Classical Formulation

The dynamics of the homicidal chauffeur game in reduced coordinates have the form

$$\begin{aligned} \dot{x}_1 &= -\frac{w^{(1)}}{R} x_2 \varphi + v_1 \\ \dot{x}_2 &= \frac{w^{(1)}}{R} x_1 \varphi + v_2 - w^{(1)}, \qquad |\varphi| \le 1, \quad v \in Q. \end{aligned} \tag{1}$$

Here $(x_1, x_2)'$ is the state vector, and $w^{(1)}$ and R are constants that define the pursuer's velocity and the minimal radius of turn, respectively. The objective of the control φ of the pursuer P is to minimize the time it takes to attain a given terminal set M. The objective of the control $v = (v_1, v_2)'$ of the evader E is to maximize this time. Thus, the payoff of the game is the time of attaining the terminal set.

The classical formulation of the homicidal chauffeur game assumes that the sets M and Q are circles of the radii l (capture radius) and $w^{(2)}$, respectively, with the centers in the origin. It is accepted that $w^{(2)} < w^{(1)}$. With the proposed algorithm, level sets of the value function can be computed for a sufficiently wide class of sets M and Q.

In this paper, we assume that Q is a convex polygon in the plane such that $(0, 0) \in \operatorname{int} Q$ and $\max\{v_2 : (v_1, v_2)' \in Q\} < w^{(1)}$. In particular, Q can be a polygon approximating the circle of radius $w^{(2)} < w^{(1)}$ with the center in the origin.

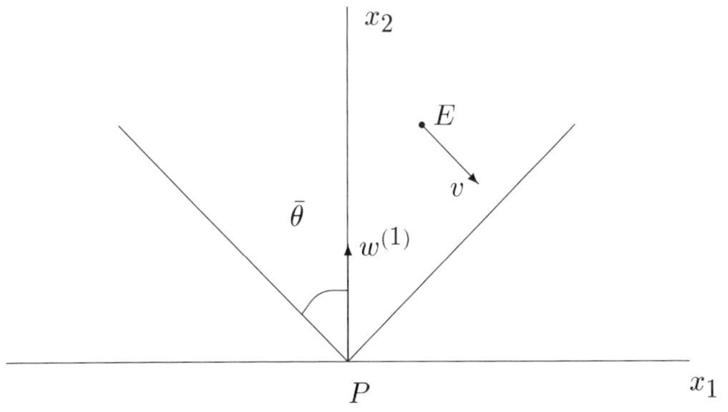

Figure 1: Detection cone.

2.2 Conic Surveillance–Evasion Game

The statement of the problem is given in [11]. The dynamics of the game are described by (1) with $R = 1$ (normalized). The terminal set M is the complement of the open detection cone depicted in Figure 1. The objective of the control $v = (v_1, v_2)'$ of the evader E is to minimize the time of attaining M. The objective of the control φ of the pursuer P is to maximize this time. Hence, in contrast to Isaacs' homicidal chauffeur game, the roles of the players change: the evader is the "minimizing" player and the pursuer is the "maximizing" one. We will assume that the convex polygon Q satisfies the same conditions as in Section 2.1.

2.3 Level Sets of the Value Function

We restrict ourselves to a conceptual definition of level sets of the value function. The precise definition can be found in [8] and [9].

Let $T \geq 0$. The level set (the Lebesgue set) of the value function is denoted by $W(T, M)$. This is the set of all points in the plane such that the minimizing player using feedback strategies can guarantee the transition of trajectories of system (1) to the terminal set M within time T.

3 The Algorithm

Here the basic idea of the algorithm for computing sets $W(T, M)$ is described.

The algorithm is based on ideas of the algorithms for linear time-optimal game problems [15], [16], [17], and [19]. The set $W(T, M)$ is formed via a step-by-step backward procedure giving a sequence of embedded sets

$$W(\Delta, M) \subset W(2\Delta, M) \subset W(3\Delta, M) \subset \ldots \subset W(i\Delta, M) \subset \ldots \subset W(T, M).$$

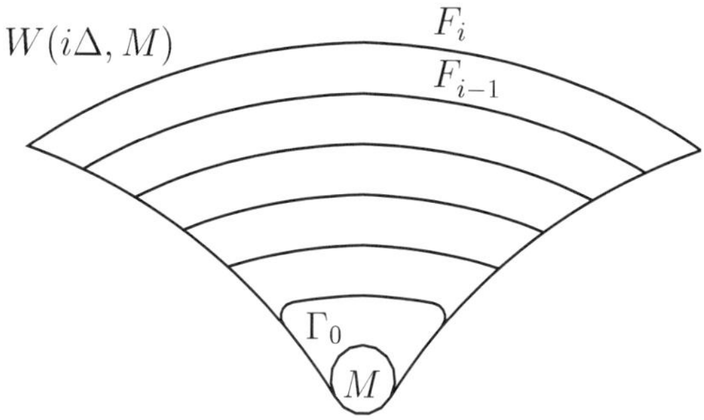

Figure 2: Construction of the sets $W(i\Delta, M)$.

Here Δ is the step of the backward procedure. Each set $W(i\Delta, M)$ consists of all initial points such that the minimizing player brings system (1) into the set $W((i-1)\Delta, M)$ within the time duration Δ. We put $W(0, M) = M$.

The crucial point of the algorithm is the computation of "fronts." The front F_i (Figure 2) is the set of all points of $\partial W(i\Delta, M)$ for which the minimum guaranteeing time of the achievement of $W((i-1)\Delta, M)$ is equal to Δ. For other points of $\partial W(i\Delta, M)$, the optimal time is less than Δ. The line $\partial W(i\Delta, M) \setminus F_i$ possesses the properties of the barrier [6]. The front F_i is designed using the previous front F_{i-1}. For the first step of the backward procedure, F_0 coincides with the usable part [6] Γ_0 of the boundary of M. It may be one or several usable parts. The computations are carried out separately from each usable part. One should take into account that the obtaining parts of the level set can collide with each other. From here on we will assume for simplicity that only one usable part is on the boundary of M.

Let's explain, assuming the problem from Section 2.1, how the fronts can be constructed. Denote $p(x) = (-x_2, x_1)' \cdot w^{(1)}/R$, $g = (0, -w^{(1)})'$. Using this notation, the equations (1) can be rewritten as follows: $\dot{x} = p(x)\varphi + v + g$. First, suppose that the front F_{i-1} is a smooth curve. Let x_* be an arbitrary point of F_{i-1} and ℓ be the normal vector to the front at x_*. Let $\varphi^\circ = \arg\min_{|\varphi|\leq 1} \ell' p(x_*)\varphi$ and $v^\circ = \arg\max_{v\in Q} \ell' v$. We call φ° and v° the extremal controls. The controls φ° and v° are chosen from the conditions of minimizing and, respectively, maximizing the projection of the velocity vector of (1) onto the direction ℓ. If the vector x_* is collinear to ℓ, then any control $\varphi \in [-1, 1]$ is extremal. If Q is a polygon in the plane, and ℓ is collinear to some normal vector to an edge $[q_1, q_2]$ of Q, then any control $q \in [q_1, q_2]$ is extremal.

After computing the extremal controls, the extremal trajectories started from the front's points in the reverse time are considered: $x(\tau) = x_* - \tau\ (p(x_*)\varphi^\circ + v^\circ + g)$. The ends of these trajectories at $\tau = \Delta$ are used to form the next front F_i. If the extremal control φ° is not unique at some point $x_* \in F_{i-1}$, then the segment $\Phi(x_*) = \{\bigcup_{\varphi^\circ\in[-1,1]}(x_* - \Delta(p(x_*)\varphi^\circ + v^\circ + g))\}$ is considered instead of the single point. Similarly, if the extremal control v° is not unique, the segment $\Xi(x_*) = \{\bigcup_{v^\circ \in [q_1, q_2]} (x_* - \Delta(p(x_*)\varphi^\circ + v^\circ + g))\}$ is considered.

For each front, we distinguish points of the local convexity and points of the local concavity. In Figure 3, d is a point of the local convexity, and e is a point of the local concavity. If x_* is a point of the local convexity, and the extremal control φ° is not unique, one obtains a local picture like that shown in Figure 4(a) after computing the extremal trajectories from the point x_*. Here, the additional segment $\Phi(x_*)$ appears on the new front F_i. If the extremal control v° is not unique, a local picture similar to that shown in Figure 4(b) is obtained: the "swallow tail" $\beta_1 \xi \beta_2$ does not belong to the new front F_i, and it is taken away. For points of the local concavity, there is an inverse situation: if the extremal control φ° is not unique, a swallow tail that should be removed appears [Figure 5(a)]; if the extremal control v° is not unique, an additional segment $\Xi(x_*)$ appears on the new front F_i [Figure 5(b)]. If both φ° and v° are non-unique, the insertion or the swallow tail arises depending on which of segments $\Phi(x_*)$ or $\Xi(x_*)$ is greater.

In the course of numerical computations, we operate with polygonal lines instead of smooth curves. Let the front F_{i-1} be computed. One can divide it into regular parts so that only one extremal trajectory emanates from each point of the regular part. The notion of regularity can be generalized for the case when a bundle of extremal trajectories, which does not destroy the regularity of the field of trajectories, emanates from some point. Thus, each regular part gives an ordered collection of points consisting of the ends of the extremal trajectories. Let's call this ordered collection the *secondary arc*. The new front F_i is obtained after processing secondary arcs, and this processing is reduced to the intersection of secondary arcs. In Figure 6, the front F_{i-1} consists of two regular parts: $[z_1, z_\omega]$ and $[z_\omega, z_r]$. The ends of the extremal trajectories computed at $\tau = \Delta$ give two secondary arcs $[\xi_1, \xi_2, \ldots, \xi_s]$ and $[\xi_{s+1}, \ldots, \xi_m]$. The front $F_i = [\xi_1, \xi_2, \ldots, \xi_\alpha, \ldots, \xi_m]$ is obtained after removing the swallow tail $\xi_s \xi_\alpha \xi_{s+1}$.

Unfortunately, very often it is not sufficient to intersect the neighboring secondary arcs only. Figure 7 gives an example where the secondary arcs S_1, S_2, and S_3 are sequentially computed, and the next front is obtained due to the intersection of S_1 and S_3.

The decomposition of the front F_{i-1} into regular parts is being done when processing its vertices. Two normal vectors to the links $[a, b]$, $[b, c]$ of the polygonal line are considered at each vertex b (Figure 8). In the endpoints of the front, the missing normals are computed from special relations [15]. The algorithm treats

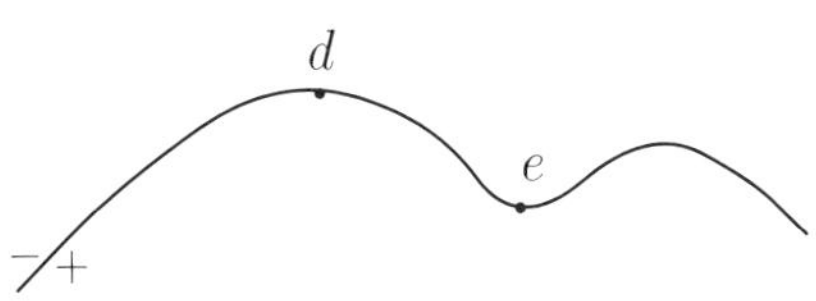

Figure 3: Local convexity and concavity.

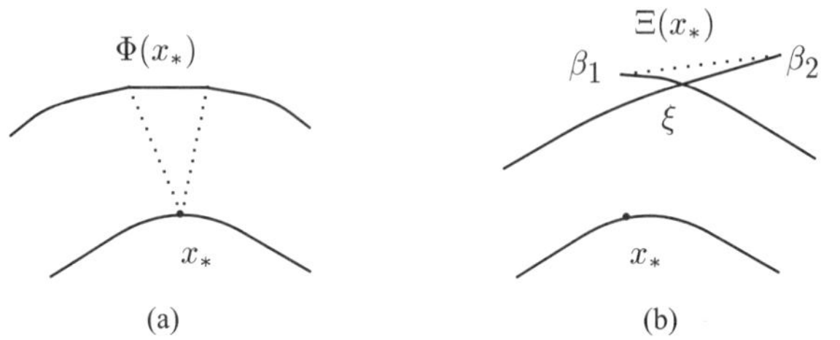

Figure 4: Nonuniqueness of extremal controls in the case of local convexity.

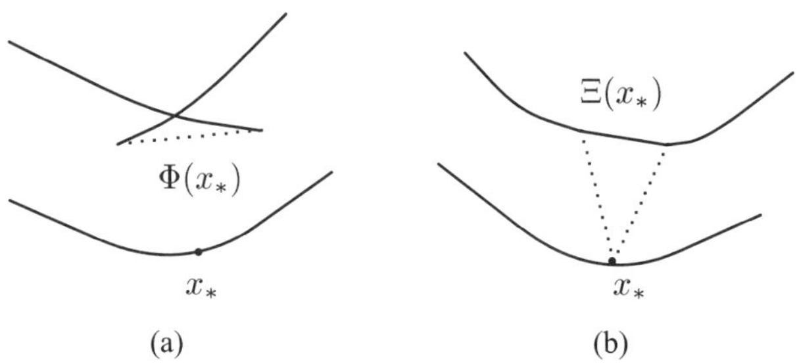

Figure 5: Nonuniqueness of extremal controls in the case of local concavity.

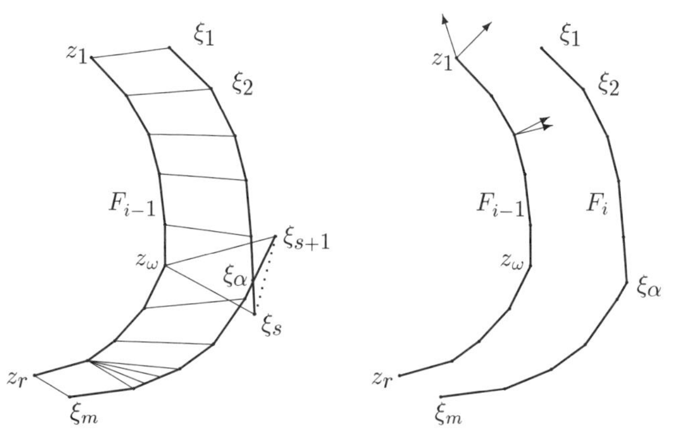

Figure 6: Removing the swallow tail.

all possible variants of disposition of the normal vectors $\ell_{[ab]}$, $\ell_{[bc]}$ to the edges of F_{i-1}, normals to the edges of Q, and normals to the segment $\bigcup_{\varphi\in[-1,1]} p(x)\varphi$. The vectors b (from the origin to the point b) and $-b$ are used as the normal vectors to the last segment. In Figure 8, for instance, the vector b is between the vectors $\ell_{[ab]}$ and $\ell_{[bc]}$; the normals n_1, n_2 to the set Q are between the vectors b and $\ell_{[bc]}$. Since b is the point of the local convexity, the location of normals to the set Q between the vectors $\ell_{[ab]}$ and $\ell_{[bc]}$ means that b is one of the decomposition points that separate the front F_{i-1} into regular parts. The ends of the extremal trajectories

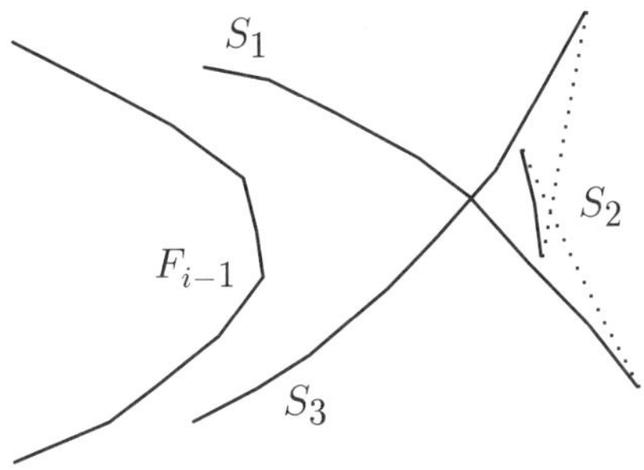

Figure 7: Processing of secondary arcs; complicated case.

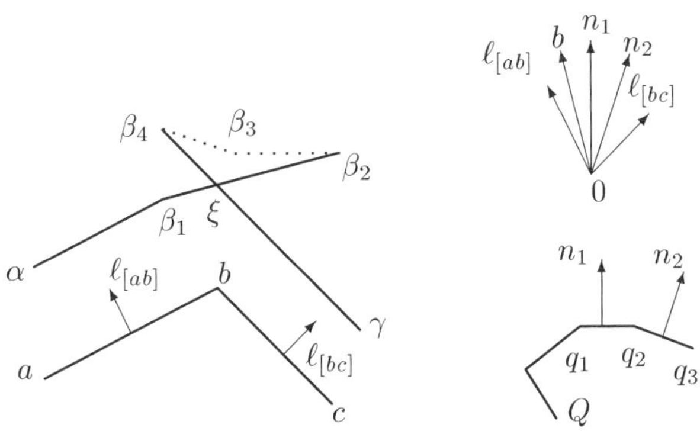

Figure 8: Example of local constructions.

computed at $\tau = \Delta$ give a local picture shown in Figure 8. Here four extremal trajectories emerge from the point b. Their ends are β_1, β_2, β_3, and β_4. The segment $[\beta_1, \beta_2]$ appears due to nonuniqueness of the extremal control φ° for the vector b. The segments $[\beta_2, \beta_3]$ and $[\beta_3, \beta_4]$ arise due to nonuniqueness of the extremal control v° for the vectors n_1 and n_2. The curve $\alpha\beta_1\beta_2$ is assigned to one secondary arc, and the curve $\beta_4\gamma$ is assigned to another secondary arc. After removing the swallow tail $\beta_2\xi\beta_4$, the polygonal line $\alpha\beta_1\xi\gamma$ becomes a fragment of the next front F_i.

Some additional details of such local constructions are given in [17]. The main difference from the case of the linear dynamics [17] is that the extremal control of player P can change its value not only at the front's vertices but also at some interior points of the front's links. In the game considered, such a switching may occur only once for each front's link.

Let's explain the last assertion. Let K be a straight line and ℓ its normal vector. Since the restriction of the scalar function $\ell' p(x)$ to K is a linear function in x, the expression $\ell' p(x)$ can change its sign only once for $x \in K$. Therefore, the extremal control of player P, which is determined by the formula $\arg\min_{|\varphi|\leq 1} \ell' p(x)\varphi$, can switch only once if x runs through K. The switching happens at the point $\bar{x} \in K$ for which the vector $\bar{x}$ is orthogonal to K. If $0 \in K$, then $\bar{x} = 0$.

In the case of the surveillance–evasion game from Section 2.2, the players change their roles: the extremal controls of the pursuer P and the evader E are determined via the relations $\varphi^\circ = \arg\max_{|\varphi|\le 1} \ell' p(x_*)\varphi$ and $v^\circ = \arg\min_{v\in Q} \ell' v$, respectively. Here $p(x) = (-x_2, x_1)' \cdot w^{(1)}$. So the local constructions described earlier for the points of the local convexity are now true for the points of the local concavity, and vice versa.

4 Semipermeable Curves in Differential Games with the Homicidal Chauffeur Dynamics

This section gives the results of some analysis of the families of smooth semipermeable curves in the differential games with the homicidal chauffeur dynamics. The semipermeable curves can be helpful for checking the computation of level sets of the value function.

The families of semipermeable curves are determined through the dynamics of the system and the bounds on controls of the players only. Let's now review what semipermeable curves mean [6]. Let

$$H(\ell, x) = \min_{\varphi} \max_{v} \ell' f(x, \varphi, v) = \max_{v} \min_{\varphi} \ell' f(x, \varphi, v), \tag{2}$$

for $x \in R^2$, $\ell \in R^2$ be the Hamiltonian of the game. Here $f(x, \varphi, v) = p(x)\varphi + v + g$. Fix $x \in R^2$ and consider ℓ such that $H(\ell, x) = 0$. Denote $\varphi^* = \arg\min_\varphi \ell' f(x, \varphi, v)$, $v^* = \arg\max_v \ell' f(x, \varphi, v)$. It holds that $\ell' f(x, \varphi^*, v) \le 0$ for any $v \in Q$, and $\ell' f(x, \varphi, v^*) \ge 0$ for any $\varphi \in [-1, 1]$. This means that the direction $f(x, \varphi^*, v^*)$, which is orthogonal to ℓ, separates the vectograms $U(v^*) = \bigcup_{\varphi\in[-1,1]} f(x, \varphi, v^*)$ and $V(\varphi^*) = \bigcup_{v\in Q} f(x, \varphi^*, v)$ of players P and E (Figure 9). Such a direction is called *semipermeable*. A smooth curve is called the semipermeable curve if the tangent vector at any point of this curve is a semipermeable direction.

Let's describe how the families of semipermeable curves can be constructed.

The number of semipermeable directions depends on the form of the function $\ell \to H(\ell, x)$ at the point x. In the case considered, the function $H(\cdot, x)$ is composed of two convex functions:

$$H(\ell, x) = \begin{cases} \max\limits_{v\in Q} \ell' v + \ell' p(x) + \ell' g & \text{if } \ell' p(x) < 0, \\ \max\limits_{v\in Q} \ell' v - \ell' p(x) + \ell' g & \text{if } \ell' p(x) \ge 0. \end{cases}$$

The semipermeable directions are defined by the roots of the equation $H(\ell, x) = 0$. We will distinguish the roots "$-$" to "$+$" and the roots "$+$" to "$-$". When defining these roots, we will exploit that $\ell \in C$ where C is a closed line that bounds a convex polygon containing the origin. We say that ℓ_* is a root "$-$" to "$+$" if the following conditions hold: (1) $H(\ell_*, x) = 0$; (2) $H(\ell, x) < 0$,

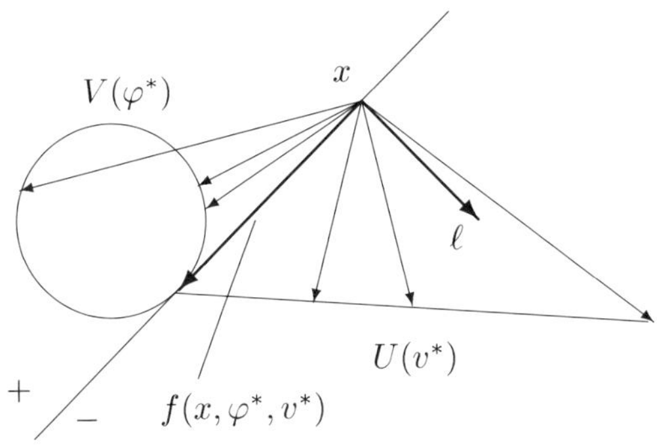

Figure 9: Semipermeable direction.

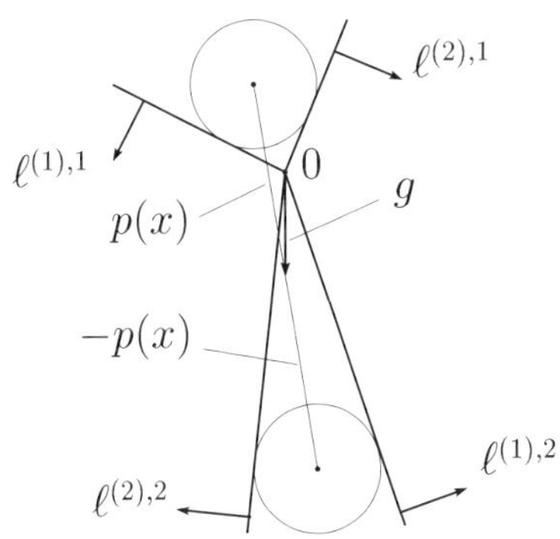

Figure 10: Appearance of four roots.

$\ell < \ell_*$; (3) $H(\ell, x) > 0$, $\ell > \ell_*$, for any ℓ sufficiently close to ℓ_*. The notation $\ell < \ell_*$ means that the direction of the vector ℓ can be obtained from the direction of the vector ℓ_* using the counterclockwise rotation by the angle not exceeding π. The roots "$-$" to "$+$" and the roots "$+$" to "$-$" are called the roots of the first and second type, respectively.

The equation $H(\ell, x) = 0$ has at least one root of the first type and one root of the second type. To prove this, it is sufficient to show that, for any x, there exist the vectors $\overline{\ell}$ and $\widetilde{\ell}$ such that $H(\overline{\ell}, x) < 0$ and $H(\widetilde{\ell}, x) > 0$. Consider $\overline{\ell}$ with $\overline{\ell}_1 = 0$ and $\overline{\ell}_2 > 0$. Taking into account the assumption $\max\{v_2 : v \in Q\} < w^{(1)}$, one obtains

$$H(\overline{\ell}, x) = \min_{|\varphi| \le 1} \overline{\ell}' p(x)\varphi + \overline{\ell}_2 \max\{v_2 \colon v \in Q\} - \overline{\ell}_2 w^{(1)} < 0.$$

Consider $\widetilde{\ell} \neq 0$ such that $\widetilde{\ell}' p(x) = 0$ and $\widetilde{\ell}' g \ge 0$. With the assumption $0 \in \operatorname{int} Q$, one derives

$$H(\widetilde{\ell}, x) = \max_{v \in Q} \widetilde{\ell}' v + \widetilde{\ell}' g > 0.$$

It follows from the above-mentioned property of the piecewise convexity of the function $H(\cdot, x)$ that the equation $H(\ell, x) = 0$ has at most two roots of the first type and two roots of the second type.

We denote the roots of the first type by $\ell^{(1),i}(x)$ and the roots of the second type by $\ell^{(2),i}(x)$. The index $i = 1$ $(i = 2)$ stands for the half-plane $\{\ell \in R^2 : \ell' p(x) < 0\}$ $(\{\ell \in R^2 : \ell' p(x) \geq 0\})$. Figure 10 explains the appearance of four roots in terms of vectograms. The upper and lower circles are the vectograms of player E for $\varphi = 1$ and $\varphi = -1$, respectively.

One can find the domains of the functions $\ell^{(j),i}(\cdot)$, $j = 1, 2$, $i = 1, 2$. The form of these domains for the set Q approximating a circle of some radius $w^{(2)}$ with the center in the origin is shown in Figure 11. Two symmetric cones with the joint apex in the origin are spanned onto polygonal approximations of the circles of the same radius $w^{(2)}R/w^{(1)}$ with the centers at the points $(-R, 0)$ and $(R, 0)$. The subsets A and B of the cones are considered. The boundaries of A and B are drawn with the thick dashed lines. There exist two roots of the first type and two roots of the second type at each internal point of the sets A and B. For any point in the exterior of A and B, there exist one root of the first type and one root of the second type.

The sets analogous to that shown in Figure 11 are used in [4], [6], [10], [11], and [12] to construct semipermeable curves. Here the modification is that the roots of the first and second types are being distinguished and the corresponding semipermeable curves of the first and second types are considered.

The function $\ell^{(j),i}(\cdot)$ is Lipschitz for any closed, bounded subset of the interior of its domain. One can consider the differential equation

$$dx/dt = \Pi \ell^{(j),i}(x), \tag{3}$$

where Π is the matrix of rotation by the angle $\pi/2$. If $j = 1$, the clockwise rotation is assumed; if $j = 2$, the counterclockwise rotation is supposed. Since the tangent vector at each point of phase trajectories of this equation is a semipermeable direction, the trajectories are semipermeable curves. It means that player P can keep one side of the curve (positive side) and player E can keep another side (negative side). So the equation (3) specifies a family $\Lambda^{(j),i}$ of smooth semipermeable curves.

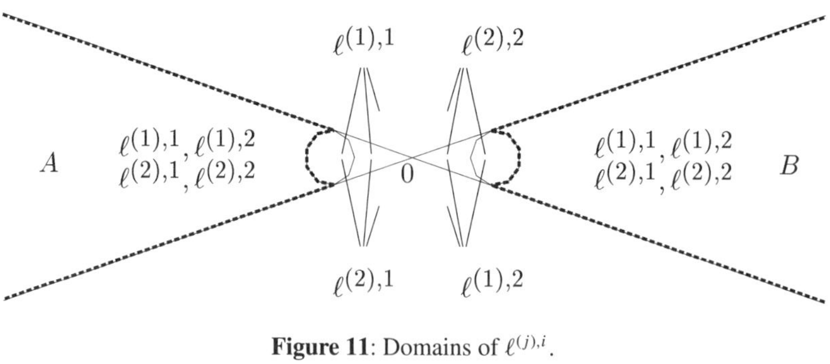

Figure 11: Domains of $\ell^{(j),i}$.

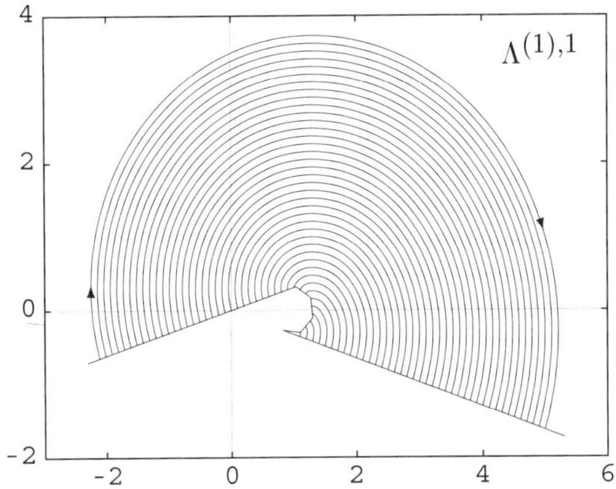

Figure 12: Family $\Lambda^{(1),1}$ of semipermeable curves.

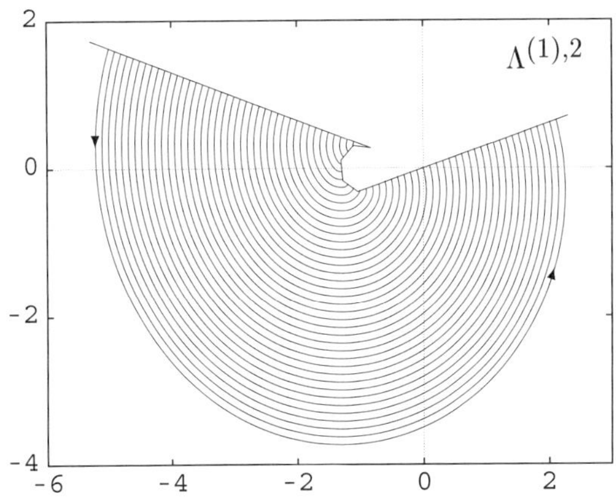

Figure 13: Family $\Lambda^{(1),2}$ of semipermeable curves.

The families $\Lambda^{(j),i}$, $j = 1, 2, i = 1, 2$, for the games from Sections 2.1 and 2.2 are depicted in Figures 12 to 15. Each smooth semipermeable curve is a trajectory of system (1) for controls of the players that deliver minimum and maximum in (2). The direction of the motion in the reverse time is shown by arrows.

The curves of families belonging to the same type can be sewed so that the semipermeability property is preserved for obtaining a composite [3], [4], [13] curve obtained. Such a sewing for semipermeable curves of the first (second) type is only possible if the sewing point belongs to certain parts of the boundaries of the sets where the families $\Lambda^{(1),1}$ and $\Lambda^{(1),2}$ ($\Lambda^{(2),1}$ and $\Lambda^{(2),2}$) are defined. The composite curve can be smooth in some cases.

Since any barrier line arising in the problem is a semipermeable curve, one can inspect the computed barrier lines comparing them with corresponding semiper-

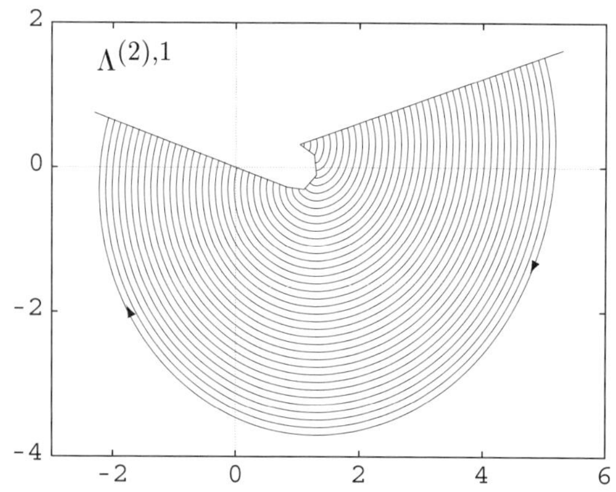

Figure 14: Family $\Lambda^{(2),1}$ of semipermeable curves.

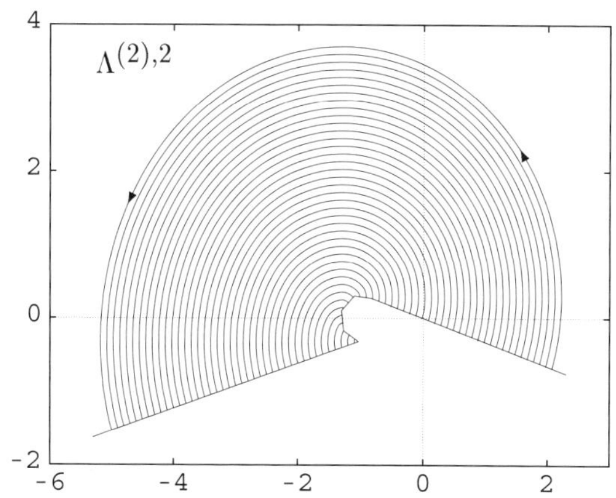

Figure 15: Family $\Lambda^{(2),2}$ of semipermeable curves.

meable curves. The procedure for finding the solvability set of the game of kind can be based [13], [14], [15], and [20] on the emitting in the reverse time two semipermeable curves of the first and second types (they are faced each to other with sides of the same sign) from the endpoints of M's usable part, on the sewing semipermeable curves of two different families belonging to the same type, and on the analysis of mutual dispositions of composite curves. Since the set $W(T, M)$ converges to the solvability set of the corresponding game of kind as $T \to \infty$, solutions to the game of kind can be used for the verification of the computation of the sets $W(T, M)$.

5 Analysis of Computation Results

We compute the level sets of the value function. The optimal time for a given state x is the minimal time τ subject to $x \in W(\tau, M)$.

We do not have a priori estimates of the accuracy of the algorithm. The correctness of computations is verified via comparison of computation results with the qualitative results of [11] and [12]. Additionally, the correctness of the construction of barriers is checked and the front's behavior after the termination of barriers is inspected. An analytical description of fronts is absent for the examples considered.

The computation step Δ in the examples below is selected experimentally. For the examples of Section 5.1, a smaller time step than the one for Section 5.2 is required to find all fine details of the evolution of the fronts. The time step depends also on the size of the terminal set M and on the length of the reverse time interval considered.

5.1 Level Sets of the Value Function in the Homicidal Chauffeur Game

In this section, the results of computing the sets $W(\tau, M)$, $\tau = i\Delta$, for the problem from Section 2.1 are discussed.

For two examples shown in Figures 16 and 17, the following values of parameters of the problem are used: $w^{(1)} = 2$, $w^{(2)} = 0.6$, $R = 0.2$. The set Q is a 25-polygon inscribed into the circle of radius $w^{(2)}$ with the center at $(0, 0)$. The set M is a regular polygon inscribed into the circle of radius 0.015. The center of the circle is $(0.2, 0.3)$, $(0, -0.45)$ for Figures 16 and 17, respectively. The step Δ is 0.001. The sets $W(8k\Delta, M)$, $k = 1, 2, \ldots$, are depicted.

Let us explain the constructions presented in Figure 16. The right barrier line terminates on the lower boundary of the auxiliary set B (here and below, the boundaries of the sets A and B are drawn with dashed lines). The front begins to bend around this barrier line. After some time, the left barrier line ends on the lower boundary of the set A, and the left part of the front bends around the left barrier. The left and right parts of the front go toward each other until the first self-intersection of the front happens at $\tau = 0.725$. The front is divided into two parts (internal and external). For $\tau > 0.725$, only internal fronts that propagate into the "region of turn" are drawn. Here a very complicated structure of fronts arises. At $\tau = 0.904$, the second self-intersection of the front (which is drawn with the thick dashed line) produces two gaps that are filled out afterwards separately. The next front consists of three parts: one exterior part (which is not shown), and two interior parts (two loops inside the dashed contour). As a result, the sets $W(\tau, M)$ for $0.904 < \tau < 0.95$ are triply connected.

In Figure 17, the fronts are symmetric with respect to the x_2-axis. The self-intersection of the front happens at $\tau = 0.355$. The gap that arises after $\tau = 0.355$, and whose center has coordinates $(0, 0.22)$, is filled out with level lines separately. Filling out the gap finishes at $\tau = 0.38$. The sets $W(\tau, M)$ for $0.355 < \tau < 0.38$

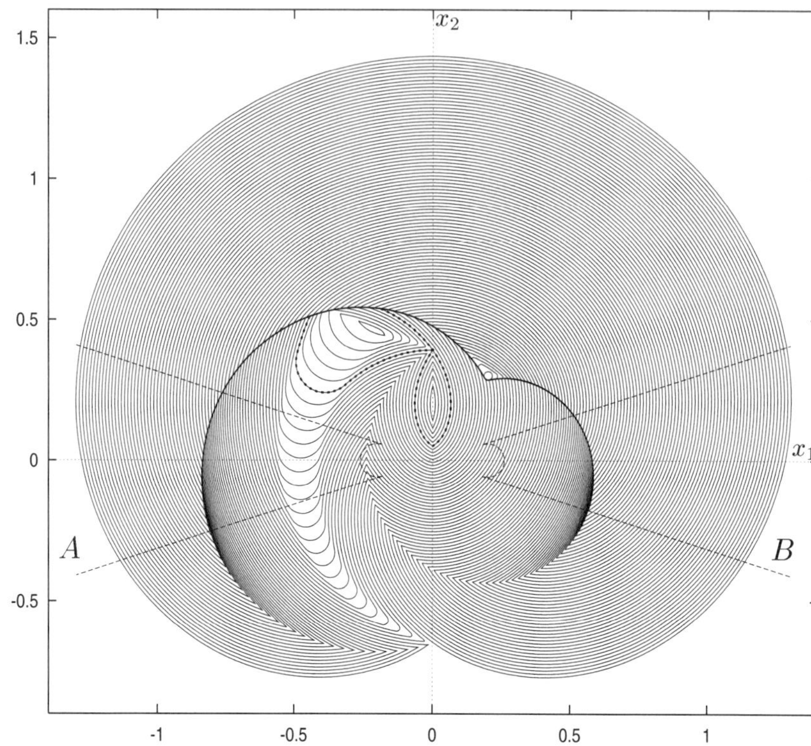

Figure 16: Complicated structure of level sets in the "region of turn."

are doubly connected. At $\tau = 0.766$, the front collides with the barrier lines. Two symmetric gaps arise. The structure of the fronts within the right gap is shown in enlarged scale in Figure 18. Filling out the gap ends at $\tau = 0.85$. The sets $W(\tau, M)$, $0.766 < \tau < 0.85$, are triply connected. The computations are carried out up to $\tau = 0.872$. The value function is discontinuous on two symmetric barrier lines that emerge tangentially from the boundary of M and terminate on the upper boundaries of auxiliary sets A and B. In regions of accumulations of fronts, the value function changes very rapidly, but it is continuous: see Figure 19 for the enlarged accumulation region within the rectangle in Figure 18.

Let's explain more precisely what happens when fronts begin to bend around the accumulation region. After the front comes onto the lower boundary of set B, a new corner point $k(\tau)$ arises on the front [Figure 20(a)]. The point $k(\tau)$ moves up as τ increases. The point $b(\tau)$ of the front moves down along the outer side of the barrier line dg at a low rate. The part of the front between the points $b(\tau)$ and $k(\tau)$ propagates to the outside very slowly. The length of the curve $b(\tau)k(\tau)$ becomes smaller as τ increases. The part of the front between points $k(\tau)$ and $c(\tau)$

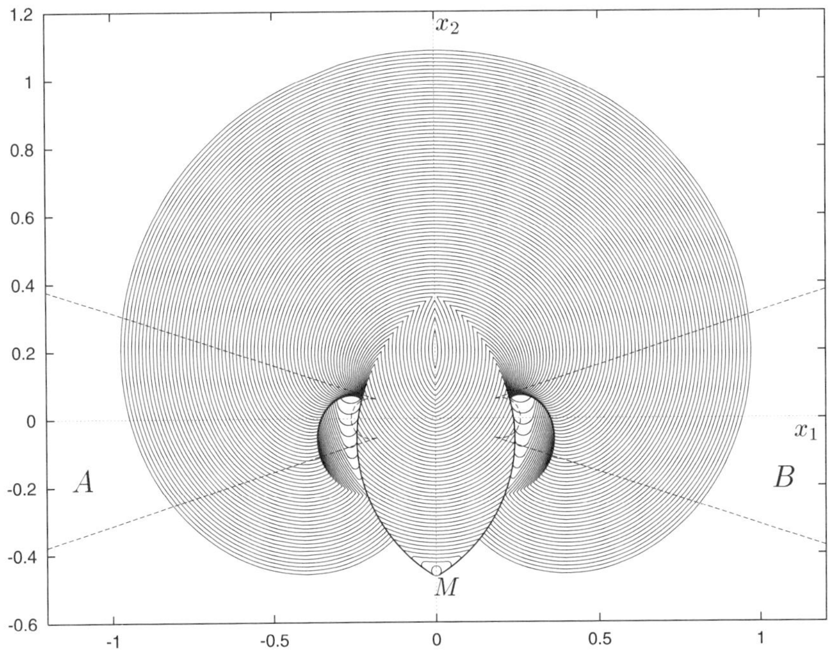

Figure 17: Level sets for the terminal set M on x_2-axis.

moves to the left rapidly enough, and the front collides with the barrier line dg at $\tau = 0.766$.

It is useful to compare this evolution of the front with the one from Figure 16, where the ends of the front move down along the corresponding barrier lines and then bend around them. Figure 20(b) shows the bending of the right barrier line. Here the point $k(\tau)$ is the endpoint of the front. It moves up along the outer side of the barrier line fe.

For the example in Figure 21, the following values of parameters are used: $w^{(1)} = 2$, and $R = 0.2$. The terminal set M is a regular 25-polygon inscribed into the circle of the radius 0.015 with the center at $(0.2, -0.4)$. The set Q is the triangle with the vertices $(-0.96, 0.8)$, $(0.96, 0.8)$, and $(0, -4.8)$. The step Δ is 0.001. The sets $W(20k\Delta, M), k = 1, 2, \dots,$ are depicted. In this example, the left barrier line finishes on the lower part of the boundary of set A, and the right barrier line finishes on the upper part of the boundary of set B. The structure of fronts near the end of the right barrier line is shown in Figure 22. The accumulation of fronts means very rapid changing of the value function. The twisted part of the left barrier line is depicted in Figure 23.

In all examples presented in this section, the barriers finish on the boundaries of the auxiliary sets A and B. This is consistent with the fact that every barrier line is a smooth, semipermeable curve of one of the families described in Section 4,

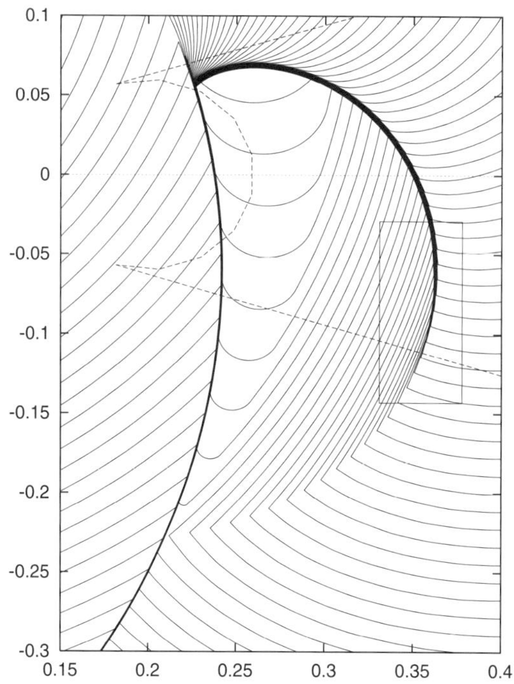

Figure 18: Enlarged fragment of Figure 17.

or it is composed of such smooth semipermeable curves. All barriers constructed approximate corresponding semipermeable curves very well.

In each concrete example, the terminal set generates a system of semipermeable curves of the first and second types, which determines the solvability set of the game of kind. Let us explain this with the help of the example from Figure 21. The semipermeable curves $p^{(2),1} \in \Lambda^{(2),1}$ and $p^{(1),2} \in \Lambda^{(1),2}$ emanated in the reverse time from the endpoints of the usable part of M do not intersect each other before they finish on the boundaries of the corresponding domains (Figure 24). The conjunction of $p^{(2),1}$ and $p^{(2),2}$ at the point b is smooth. It provides the semipermeability property of $p^{(2),1} \bigcup p^{(2),2}$ at b. The composite semipermeable curve $p^{(2),1} \bigcup p^{(2),2}$ does not intersect $p^{(1),2}$. Though the conjunction of the arc $as \subset p^{(1),2}$ and the curve $p^{(1),1}$ is not smooth, the semipermeability property is fulfilled (see [13], [14], [15], and [20] for analogous situations) at the conjunction point s. So the curve asr is the composite semipermeable curve of the first type. The composite semipermeable curves of the first and second types do not intersect each other. Further semipermeable curves are not being produced. One can prove, using the described mutual disposition of the semipermeable curves, that the solvability set of the game of kind is the whole plane. Numerical computation of the level sets of the value function confirms this fact.

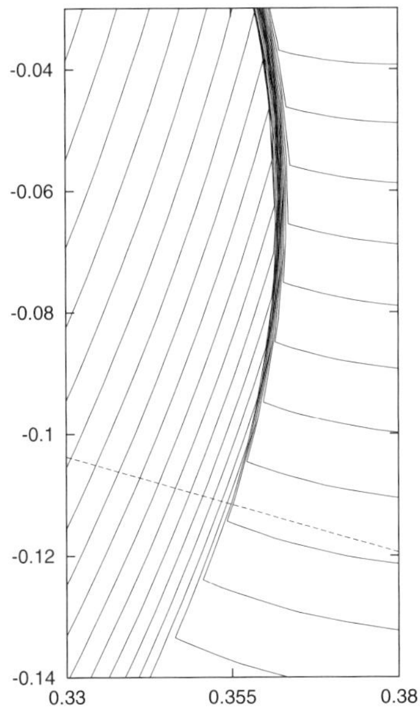

Figure 19: Enlarged fragment of Figure 18.

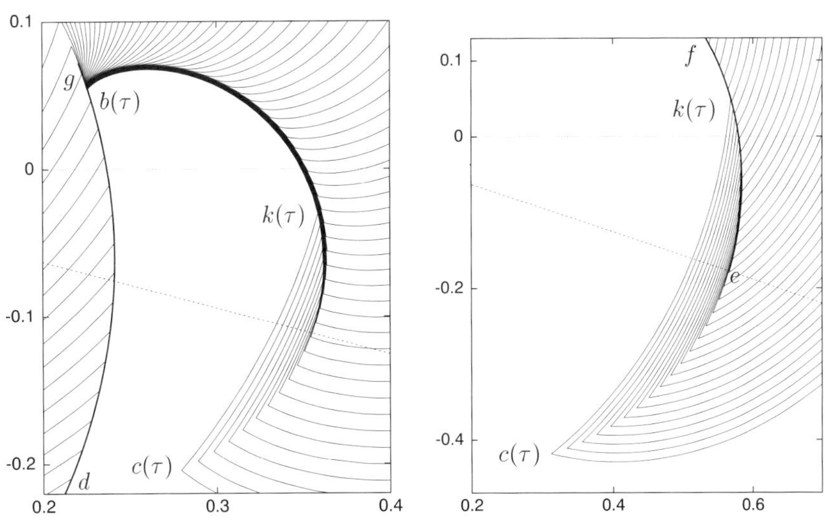

Figure 20: The evolution of fronts in regions of continuous and discontinuous change of the value function.

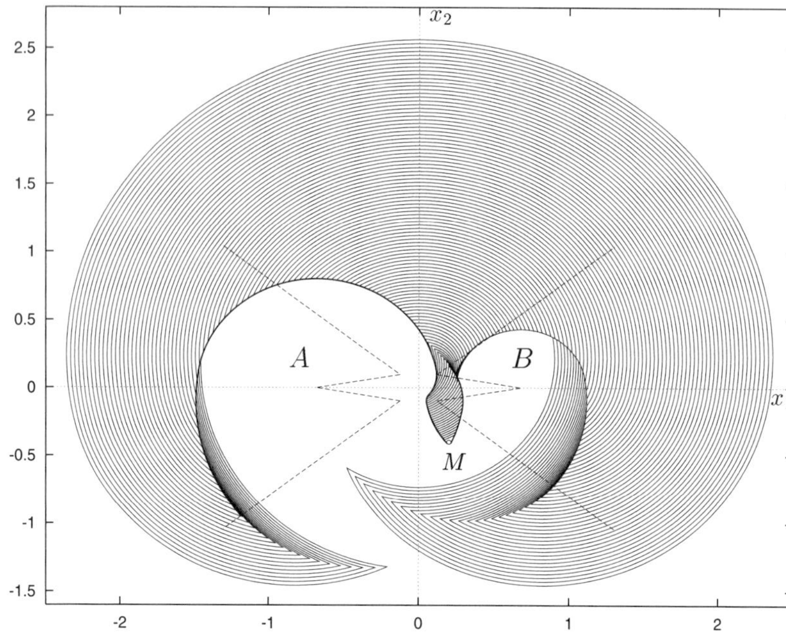

Figure 21: Level sets for a triangular constraint Q.

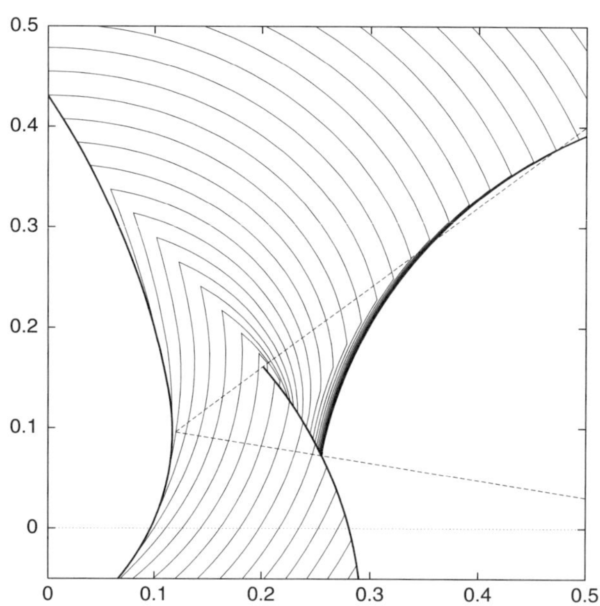

Figure 22: Enlarged fragment of Figure 21.

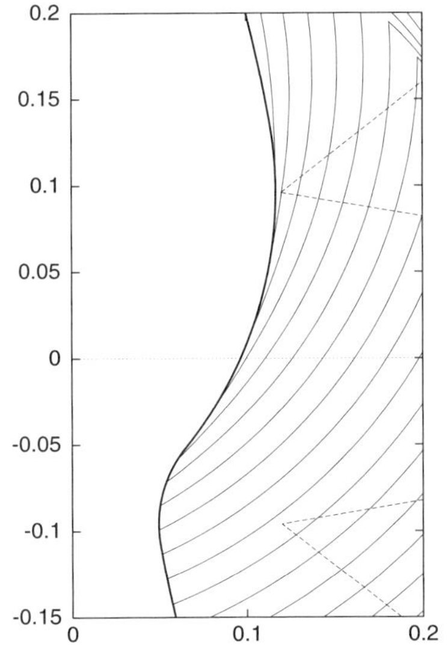

Figure 23: Enlarged fragment of Figure 21 near the left barrier.

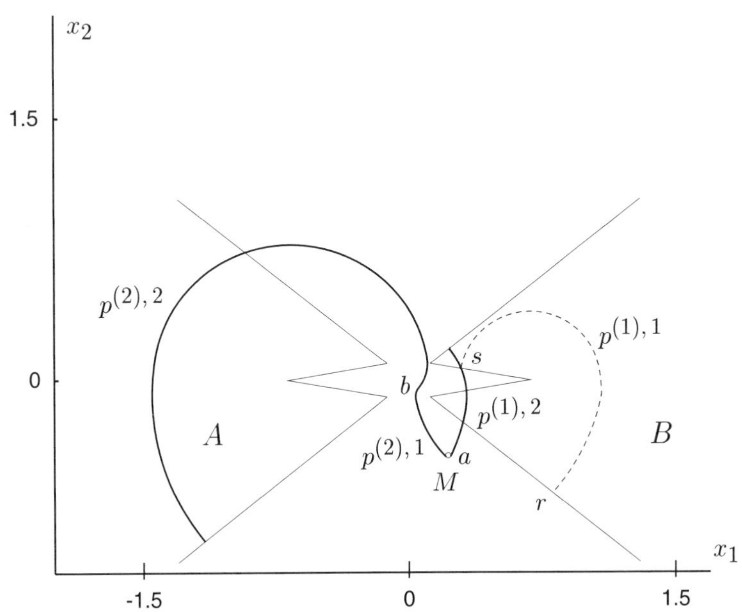

Figure 24: The system of semipermeable curves for the example with the triangle set Q.

The similar location of the accumulation region of fronts can be established for the example from Figure 17. One can see that the curve $p^{(2),1} \bigcup p^{(2),2}$ is the left barrier line in Figure 21. The curve $p^{(1),2}$ is the other barrier. It is interesting to observe that the curve $p^{(1),1}$ is not a barrier. The value function is continuous near this curve. Since *asr* is the composite semipermeable curve, the fronts cannot penetrate the curve $p^{(1),1}$ from above. Therefore, the arcs of the fronts that form the accumulation region remain above the curve $p^{(1),1}$. The fulfillment of this property for computed fronts can be considered as an additional verification of the algorithm.

5.2 Level Sets of the Value Function in the Surveillance–Evasion Game

Figures 25 to 32 present the computational results for $w^{(1)} = 1.7$. For figures 25 to 31 the set Q is a regular hexagon inscribed into the unit circle with the center in the origin. The vertices are (0, 1), (0.5, 0.87), (0.5, −0.87), (0, −1), (−0.5, −0.87), and (−0.5, 0.87). The half-angle $\bar{\theta}$ of the detection cone is 45^0, 143^0, 136.3^0, 130^0, 125.6^0, and 121^0 for Figures 25, 26, 27, 28, 29, and 30, respectively. According to the classification of [11], Figures 25, 26, 28, and 30 correspond to the cases I, VI, V, and IV, respectively. Figures 27 and 29 correspond to transient states. The step Δ is 0.01. In Figure 25, the sets $W(\tau)$ for $\tau = 2\Delta i$, $i = \overline{1, 65}$, are shown. In Figures 26 to 30, the sets $W(\tau)$ for $\tau = 10\Delta i$, $i = \overline{1, k}$, are depicted ($k = 16$, 18, 58 for Figures 26, 27, 28 to 30, respectively).

In Figure 25, player E escapes from all initial positions within the detection set. The computations are done up to $\tau = 1.3$.

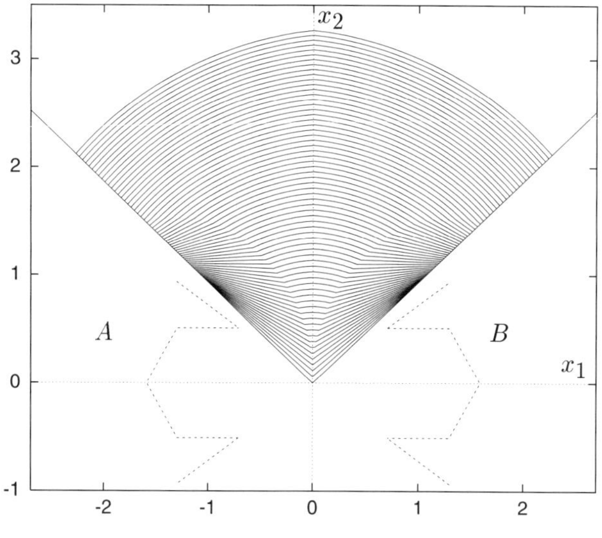

Figure 25: Level sets of the surveillance-evasion game for $\bar{\theta} = 45^\circ$.

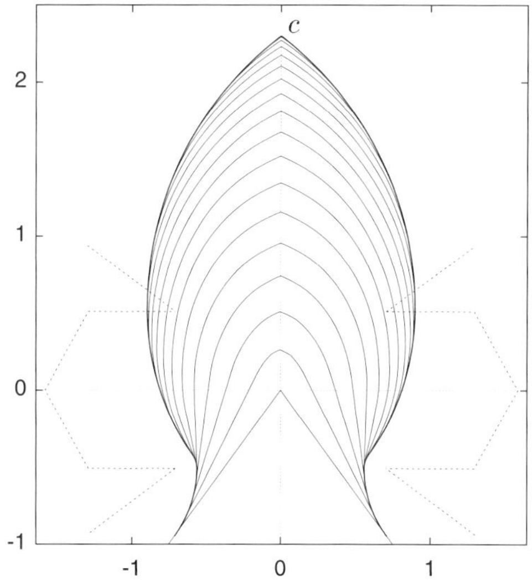

Figure 26: Level sets for $\bar{\theta} = 143°$.

In Figure 26, the escape zone is bounded by two symmetric barrier lines that emanate from the ends of the usable part and intersect at point c. The maximal escaping time is $\tau = 1.6$.

In Figures 27 to 30, the gradual change of the escape zone depending on the value of $\bar{\theta}$ is shown. Only the right half of each picture is presented.

The escape zone in Figure 27 is similar to the one in Figure 26. The barrier lines that bound the escape set touch the auxiliary sets A and B in the cusp points. The maximal escaping time is $\tau = 1.8$.

Further decreasing of the angle $\bar{\theta}$ yields Figure 28. The escape zone is bounded by mutually symmetric (with respect to the vertical axis) curves aef and $a'e'f'$. The arcs ae and $a'e'$ are the parts of the barrier lines ab and $a'b'$ that finish on the boundary of auxiliary sets B and A, respectively. The arc efe' is a limit of fronts as $\tau \to \infty$.

In Figure 30, player E can escape from any point within the detection set. The fronts go around the endpoint b of the barrier ab. After passing the point a, the end of the front propagates along the line $\theta = \bar{\theta}$ with a constant velocity. So the detection cone can be completely filled out with the fronts.

The case shown in Figure 29 is a transient state from the case in Figure 28 to the case in Figure 30. The boundary of the escape zone is a limit of fronts as $\tau \to \infty$. Here the point e (see Figure 28) coincides with the point a.

In Figure 31, the fronts for an asymmetric detection set are presented. The left inclination angle corresponds to case IV in the classification of [11]; the right one corresponds to case VI.

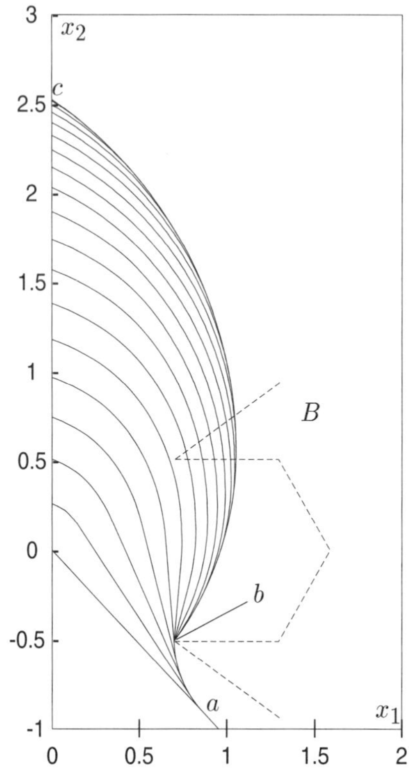

Figure 27: Level sets for $\bar{\theta} = 136.3°$.

The escape zone in Figure 32 is similar to the one in Figure 26, but a 25-polygon inscribed into the unit circle is used as set Q instead of the 6-polygon.

In [11], the solution to the surveillance–evasion game posed as a game of kind is given for the case where the set Q is a circle. This solution is used here to verify the correctness of our computations. The verification is based on the geometric form of solutions presented on pictures in [11]. The geometric classification of solutions is actually based on the type of disposition of semipermeable curves emanated from endpoints of the usable part with respect to the sets A and B. This gives us the possibility to obtain solutions to the game of kind for arbitrary sets Q.

When solving the game of kind, semipermeable curves are emitted in the reverse time from endpoints of the usable part so that the semipermeability property (the property of barrier) holds at these points. In the game considered, the negative sides of the semipermeable curves emitted should face the usable part. To provide this, the curve of the second type must emanate from the right endpoint of the usable part, and the curve of the first type must emanate from the left endpoint. The semipermeable curves with such properties are absent for the example in Figure 25. For the examples in Figures 26 to 30 and 32, the semipermeable curve

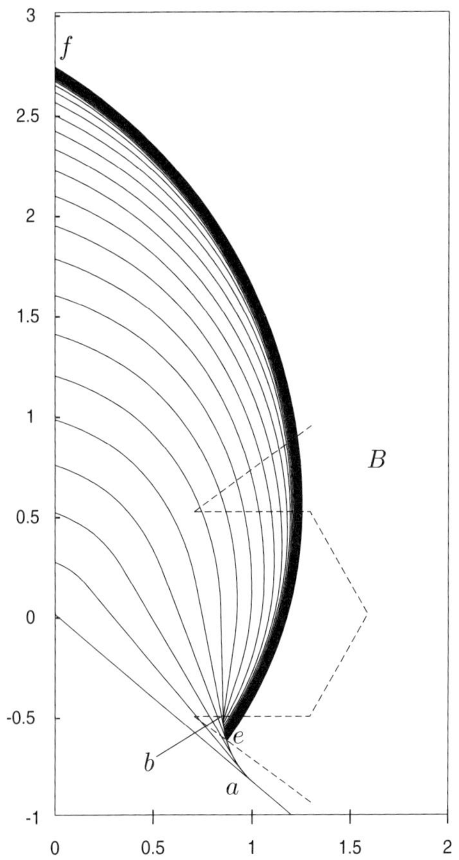

Figure 28: Accumulation of fronts; $\bar{\theta} = 130°$.

of the family $\Lambda^{(2),1}$ emanated in the reverse time from the right endpoint a of the usable part is proper. Let's denote it by $p^{(2),1}$. The curve $p^{(2),1}$ is extended to the boundary of the domain of $\Lambda^{(2),1}$. In Figures 26 to 30 and 32, the curve ab coincides practically with $p^{(2),1}$.

For the examples in Figures 26 and 32, the extended curve can be smoothly continued by a curve $p^{(2),2}$ of the family $\Lambda^{(2),2}$. After this continuation, a smooth, composite semipermeable curve of the second type is obtained. Let's denotc it by $g^{(2)}$. Because of the symmetry, there is a smooth, composite semipermeable curve $g^{(1)}$ of the first type that emanates from the left endpoint a' of the usable part. The curves $g^{(2)}$ and $g^{(1)}$ intersect at a point that lies on the x_2-axis. This yields a closed barrier that bounds the solvability set of the game of kind (the escape set). The optimal guaranteeing time for attaining the terminal set (escaping time) is finite for any initial point on the closed barrier. The above-mentioned facts are completely

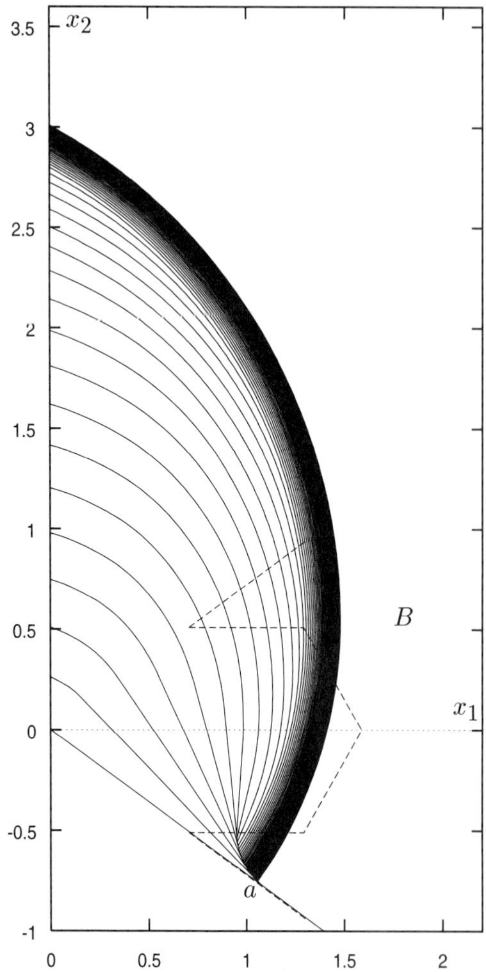

Figure 29: Accumulation of fronts; $\bar{\theta} = 125.6°$.

confirmed by the computation of the fronts shown in Figures 26 and 32. The curve bc lies on the curve $p^{(2),2}$.

For the examples in Figures 27 to 30, the semipermeable curve $p^{(2),1}$ can not be smoothly continued beyond the endpoint. Moving along the curve $p^{(2),1}$ from the endpoint toward the point a, one looks for a point from which a semipermeable curve $p^{(2),2}$ of the family $\Lambda^{(2),2}$ emanates so that the composite curve $g^{(2)}$ formed by the initial part of $p^{(2),1}$ and the curve $p^{(2),2}$ would possess the semipermeability property at the sewing point. One can establish that such sewing points can only lie on the boundary of the domain of $\Lambda^{(2),2}$. A sewing point with the above property exists for the case shown in Figures 27 to 29 and is absent for the case in Figure 30.

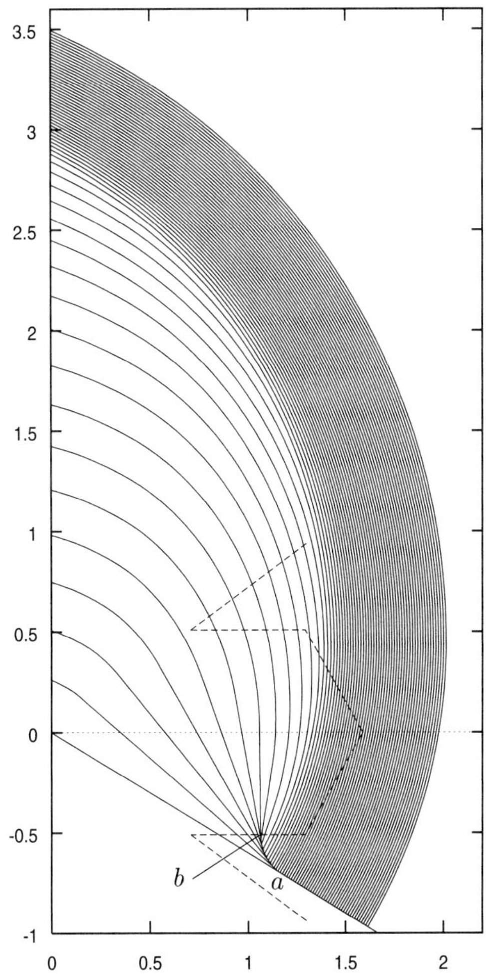

Figure 30: Unbounded escape zone; $\bar{\theta} = 121^\circ$.

One can see that point e in Figure 28 is actually the sewing point discussed, and the curve aef is close to the piecewise-smooth, composite semipermeable curve $g^{(2)}$. The solution to the game of kind is determined by the closed barrier. However, in contrast to the examples in Figures 26, 27, and 32, the escaping time is infinite for any point belonging to the common part of the barrier and symmetric curves $p^{(2),2}$ and $p^{(1),1}$. This nontrivial fact was mentioned in [11], and it is confirmed by the accumulation of fronts as $\tau \to \infty$ in Figure 28.

It is noted in [11] that the solvability set of the game of kind coincides with the whole detection cone for cases similar to that shown in Figure 30.

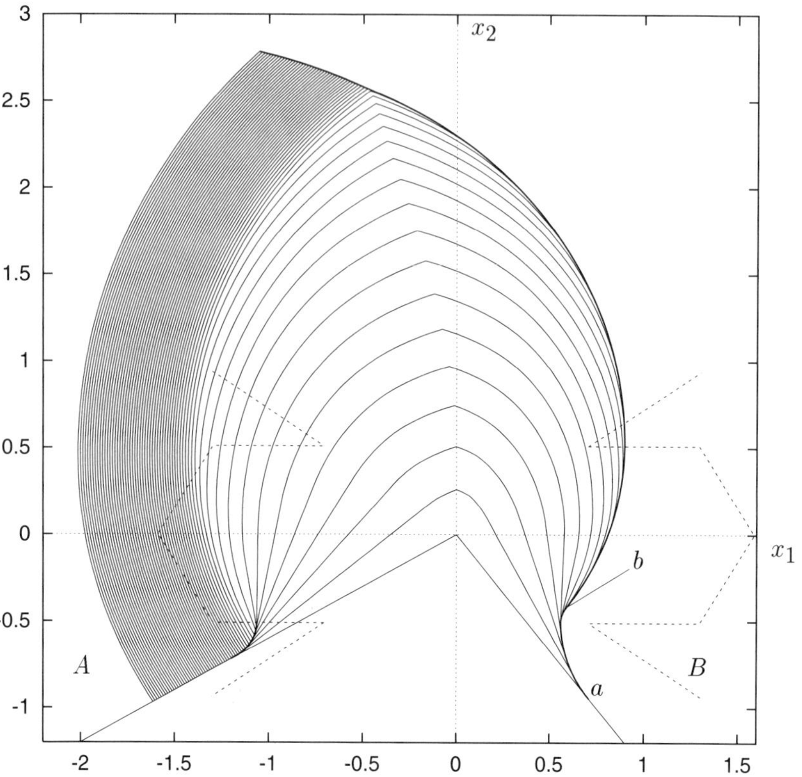

Figure 31: Level sets for an asymmetric detection cone.

The case presented in Figure 29 is degenerate. Here, the point of nonsmooth sewing of two semipermeable curves coincides with point a. The boundary of the solvability set of the game of kind is determined by two curves that are symmetric to each other: the curve $p^{(2),2}$ of the family $\Lambda^{(2),2}$ emanated from point a and the curve $p^{(1),1}$ of the family $\Lambda^{(1),1}$ emanated from point a'. This is consistent with the computed fronts in Figure 29.

In the case corresponding to Figure 31, a semipermeable curve of the family $\Lambda^{(2),1}$ emanates from the point a, and it is smoothly continued by a curve of the family $\Lambda^{(2),2}$. The resulting smooth, composite curve determines the boundary of the solvability set of the game of kind.

Acknowledgments

This research was partially supported by the Russian Foundation for Basic Researches under Grant No. 00-01-00348.

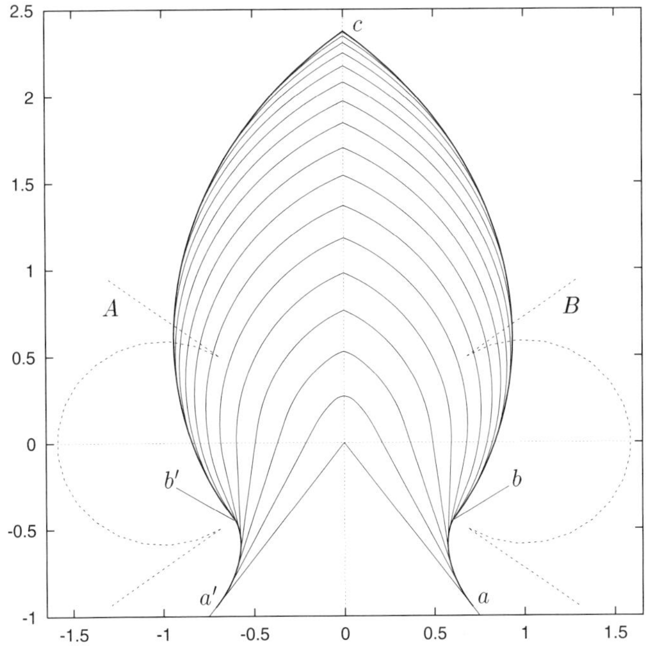

Figure 32: Level sets for a circular constraint Q.

REFERENCES

[1] Bardi, M. and Falcone, M. An approximation scheme for the minimum time function, *SIAM J. Contr. Opt.*, **28**, pp. 950–965, 1990.

[2] Bardi, M., Falcone, M., and Soravia, P. *Numerical Methods for Pursuit–Evasion Games via Viscosity Solutions*, Preprint no. 38, Universita degli di Roma "La Sapienza," Dipartimento di Matematica "Guido Castelnuovo," 1997.

[3] Bernhard, P. Conditions de coin pour les jeux différentiels, *Seminaire sur les Jeux Differentiels*, Centre d'Automatique, Paris, 1971.

[4] Breakwell, J. V. Zero-sum differential games with terminal payoff, pp. 70–95, in *Differential Games and Applications*, P. Hagedorn, H. W. Knobloch, and G. J. Olsder, eds., Lecture Notes in Control and Information Sciences, Springer-Verlag, Berlin, 1977.

[5] Cardaliaguet, P., Quincampoix, M., and Saint-Pierre, P. Some algorithms for differential games with two players and one target, *RAIRO-Modelisation-Mathematique-et-Analyse-Numerique*, **28**(4):441–461, 1994.

[6] Isaacs, R. *Differential Games*, John Wiley, New York, 1965.

[7] Ivanov, V. A., Taras'yev, A. M., Ushakov, V. N., and Khripunov, A. P. The toreador problem, *J. Appl. Math. Mech.*, **57**(3):419–425, 1993.

[8] Krasovskii, N. N. and Subbotin, A. I. *Positional Differential Games*, Nauka (in Russian), Moscow, 1974.

[9] Krasovskii, N. N. and Subbotin, A. I. *Game-Theoretical Control Problems*, Springer-Verlag, New York, 1988.

[10] Lewin, J. and Breakwell, J. V. The surveillance–evasion game of degree, *J. Opt. Th. Appl.*, **16**(3–4):339–353, 1975.

[11] Lewin, J. and Olsder, G. J. Conic surveillance evasion, *J. Opt. Th. Appl.*, **27**(1):107–125, 1979.

[12] Merz, A. W. The homicidal chauffeur—A differential game, Ph.D. dissertation, Stanford University, 1971.

[13] Patsko, V. S. Problem of Kind in Linear Differential Games of the Second Order, Ph.D. Dissertation (in Russian), Ural State University, Sverdlovsk, 1973.

[14] Patsko, V. S. The problem of quality in second-order linear differential games, pp. 167–227, *Differential Games and Control Problems*, A. B. Kurzhanskii, ed., Institute of Mathematics and Mechanics (in Russian), Sverdlovsk, 1975.

[15] Patsko, V. S., and Turova, V. L. *Numerical Solution of Two-dimensional Differential Games*, Preprint, Institute of Mathematics and Mechanics, Ekaterinburg, 1995.

[16] Patsko, V. S. and Turova, V. L. Minimum-time problem for linear second-order conflict-controlled systems, *Proc. UKACC Intl. Conf. CONTROL'96*, Exeter, UK, September 2–5, 1996, Conf. Pub. no. 427, pp. 947–952, IEE 1996.

[17] Patsko, V. S. and Turova, V. L. Numerical solutions to the minimum-time problem for linear second-order conflict-controlled systems, pp. 327–338, *Proc. Seventh Intl. Colloquium on Differential Equations*, D. Bainov, ed., Utrecht, the Netherlands, 1997.

[18] Subbotin, A. I. *Generalized Solutions of First-Order PDEs: The Dynamical Optimization Perspective*, Birkhäuser, Boston, 1995.

[19] Subbotin, A. I. and Patsko, V. S., eds., *Algorithms and Programs for Solving Linear Differential Games*, Institute of Mathematics and Mechanics (in Russian), Sverdlovsk, 1984.

[20] Turova, V. L. Nonlinear differential game of kind in the plane, pp. 91–116, in *Investigations of Minimax Control Problems*, A. I. Subbotin and V. S. Patsko, eds., Institute of Mathematics and Mechanics, (in Russian), Sverdlovsk, 1985.

[21] Ushakov, V. N. On the problem of constructing stable bridges in a differential game of approach and avoidance, *Eng. Cybernetics*, **18**(4):16–23, 1981.

“The Tragedy of the Commons” Modelled by Large Games

Agnieszka Wiszniewska-Matyszkiel
Institute of Applied Mathematics and Mechanics
Warsaw University
Warsaw, Poland

Abstract

This paper deals with a dynamic game of extraction of common, renewable resources by players constituting a nonatomic measure space.

The main issue is the existence and properties of equilibria. Another issue is a taxation system, with socially acceptable properties, enforcing an equilibrium that is Pareto optimal and at least as good as the one arising after the introduction of private property.

1 Introduction

The expression *tragedy of the commons* was first used by Hardin in [10] to express the fact that when common resources are extracted, the market mechanism inevitably fails. When an ecological system is exploited without precisely defined property rights, individually rational decisions do not lead to a collectively rational result but may cause a common disaster. This applies to common pastures, fisheries, air and water pollution, and even demography.

However, in [6] Coase claimed that, if the property rights are clearly defined and there is a possibility of bargaining, the problem of externalities is automatically solved. His reasoning was proven to be true in some special cases involving two players. The statement of this, equipped with some assumptions, is now known as the Coase theorem.

From that time, management of common resources became one the main issues of environmental economics. Many papers on this subject and related to the framework of the present paper are included, for example, in [9] and [17].

Although the game-theoretic approach has been applied to this problem (see, e.g., Levhari and Mirman [14], Fisher and Mirman [8], Dutta and Sundaram [7], the survey by Kaitala [12] and articles contained in the volume edited by Carraro and Filar [5]), the games considered involved only a finite number of players, usually two. Games with a nonatomic measure space of players (introduced by

Schmeidler in [16]; continued, e.g., by Balder [3], Wiszniewska-Matyszkiel [22], and in an anonymous version by Mas-Colell [15] and Wieczorek [18]) were not applied to the tragedy of the commons.

It seems that dynamic games with a nonatomic space of players have been examined only in [20], [13], and [19].

One aspect of games with a nonatomic measure space of players is especially essential in games modelling the effect of human activity on ecosystems, and that does not occur in finite games: *negligibility* of the players. This aspect reflects some real situations: the sense of negligibility in the case of global ecology is, together with natural selfishness, one of the most important generators of the *tragedy*. This negligibility makes real sense. For instance, one person's decision to refrain from using deodorants containing freon has, in fact, no impact on the shrinking of the ozone layer, but an aggregate of such decisions does have an impact. We can find many similar relations: between house owners' decisions to use environment-friendly energy instead of coal and the greenhouse effect, between fishermen's decisions to decrease catch and the threat of extinction of the exploited species of fish, or between factories' decisions to use sulfer dioxide filters and acid rains. In all these cases one player's decision is negligible while their aggregate is not. In finite games such an effect is hard to model.

The present paper contains some results of the thesis of the author [20] and is a continuation of it. The objective of this paper is to present a model describing the exploitation of a common ecosystem by a large number of players, formally constituting a measure space, which allows the phenomena of negligibility of single players to be captured, which is otherwise hard or even impossible to model. In this context the existence and properties of various types of equilibria are investigated, as well as ways and means to enforce optimality. Those means are interpreted in this paper as taxation systems also allowing for the subsidy of certain socially desired activities.

The paper is constructed as follows. The model is defined in Section 2. Section 3 is devoted mainly to the existence and properties of equilibria: a relation between a dynamic equilibrium and a family of static equilibria is established in Theorem 3.1, while Theorem 3.2 gives a sufficient condition for the existence of a static equilibrium. Since we cannot expect a dynamic equilibrium to be optimal, we solve a related problem by finding an optimal control function (aggregate extraction) and decomposing the primary dynamic game into a family of static games. The optimal control function is one maximizing a functional called a dynamic social welfare function (Section 4). The static games are designed to enforce the result of this maximization with tax-subsidy systems changing the static payoffs (Section 5). We prove the existence of equilibria in those modified games. In Section 6 we examine three optimality concepts in the dynamic game: usual Pareto optimality, specific to our problem optimality in the sense of property, and optimality in the sense of maximizing the aggregate payoff. Two optimality theorems are formulated. Two examples illustrating the theoretical results are presented in Section 7; both describe an exploitation of a rainforest: stable or unstable.

2 Formulation of the Model

A *static game* is defined as a system

$$\mathfrak{G} = ((\Omega, \Im, \mu), \mathbb{S}, s, r)$$

consisting of a measure *space of players* $(\Omega, \Im, \mu)$, a *set of strategies* $\mathbb{S}$, a *correspondence* $s : \Omega \multimap \mathbb{S}$ *of* the players' *available strategies* and *the players' payoff function* $r : \mathrm{Gr}(s) \times \mathcal{S} \to \mathbb{R} \cup \{-\infty\}$, where $\mathcal{S}$ denotes the set of all measurable a.e. selections of s, in this paper called *static profiles*.

A *dynamic game* is defined in a more complex way: it is a system

$$\mathfrak{G} = (\mathbb{X}, x_0, T, (\Omega, \Im, \mu), \mathbb{S}, S, R, \varphi, \Psi)$$

consisting of a set of states, an initial state, a time horizon, a measure space of players, a set of (static) strategies, a correspondence of the players' (static) strategies, a players' instantaneous utility function, a regeneration function, and a discounting function.

The elements of the game are defined and interpreted as follows:

The set $\mathbb{X} \subset \mathbb{R}^n_+$ will consist of all *possible states*; here the dimension n of the space is understood as the number of natural *resources* existing in nature and relevant to the model.

The *initial state* of the system is $x_0 \in \mathbb{X}$.

The *time horizon* will be $0 < T \leq +\infty$.

Every function $X : [0, T) \to \mathbb{X}$ such that $X(0) = x_0$ will be called a *trajectory of the system* [$X(t)$ will denote the state of the system at time t].

The *players* are assumed to form a nonatomic normed measure space $(\Omega, \Im, \mu)$.

The bounded set $\mathbb{S} \subset \mathbb{R}^m$ will consist of all *static strategies*. In the simplest model $m = n$ and the ith coordinate of a strategy denotes intensity of extraction of the ith resource. In more complex models m might be the number of activities available to the players and affecting the system: then the ith coordinate represents intensity of the ith activity. Whatever interpretation is chosen, an element of $\mathbb{S}$ will be called *extraction*.

A nonempty-valued correspondence $S : \Omega \times \mathbb{X} \multimap \mathbb{S}$ is a *correspondence of the players' strategies*. Instead of $S(\omega, \cdot)$, we shall write S_ω. Then $S_\omega(x)$ is understood as the *set of strategies available to player* ω *at* (state of the system) x. If the state of the system is x, then every $d \in S_\omega(x)$ is an *individual static extraction* (or strategy) *available to player* ω *at* (state) x. If the trajectory of the system is X, then any measurable function $D : [0, T) \to \mathbb{S}$ such that $D(t) \in S_\omega(X(t))$ is *ω's dynamic strategy available to him at X*.

Before defining the remaining elements of the game, we shall specify, for convenience, the σ-fields associated with the defined objects: for subsets of $\mathbb{R}^k$ (set of states, set of strategies, and set of aggregate strategies—the last not defined yet) we shall cope with Borel σ-fields, for the time interval $[0, T)$—the Lebesgue-measurable sets, $\Im$ for the space of players, and product σ-fields for products. Therefore, measurability will always be understood with respect to the corresponding σ-fields, as stated above.

Any measurable function $\delta : \Omega \to \mathbb{S}$ such that for almost every ω, $\delta(\omega) \in S_\omega(x)$ will be called a *static (extraction) profile* (*profile* for short) *available at* x; $\delta(\omega)$ is the *player* ω*'s strategy at* (profile) δ.

The set of all measurable functions $\delta : \Omega \to \mathbb{S}$ will be denoted by Σ.

If $u_\delta = \int_\Omega \delta(\omega) d\mu(\omega)$ for a static profile δ, then u_δ is called *aggregate (total) extraction* of δ (we omit the subscript if it does not lead to confusion).

For a given state x, the set $Y(x)$ is equal to $\int_\Omega S_\omega(x) d\mu(\omega)$ (the Aumann [1] integral): it is the *set of amounts of aggregate extraction available at* (state) x, so the correspondence $Y : \mathbb{X} \multimap \text{co}\,\mathbb{S}$ may be called the *correspondence of available aggregate extraction*. For simplicity of notation, we will denote $\text{co}\,\mathbb{S}$ by $\mathbb{Y}$.

The next element of the game is a function $R : \Omega \times \mathbb{Y} \times \mathbb{S} \times \mathbb{X} \times [0, T) \to \mathbb{R} \cup \{-\infty\}$ called the *players' instantaneous payoff function*. For any fixed $\omega \in \Omega$, R_ω, as a function of the remaining arguments y, d, x, t, is called the *individual instantaneous payoff function of player* ω. The dependence of the instantaneous payoff function on time does not include any discounting, which is reflected by an additional element of the model, the discounting function.

A function $\varphi : \mathbb{X} \times \mathbb{Y} \to \mathbb{R}^n$ is called a *regeneration function* and its meaning is as usual: the behavior of the environment is ruled by the equation $\dot{X}(t) = \varphi(X(t), U(t))$.

Throughout this paper we shall assume that the regeneration function fulfills a Lipshitz condition. This is only necessary for the existence and uniqueness of a solution of the above equation. The results of this paper remain valid without this condition if we have only the existence of a solution, but it seems unnatural and inconvenient to have to cope with nonunique trajectories.

Any function $U : [0, T) \to \mathbb{Y}$ will be called a *control function* (in fact, it is a control function of Ω as a whole).

Given a measurable control function U, we shall call the unique absolutely continuous trajectory fulfilling the equation $\dot{X}(t) = \varphi(X(t), U(t))$ *corresponding to* U and it will be denoted by X_U. A measurable control function U is called *admissible* if there exists the absolutely continuous corresponding trajectory X_U and, for almost all t, $U(t) \in Y(X_U(t))$ [i.e., $U(t)$ is available at $X_U(t)$].

An integrable function $\Psi(t) : [0, T) \to (0, +\infty)$ will be a *discounting function*.

A measurable function $\Delta : \Omega \times [0, T) \to \mathbb{S}$ is called a *dynamic extraction profile* (*dynamic profile* for short) if, for almost every t, the function $\Delta(\cdot, t)$ is a static extraction profile available at $X^\Delta(t)$, where X^Δ denotes the trajectory corresponding to the control function U_Δ defined by $U_\Delta(t) = u_{\Delta(\cdot,t)}$.

Finally we describe the player's actual objective.

The *payoff* function of player ω is a function of two statistics of a dynamic profile: ω's own strategy and the aggregate strategy. The payoff is equal to the discounted utility integrated over time:

$$\Pi_\omega(U, D) = \int_0^T R_\omega(U(t), D(t), X(t))\Psi(t)dt.$$

If player ω expects some admissible function of aggregate extraction U, he maximizes $\Pi_\omega(U, D)$.

A *static game at* time $\bar{t}$ *and* state $\bar{x}$, *associated with* a dynamic game $\mathfrak{G}$, is a static game $G_{\bar{t},\bar{x}}$ defined by

$$G_{\bar{t},\bar{x}} = ((\Omega, \Im, \mu), \mathbb{S}, s, r),$$

with the same space of players, the set of strategies equal to the set of static strategies of $\mathfrak{G}$, the correspondence of available strategies s defined by $s(\omega) = S(\omega, \bar{x})$ (the correspondence of available strategies the same as the correspondence of strategies available in $\mathfrak{G}$ at $\bar{x}$), and r defined by $r(\omega, d, \delta) = R(\omega, d, u_\delta, \bar{x}, \bar{t})$ (the players' payoff function equal to the instantaneous payoff function at time $\bar{t}$ and state $\bar{x}$ in $\mathfrak{G}$).

We need the definition of a general dynamic game to cope with two kinds of dynamic games: the tragedy of the commons game—the original game of our model (defined below) and its modifications resulting from introduction of tax-subsidy systems (to be defined in Section 5).

A *tragedy of the commons game* is a dynamic game $\mathfrak{G} = (\mathbb{X}, x_0, T, (\Omega, \Im, \mu), \mathbb{S}, S, R, \varphi, \Psi)$ with the instantaneous payoff function independent of time. Hence, we may introduce the notation $P(\omega, u, d, x) = R(\omega, u, d, x, t)$ for any t. The so-defined function P acts from $\Omega \times \mathbb{Y} \times \mathbb{S} \times \mathbb{X}$ into $\mathbb{R} \cup \{-\infty\}$ and is called, for distinction, the *players' instantaneous utility function* or, for short, *utility*.

3 Equilibrium

In this section we deal with a dynamic game and static games associated with it, so we are going to deal with two kinds of equilibria: dynamic and static (in static games associated with the original dynamic game).

A *Cournot–Nash equilibrium* is, in general, a profile such that almost no player has an incentive to change his strategy unless the remaining players have changed theirs. In our framework this is equivalent to saying "unless the aggregate has changed." Formally, we have a sequence of definitions.

Definition 3.1. A static equilibrium at (time) $\bar{t}$ and (state) $\bar{x}$ is a static profile δ such that

$$\text{for a.e. } \omega, \quad \delta(\omega) \in \text{Argmax}_{d\in S_\omega(\bar{x})} R_\omega\left(\int_\Omega \delta(\omega)d\mu(\omega), d, \bar{x}, \bar{t}\right).$$

Definition 3.2. The set of static best responses of player ω to $u \in \mathbb{Y}$ (representing the aggregate extraction) at (time) $\bar{t}$ and (state) $\bar{x}$ is defined by

$$B_\omega(u, \bar{x}, \bar{t}) = \text{Argmax}_{d\in S_\omega(\bar{x})} R_\omega(u, d, \bar{x}, \bar{t}),$$

and the set of static aggregate best responses to u at (time) $\bar{t}$ and (state) $\bar{x}$ is defined by

$$\overline{B}(u, \bar{x}, \bar{t}) = \int_\Omega B_\omega(u, \bar{x}, \bar{t}) d\mu(\omega).$$

Thus, so we have defined the *correspondence of static best responses* and the *correspondence of static aggregate best responses.*

The static equilibrium profile δ may be equivalently defined by the condition

$$\int_\Omega \delta(\omega) d\mu(\omega) \in \overline{B}\left(\int_\Omega \delta(\omega) d\mu(\omega), \bar{x}, \bar{t}\right),$$

that is, it is a profile whose aggregate is a fixed point of the correspondence $\overline{B}(\cdot, \bar{x}, \bar{t})$.

Definition 3.3. The dynamic best-response correspondence is defined by: $\mathbf{B}_\omega(U) = \mathrm{Argmax}\left\{\Pi_\omega(U, D) \mid D : \mathbb{T} \to \mathbb{S} \text{ measurable}, D(t) \in S_\omega(X_U(t))\right\}$ (where U is any admissible control function). The dynamic aggregate best-response correspondence is defined by

$$\overline{\mathbf{B}}(U) = \left\{\zeta : \mathbb{T} \to \mathbb{Y} \text{ measurable} \mid \text{for a.e. } t, \zeta(t) \in \int_\Omega \mathbf{B}_\omega(U)(t)\, d\mu(\omega)\right\}.$$

Definition 3.4. A *dynamic equilibrium* is a dynamic profile Δ such that $\int_\Omega \Delta(\omega, t) d\mu(\omega)$ is a fixed point of the correspondence $\overline{\mathbf{B}}$.

3.1 Relations between Static and Dynamic Equilibria

There are some relations between dynamic and static best-reply correspondences, and between dynamic and static equilibria, which do not usually appear in games with a finite space of players.

We can state the following theorem, which holds not only for the tragedy of the commons game, but for a wider class of similar dynamic games with a continuum of players (see [23] for a more general theorem and [24] for a discrete-time version).

Theorem 3.1. *Let's assume that almost every function P_ω is upper semicontinuous in player's ω own strategy, bounded from above, and measurable. Further, assume that almost every correspondence S_ω is measurable and has compact values.*

(a) *Let U be any admissible measurable control function. Let X be the corresponding absolutely continuous trajectory, and let Δ be a dynamic profile for which the payoff of almost every player is finite. For almost every ω, the following statements (*) and (**) are equivalent:*

(*) *For almost every t, the vector $\Delta(\omega, t)$ is an optimal strategy of player ω in the static game with the state of system $X(t)$ given $u = U(t)$.*

(**) *The function $\Delta(\omega, \cdot)$ is ω's optimal strategy in the dynamic game given the measurable control function U and the corresponding absolutely continuous trajectory of system X.*

(b) *Moreover, for every dynamic profile Δ for which the payoff is finite, we have the following equivalence:*
The dynamic profile Δ constitutes a dynamic equilibrium if and only if for almost every t the static profile $\Delta(\cdot, t)$ is an equilibrium in the static game at time t and the state of system $X(t)$ (where X is the trajectory corresponding to U_Δ defined by $U_\Delta(t) = u_{\Delta(\cdot,t)}$).

Proof.

(a) Since the players are negligible, one player's strategy affects neither the aggregate nor the system. This implies that a player cannot transfer utility in time, so the optimization at every time can be done independently, with $U(t)$ and $X(t)$ being parameters. Hence, a dynamic profile whose static profiles are optimal in static games is optimal in the dynamic game.
Let us fix any ω for whom the assumptions hold. Let's take an optimal dynamic profile Δ. Almost every static strategy of player ω, $\Delta(\omega, t)$ is optimal if there exists a measurable function $D : [0, T) \to \mathbb{S}$ such that almost every $D(t)$ is in ω's static best response to $U(t)$ at the state of system $X(t)$. Since the trajectory X is continuous and the correspondence S_ω is measurable, $S_\omega \circ X$ is measurable.
The correspondence $t \mapsto \operatorname{Argmax}_{(u,d,x)\in\{U(t)\}\times S_\omega(X(t))\times\{X(t)\}} P_\omega(u, d, x)$ is measurable (i.e., has measurable graph), for the correspondence $\Phi : [0, T) \multimap \mathbb{Y} \times \mathbb{S} \times \mathbb{X}$ defined by $\Phi(t) = \{U(t)\} \times S_\omega(X(t)) \times \{X(t)\}$ is a measurable correspondence of complete measure space into a complete separable metric space and P_ω is a measurable functional of this metric space ([11], p. 60).
Since P_ω is upper semicontinuous in d and S_ω has nonempty compact values, the maximum is attained, so the values of the defined Argmax correspondence are nonempty. Therefore, by Aumann's theorem (see [2]), there exists a measurable a.e. selection, which ends this part of the proof.

(b) Let Δ be a dynamic profile for almost every t equal to a dynamic equilibrium. It means that for U almost everywhere equal to U_Δ, $\Delta(\omega, \cdot)$ is a dynamic best response to U. Let X be the trajectory corresponding to U.
By (a), the latter condition is equivalent to the following: for a.e. t, $\Delta(\omega, t)$ is ω's static best response to $U(t)$. Hence, both conditions are equivalent to the following: for a.e. t, $\Delta(\cdot, t)$ is an equilibrium in the static game at t and $X(t)$. □

Theorem 3.1 enables us to neglect the dynamic structure of the game while searching for equilibria and best-response sets. We have reduced the problem of finding a dynamic equilibrium to finding some static equilibria corresponding to the state of the system.

Moreover, this theorem obviously implies that the dynamic equilibrium is independent of the discounting function.

Remark 3.1. We actually proved that if the assumptions of Theorem 3.1 are fulfilled, then for every measurable function U there exists ω's dynamic best response to U.

We have the following existence and uniqueness result.

Corollary 3.1. *If the function P_ω is upper semicontinuous and strictly quasi-concave in the player's own strategy, bounded from above, the values of the correspondence S_ω are compact convex, and both P_ω and S_ω are measurable, then for every measurable function U for which ω has a dynamic strategy with finite payoff, ω has exactly one dynamic best response to U (up to t in a negligible set), yielding finite payoffs.*

Proof. Static best-response sets are singletons for all assumed values of $X(t)$ and $U(t)$. By Theorem 3.1, values of a dynamic best response to U are almost everywhere equal to some static best response at t and $X(t)$ to $U(t)$, so there is at most one function D that belongs to the set of dynamic best responses to U. By Remark 3.1, such a D does exist. □

This fact is a simple corollary of Theorem 3.1 and Remark 3.1. It cannot be derived from a Hamiltonian-type argument, like strict diagonal concavity-convexity of [4], since those concepts do not fit our framework: we do not even assume continuity of the payoff functions.

Strategies yielding payoff equal to $-\infty$ are dominated, so we can remove them from our considerations. The only problem appears when every possible payoff is $-\infty$.

From now on we shall assume that even if the only possible payoff of player ω is $-\infty$, for some aggregate extraction function U and the corresponding trajectory X, he maximizes his static utility at almost every time t.

3.2 Existence of Static Equilibria

Now we shall consider a static game at some fixed moment $\bar{t}$ and state of the system $\bar{x}$ associated with the tragedy of the commons game.

Theorem 3.2. *If $P_\omega(u, d, \bar{x})$ is for a.e. ω jointly upper semicontinuous in (u, d), for a.e. ω and every d continuous in u, and for every u jointly measurable in (ω, d), the correspondence $S_{\cdot}(\bar{x})$ is measurable (i.e., has measurable graph) and has closed values, then there exists a static equilibrium.*

Proof. This follows directly by Balder's theorem 3.4.1 from [3] (with a little modification) or extension by Wiszniewska-Matyszkiel, [22], Theorem 3.1. □

4 Social Welfare Approach

The results of Section 3, especially Theorem 3.1, allow a decomposition of a dynamic game as formulated in Section 2 into a family of static games associated with it. Of course, we cannot expect such a dynamic equilibrium to be optimal.

Our model describes the exploitation of a common ecosystem; therefore, we can add some interpretation: an equilibrium destroys the system—the trajectory corresponding to the equilibrium profile is often nonoptimal, or even disastrous, lethal to the whole population (modelled with utility equal to $-\infty$). Such a situation is illustrated in Example 7.2.

To solve the problem of nonoptimality, we first specify the meaning of optimality of a control function. After finding such an optimal control, we enforce it in static games. The problem of enforcement is examined in Section 5. The way to define optimality of a control function is by introduction of an additional player: a social planner.

The social planner is optimizing his utility function, called a *social welfare function*, which is assumed to reflect the well-being of all players. Decisions of the social planner are supposed to be Pareto optimal.

The social welfare approach is very controversial, because different choices of the way to aggregate individual utility functions may yield different equilibria.

In this paper a *dynamic social welfare function* will be defined by means of a *static social welfare function.*

The *static social welfare function* $W_{h,\Gamma} : \mathbb{Y} \times \mathbb{X} \to \mathbb{R} \cup \{-\infty\}$ considered in this paper is a function of an aggregate extraction and of a state of the system, with $W_{h,\Gamma}(u, x)$ defined as a supremum of an integral of players' utilities weighted by a function h, over a certain set $\Gamma(u, x)$ of profiles available at x and with an aggregate equal to u. Formally:

Definition 4.1. Parameters of the social planner's optimization consist of: an essentially bounded function $h : \Omega \to (0, +\infty)$ such that $\int_\Omega h(\omega)d\mu(\omega) = 1$, called a social weighting function and a correspondence $\Gamma : \mathbb{Y} \times \mathbb{X} \multimap \Sigma$ with values fulfilling the inclusion $\Gamma(u, x) \subset \{\delta \in \Sigma \mid \delta(\omega) \in S_\omega(x), u_\delta = u\}$ (profiles the social planner takes into consideration).

Definition 4.2. For the parameters of the social planner's optimization h and Γ, the *static social welfare function* (associated with h and Γ) is defined by

$$W_{h,\Gamma}(u, x) = \sup_{\delta \in \Gamma(u,x)} \int_\Omega P_\omega(u, \delta(\omega), x)h(\omega)d\mu(\omega).$$

We shall omit the subscripts h, and Γ if doing so does not lead to confusion.

The parameters of social planner's optimization Γ and h may be arbitrarily chosen by the social planner or restricted by the rules of the game (e.g., by law). The question of how these two parameters might be chosen is partly answered in Section 6.

Proposition 4.1. *If for almost every ω, the expression $h(\omega) \cdot P_\omega(\overline{u}, d, \overline{x})$ does not depend on ω, so it can be written as $hP(\overline{u}, d, \overline{x})$ with hP concave as a function of d, and the set $\Gamma(\overline{u}, \overline{x})$ contains a profile $\overline{\delta}$ almost everywhere equal to $\overline{u}$, then the supremum from the definition of the social welfare function is attained at $\overline{\delta}$. Moreover, we have $W(\overline{u}, \overline{x}) = hP(\overline{u}, \overline{u}, \overline{x})$.*

Proof. This is a simple corollary of the generalized Jensen's inequality. The function hP is concave in d, Hence, for every $\delta \in \Gamma(\overline{u}, \overline{x})$, we have $\int_\Omega hP(\overline{u}, \delta(\omega), \overline{x})d\mu(\omega) \leq hP(\overline{u}, \int_\Omega \delta(\omega)d\mu(\omega), \overline{x}) = hP(\overline{u}, \overline{u}, \overline{x})$. □

Definition 4.3. The social planner is looking for an admissible control function U maximizing the extended-valued functional called a *dynamic social welfare function*

$$I_{h,\Gamma}(U) = \int_0^T W_{h,\Gamma}(U(t), X_U(t))\Psi(t)\,dt.$$

5 Tax-Subsidy Systems

The optimal control of the social planner is usually different from the aggregate resulting from the players' optimization. In this paper the social planner is assumed to enforce socially optimal strategies by modifying the players' payoff functions and without restricting their strategy sets. The only profiles enforced this way are those of finite instantaneous utility for almost every individual.

Definition 5.1. A *tax-subsidy system* is a function $\tau : \Omega \times \mathbb{Y} \times \mathbb{X} \times \mathbb{S} \to \mathbb{R}$ [we shall, as usual, put the player index ω into the subscript and write $\tau_\omega(u, x, d)$ rather than $\tau(\omega, u, x, d)$].

Definition 5.2.

(a) For an aggregate $\overline{u}$ and a state $\overline{x}$, a tax-subsidy system τ is said to *enforce* $\overline{u}$ at $\overline{x}$ if

$$\overline{u} \in \int_\Omega \operatorname{Argmax}_{d \in S_\omega(\overline{x})} (P_\omega(\overline{u}, d, \overline{x}) - \tau_\omega(\overline{u}, \overline{x}, d))\,d\mu(\omega).$$

(b) For an admissible control function $\overline{U}$, a tax-subsidy system τ is said to enforce $\overline{U}$ whenever τ enforces $\overline{U}(t)$ at $X_{\overline{U}}(t)$ [in the sense of (a)] for almost every t.

A natural assumption is that a tax paid or a subsidy obtained will be identical for players with the same characteristics (i.e., the available strategies' correspondence and the utility function) who choose the same strategy.

Definition 5.3. A tax-subsidy system is called anonymous if $\tau_\omega \equiv \tau_\varpi$ whenever $S_\omega \equiv S_\varpi$ and $P_\omega \equiv P_\varpi$.

We shall also investigate the following properties of tax-subsidy systems:

Definition 5.4. A tax-subsidy system enforcing $\overline{u}$ at $\overline{x}$ is called *self-financing* if, for every profile δ available at $\overline{x}$ and such that $\int_\Omega \delta(\omega)d\mu(\omega) = \overline{u}$,

$$\int_\Omega \tau_\omega\,(\overline{u}, \overline{x}, \delta(\omega))\,d\mu(\omega) \geq 0.$$

Definition 5.5. A tax-subsidy system enforcing $\overline{u}$ at $\overline{x}$ is called *nonprofit* if, for every profile δ available at $\overline{x}$ and such that $\int_{\Omega} \delta(\omega) d\mu(\omega) = \overline{u}$,

$$\int_{\Omega} \tau_{\omega}(\overline{u}, \overline{x}, \delta(\omega))\, d\mu(\omega) = 0.$$

Definition 5.6. For a tax-subsidy system τ, the *aftertax/subsidy utility* is a function $P^{\tau} : \Omega \times \mathbb{Y} \times \mathbb{Y} \times \mathbb{X} \times \mathbb{S}$ defined by $P^{\tau}_{\omega}(u, \overline{u}, \overline{x}, d) = P_{\omega}(u, d, \overline{x}) - \tau_{\omega}(\overline{u}, \overline{x}, d)$.

Definition 5.7. For any fixed tax-subsidy system τ and any admissible control function $\overline{U}$ enforced by τ, the dynamic game

$$\mathfrak{G}^{\tau, \overline{U}} = \left(\mathbb{X}, x_0, T, (\Omega, \Im, \mu), \mathbb{S}, S, \overline{R}, \varphi, \Psi\right),$$

where $\overline{R}_{\omega}(u, d, x, t) = P^{\tau}_{\omega}(u, \overline{U}(t), x, d)$, is called a *dynamic game with tax-subsidy system τ enforcing $\overline{U}$*.

Definition 5.8. Let $\overline{U}(t) = \overline{u}$ and $\overline{x} \in \mathbb{X}$. A static game at time $\overline{t}$ and state $\overline{x}$ corresponding to $\mathfrak{G}^{\tau, \overline{U}}$ [players' instantaneous utility functions are equal to $P^{\tau}_{\omega}(\cdot, \overline{u}, \overline{x}, \cdot)$] will be, throughout this paper, called a *static game at time $\overline{t}$ and state $\overline{x}$ with tax-subsidy system τ enforcing $\overline{u}$*.

In this paper we will not concentrate on any specific tax-subsidy system. Some examples will be examined on a similar tragedy of the commons game in [21].

5.1 Existence of Static Equilibrium in a Static Game with a Tax-Subsidy System

Suppose that at some fixed time $\overline{t}$ the state of the system is $\overline{x}$ and the social planner intends to enforce $\overline{u}$. Then we have

Theorem 5.1. *If $P_{\omega}(u, d, \overline{x})$ is for a.e. ω jointly upper semicontinuous in (u, d), for a.e. ω and every d continuous in u, and for every u jointly measurable in (ω, d), the correspondence $S_{\cdot}(\overline{x})$ is measurable (i.e., has measurable graph) and has closed values, the functions $\tau_{\omega}(\overline{u}, \overline{x}, d)$ are for a.e. ω lower semicontinuous in d, and jointly measurable in (ω, d), then there exists a static equilibrium in the static game with a tax-subsidy system.*

Proof. The proof is the same as that of Theorem 3.2. □

6 Dynamic Optimality

We finally tackle the problem of whether the dynamic profiles chosen by the players in the dynamic game with a tax-subsidy system yield payoffs optimal in the tragedy of the commons game.

We shall consider two types of optimality. The first one is the usual Pareto optimality. The definition of Pareto optimality was originally introduced for models with finitely many agents. Translated into the language of this paper, it has the following form: “a profile Δ^* is optimal if there is no profile Δ yielding payoffs greater or equal for every player and greater for at least one.” Automatic substitution of this definition to games with a continuum of players like those considered in this paper does not make sense: for instance, a profile-maximizing aggregate payoff could be not optimal in this sense, since changing one player's strategy does not affect the payoffs of the remaining players.

Therefore, the definition of Pareto optimality has the following form: a profile Δ^* is optimal in the sense of Pareto if there exists no profile Δ yielding at least the same payoff to almost every player and greater payoff for players in a nonnegligible set. Let us note that in the case of finitely many players, all of whom are nonnegligible, this definition is equivalent to the classical one.

Formally, the definition of Pareto optimality in our game is equivalent to the following:

Definition 6.1. A dynamic profile $\overline{\Delta}$ in a dynamic game $\mathfrak{G}$ (as defined in Section 2) is *Pareto optimal* if, for every dynamic profile Δ, the following implication holds: if for a.e. ω,

$$\int_0^T R_\omega(U_\Delta, \Delta(\omega,t), X^\Delta(t), t)\Psi(t)dt \geq \int_0^T R_\omega(U_{\overline{\Delta}}, \overline{\Delta}(\omega,t), X^{\overline{\Delta}}(t), t)\Psi(t)dt$$

then for a.e. ω,

$$\int_0^T R_\omega(U_\Delta, \Delta(\omega,t), X^\Delta(t), t)\Psi(t)dt = \int_0^T R_\omega(U_{\overline{\Delta}}, \overline{\Delta}(\omega,t), X^{\overline{\Delta}}(t), t)\Psi(t)dt.$$

Let us proceed to another kind of optimality, specific to our problem.

Definition 6.2. A dynamic profile $\overline{\Delta}$ is said to be *optimal in the sense of property in a modified game* $\widetilde{\mathfrak{G}}$ if for a.e. player ω, the payoff at this profile in $\widetilde{\mathfrak{G}}$ is not less than the maximum of the functional $\Pi_\omega(D, D)$ over a set of measurable functions $D : [0,T) \to \mathbb{S}$ such that for a.e. t, $D(t) \in S_\omega(X_D(t))$ (Π, S from the original tragedy of the commons game $\mathfrak{G}$).

Informally, the meaning of optimality in the sense of property is that the players cannot obtain greater payoffs even if division of the common ecosystem were possible.

The game decomposes in fact into independent players' optimization problems. Obviously, such a decomposition could be impossible in the real world. However, a result of some policy's being for almost all players at least as good as such an ideal result is a justified idea of optimality.

Theorem 6.1. *Let's assume that there exists $N \in \mathbb{R}$ such that for a.e. ω and every u, x, the function $P_\omega(u,d,x)$ is concave in d and bounded from above by N, and let the function P be measurable. We assume that the social weighting function h is such that for all u, d, and x, the expression $h(\omega) \cdot P_\omega(u,d,x)$ is essentially independent of ω. Further, for every x and $u \in Y(x)$, the set $\Gamma(u,x)$ contains the profile $\delta \equiv u$ (everywhere equal to u). Let the function $\overline{U}$ be an optimal control for* *the dynamic social welfare maximization, yielding finite dynamic social welfare, and let* $\overline{X}$ *be the corresponding trajectory.*

(a) *If a dynamic profile* $\overline{\Delta}$ *almost everywhere equal to* $\overline{U}$ *can be enforced by a measurable tax-subsidy system such that for a.e.* ω *and* t, $\tau_\omega(\overline{U}(t), \overline{X}(t), \overline{U}(t)) = 0$, *then* $\overline{\Delta}$ *is optimal in the sense of Pareto and in the sense of property in* $\mathfrak{G}$.

(b) *Every dynamic profile* $\widetilde{\Delta}$ *with the aggregate* $\overline{U}$ *enforced by a tax-subsidy system* τ *as above gives the aftertax/subsidy payoff for a.e. player not less than a profile Pareto optimal in the tragedy of the commons game, and this* $\widetilde{\Delta}$ *is optimal in the sense of property in* $\mathfrak{G}^{\tau,\overline{U}}$.

Proof. (a) Without loss of generality we may assume that $\overline{\Delta}(\omega, t) \equiv \overline{U}(t)$, which implies that the players neither pay tax nor get a subsidy, so

$$P^\tau_\omega(\overline{U}(t), \overline{U}(t), \overline{U}(t), \overline{X}(t)) = P_\omega(\overline{U}(t), \overline{U}(t), \overline{X}(t)).$$

Let us assume, contrary to our claim, that $\overline{\Delta}$ is not Pareto optimal, that is, there exist a dynamic profile Δ (with U defined by $U(t) = \int_\Omega \Delta(\omega, t)d\mu(\omega)$ and the corresponding trajectory X) and a set $\overline{\Omega} \subset \Omega$ of positive measure such that for a.e. $\omega \in \Omega$,

$$\int_0^T P_\omega(U(t), \Delta(\omega, t), X(t))\Psi(t)dt \geq \int_0^T P_\omega(\overline{U}(t), \overline{U}(t), \overline{X}(t))\Psi(t)dt$$

and for $\omega \in \overline{\Omega}$,

$$\int_0^T P_\omega(U(t), \Delta(\omega, t), X(t))\Psi(t)dt > \int_0^T P_\omega(\overline{U}(t), \overline{U}(t), \overline{X}(t))\Psi(t)dt.$$

Let's use the symbol $hP(u, x, d)$ for the function independent on ω and for a.e. ω equal to $h(\omega)P_\omega(u, x, d)$. By Proposition 4.1, $hP(\overline{U}(t), \overline{U}(t), \overline{X}(t)) = W(\overline{U}(t), \overline{X}(t))$ and $hP(U(t), U(t), X(t)) = W(U(t), X(t))$, so

$$\int_0^T hP(\overline{U}(t), \overline{U}(t), \overline{X}(t))\Psi(t)dt = \int_0^T W(\overline{U}(t), \overline{X}(t))\Psi(t)dt.$$

Throughout this proof we restrict our attention to the set of ω of measure 1 such that $h(\omega)P_\omega(u, d, x) = hP(u, d, x)$.

Note that

$$\begin{aligned} h(\omega)\int_0^T P_\omega(U(t), U(t), X(t))\Psi(t)dt &= \int_0^T hP(U(t), U(t), X(t))\Psi(t)dt \\ &= \int_0^T W(U(t), X(t))\Psi(t)dt \\ &\leq \int_0^T W(\overline{U}(t), \overline{X}(t))\Psi(t)dt \\ &= \int_0^T hP(\overline{U}(t), \overline{U}(t), \overline{X}(t))\Psi(t)dt, \end{aligned}$$

for $\overline{U}$ is the optimal control function.

Since for each u the function $hP(u, \cdot, x)\Psi(t)$ is concave, we may use the generalized Jensen's inequality:

$$\begin{aligned}&\int_\Omega hP(U(t), \Delta(\omega, t), X(t))\Psi(t)d\mu(\omega)\\&\qquad \le hP\left(U(t), \int_\Omega \Delta(\omega, t)d\mu(\omega), X(t)\right)\Psi(t)\\&\qquad = hP(U(t), U(t), X(t)).\end{aligned}$$

Let's integrate both sides of this inequality over time:

$$\begin{aligned}\int_0^T \int_\Omega hP(U(t), \Delta(\omega, t), X(t))\Psi(t)d\mu(\omega)\,dt &\le \int_0^T hP(U(t), U(t), X(t))\Psi(t)dt\\&\le \int_0^T hP(\overline{U}(t), \overline{U}(t), \overline{X}(t))\Psi(t)dt.\end{aligned}$$

The function $hP \circ (U \times \Delta \times X) \cdot \Psi$ [i.e., the one defined by the equation

$$(hP \circ (U \times \Delta \times X) \cdot \Psi)(\omega, t) = hP(U(t), \Delta(\omega, t), X(t)) \cdot \Psi(t)]$$

is measurable and bounded from above by the integrable function $N \cdot \Psi$. In addition, the dynamic profile Δ is not worse than $\overline{\Delta}$, for which the integral is finite, so $hP \circ (U \times \Delta \times X) \cdot \Psi$ is bounded from below by the integrable function $hP \circ (\overline{U} \times \overline{\Delta} \times \overline{X}) \cdot \Psi$. Therefore, $hP \circ (U \times \Delta \times X) \cdot \Psi$ is integrable, and we may apply to it the general Fubini theorem:

$$\begin{aligned}&\int_\Omega \int_0^T hP(U(t), \Delta(\omega, t), X(t))\Psi(t)dt\, d\mu(\omega)\\&\qquad = \int_0^T \int_\Omega hP(U(t), \Delta(\omega, t), X(t))\Psi(t)d\mu(\omega)\,dt.\end{aligned}$$

Hence

$$\begin{aligned}&\int_\Omega \int_0^T hP(U(t), \Delta(\omega, t), X(t))\Psi(t)\,dt\, d\mu(\omega)\\&\qquad \le \int_0^T hP(\overline{U}(t), \overline{U}(t), \overline{X}(t))\Psi(t)\,dt\, d\mu(\omega).\end{aligned}$$

But by the assumption about $\Delta(\omega, t)$, we have for a.e. $\omega \in \Omega$,

$$h(\omega)\int_0^T P_\omega(U(t), \Delta(\omega, t), X(t))\Psi(t)dt \ge h(\omega)\int_0^T P_\omega(\overline{U}(t), \overline{U}(t), \overline{X}(t))\Psi(t)dt$$

and for $\omega \in \overline{\Omega}$,

$$h(\omega)\int_0^T P_\omega(U(t), \Delta(\omega, t), X(t))\Psi(t)dt > h(\omega)\int_0^T P_\omega(\overline{U}(t), \overline{U}(t), \overline{X}(t))\Psi(t)dt.$$

Thus,

$$\int_\Omega \int_0^T hP(U(t), \Delta(\omega, t), X(t))\Psi(t)dt\, d\mu(\omega)$$
$$> \int_\Omega \int_0^T hP(\overline{U}(t), \overline{U}(t), \overline{X}(t))\Psi(t)dt\, d\mu(\omega),$$

which is a contradiction.

Optimality in the sense of property is obvious, since a.e. player looks for the optimal control at the same problem: multiplying the utility function by a positive constant does not change the optimal control.

(b) Let us compare $\widetilde{\Delta}$ with $\overline{\Delta}$ from (a) almost everywhere equal to $\overline{U}$. The profile $\overline{\Delta}$ is optimal in the original game in both senses. Player ω's decision to choose $d = \widetilde{\Delta}(\omega, t)$ instead of the available $\overline{U}(t)$ shows that his after-tax/subsidy utility at d is not less than at $\overline{U}(t)$. Hence the after-tax/subsidy payoff at $\widetilde{\Delta}(\omega, \cdot)$ is not less than the payoff at $\overline{\Delta}(\omega, \cdot)$ in the tragedy of the commons game, which completes the proof. □

On the basis of this theorem, in the case of such nearly identical players, the social planner who is interested in Pareto optimality in the original game and optimality in the sense of property only, can take this "uniforming" h and restrict Γ to constant profiles.

Theorem 6.2. *Let P be measurable and essentially bounded (essentially w.r.t. ω) from above. Suppose that the parameters of the social planner's optimization are $h \equiv 1$ and*

$$\Gamma(u, x) = \left\{\delta \in \Sigma \mid \delta(\omega) \in S_\omega(x),\ \int_\Omega \delta(\omega)d\mu(\omega) = u\right\};$$

$\overline{U}$ is the optimal control yielding finite payoff, for the static social welfare maximization, with $\overline{X}$ as the corresponding trajectory. Then the following implication holds:

If the maxima from the definition of the static social welfare function for $\overline{U}(t)$, $\overline{X}(t)$ are attained at profiles constituting a dynamic profile $\widetilde{\Delta}$, which can be enforced by a nonprofit, measurable tax-subsidy system, and there exists M such that for a.e. ω and t, $\left|\tau_\omega(\overline{U}(t), \overline{X}(t), \widetilde{\Delta}(\omega, t))\right| < M$, then the aggregate aftertax/subsidy payoff for this $\widetilde{\Delta}$ is equal to the maximal aggregate payoff that can be obtained in the tragedy of the commons game.

Proof. The aggregate payoff for a profile Δ with the corresponding trajectory X is equal to:

$$\int_\Omega \int_0^T P_\omega(U_\Delta, \Delta(\omega, t), X(t))\Psi(t)\, dt\, d\mu(\omega).$$

Since the integrand is measurable and the integral is finite, we can use the Fubini theorem. We get the aggregate payoff equal to

$$\int_0^T \int_\Omega P_\omega(U_\Delta, \Delta(\omega, t), X(t))\, d\mu(\omega)\Psi(t)\, dt.$$

Let's take a supremum over all the dynamic profiles (equal to the supremum over all the profiles with a.e. finite payoff). To make reading easier we introduce the following abbreviations: $\mathbb{D}$ for the set of all dynamic profiles, $\mathbb{U}$ for the set of all admissible control functions and

$$\mathbb{D}_{U,t} = \left\{ \delta \in \Sigma \mid \delta(\omega) \in S_\omega(X_U(t)) \text{ for a.e. } \omega, \int_\Omega \delta(\omega)\, d\mu(\omega) = U(t) \right\}.$$

Now let's make a sequence of transformations:

$$\begin{aligned}
&\sup_{\Delta\in\mathbb{D}} \left(\int_\Omega \int_0^T P_\omega(U_\Delta, \Delta(\omega, t), X(t))\, \Psi(t) dt\, d\mu(\omega) \right) \\
&\quad = \sup_{\Delta\in\mathbb{D}} \left(\int_0^T \int_\Omega P_\omega(U_\Delta, \Delta(\omega, t), X(t))\, d\mu(\omega)\, \Psi(t) dt \right) \\
&\quad \le \sup_{U\in\mathbb{U}} \left(\int_0^T \sup_{\delta\in\mathbb{D}_{U,t}} \left(\int_\Omega P_\omega(U(t), \delta(\omega), X(t)) d\mu(\omega) \right) \Psi(t) dt \right) \\
&\quad = \sup_{U\in\mathbb{U}} \left(\int_0^T W(U(t), X(t)) \Psi(t) dt \right) \\
&\quad = \int_0^T W(\overline{U}(t), \overline{X}(t)) \Psi(t) dt \\
&\quad = \int_0^T \int_\Omega P_\omega(\overline{U}(t), \widetilde{\Delta}(\omega, t), \overline{X}(t))\, d\mu(\omega)\, \Psi(t) dt
\end{aligned}$$

Since the tax-subsidy system is nonprofit, the last term is equal to

$$\int_0^T \int_\Omega P_\omega^\tau(\overline{U}(t), \overline{U}(t), \overline{X}(t), \widetilde{\Delta}(\omega, t)) d\mu(\omega)\, \Psi(t) dt,$$

which by the Fubini theorem, is equal to

$$\int_\Omega \int_0^T P_\omega^\tau(\overline{U}(t), \overline{U}(t), \overline{X}(t), \widetilde{\Delta}(\omega, t)) \Psi(t) dt\, d\mu(\omega).$$

We could use Fubini's theorem, since $P_\omega^\tau \circ (\overline{U} \times \overline{U} \times \overline{X} \times \widetilde{\Delta}) \cdot \Psi$ is integrable: it is measurable and bounded from above by $(N + M) \cdot \Psi$ (where N is an essential bound of P) and from below by $\big(P_\omega \circ (\overline{U} \times \widetilde{\Delta} \times \overline{X}) - M\big) \cdot \Psi$, which are integrable. □

This theorem can be regarded as an answer to the question of how Γ and h might be chosen, since the requirement of maximizing the aggregate payoff is strong and socially desirable: there is no deadweight loss. What two Coasian players obtain due to free bargaining, we obtain by means of a nonprofit tax-subsidy system in the case of an infinite set of players.

Although the answer to the question of how to choose h and Γ remains political, avoiding deadweight loss should be one of the social planner's main objectives. Political aims can be reached, for instance by means of lump-sum taxes or subsidies, which do not affect the choice of strategies.

7 Examples

In this section we give two examples in the scheme considered in Section 2, so we feel free to use the notation introduced there, along with the notation introduced in Section 4.

Example 7.1. Let us consider what will be called a *stable rainforest problem*. The state of the system is the area of the forest. We can extract some fixed surplus part of the state without damaging the system. If we extract more, we affect the "basis" of regeneration (the soil erodes and no more trees will grow there). Even if we extract less than the surplus, the future state of the system cannot be improved. We assume that every state is sustainable and it is impossible to completely destroy the system at any finite time.

The space of players $(\Omega, \Im, \mu)$ is $[0, 1]$ with the Lebesgue measure. The space of states $\mathbb{X}$ is $[0, M]$, the initial state is $x_0 > 0$ and the time horizon is equal to $+\infty$. All the players are identical: for all ω, $S_\omega(x) = \{d : 0 \le d \le x\}$ and $P_\omega(u, d, x) = \ln(d)$ (we also let $\ln 0 = -\infty$).

The regeneration function $\varphi(x, u) = -\max(0, u - r \cdot x)$, for fixed $0 < r \le 1$. The discounting function $\Psi(t) = e^{-\xi t}$ for some fixed $\xi > 0$. Due to legal restrictions, the choice of the social planner is restricted to $h \equiv 1$ and Γ such that $\Gamma\big(u, x\big) = \big\{\delta \in \Sigma \mid \delta \equiv u\big\}$ for all u and x. Note that the model is characterized by four positive reals: M, x_0, r, and ξ.

Proposition 7.1. (a) *If $\xi \le r$, then the constant function defined by $\overline{U} \equiv r \cdot x_0$ is the unique (up to measure equivalence) optimal control of the social planner.*

(b) *If $\xi > r$, then the function defined by $\overline{U}(X(t)) = \min(\xi, 1) \cdot X(t)$ is the unique (up to measure equivalence) optimal control of the social planner.*

(c) *The profile $\overline{\Delta} \equiv \overline{U}$ is the only profile (up to measure equivalence) maximizing the aggregate payoff. If this profile is enforced by a measurable, self-financing and anonymous tax-subsidy system such that for every t, $T_\omega(\overline{U}(t), \overline{X}(t), \overline{U}(t)) \equiv 0$, then $\overline{\Delta}$ is optimal in the sense of Pareto and in the sense of property. If it is enforced by a measurable, nonprofit tax-subsidy system such that $\big|T_\omega(\overline{U}(t), \overline{X}(t), \overline{U}(t))\big|$ is essentially bounded (in ω and t), then the aftertax/subsidy payoff equals the maximal aggregate payoff in the tragedy of the commons game.*

Proof. It is obvious that the correspondence of aggregate strategies fulfills $Y(x) = \{u : 0 \le u \le x\}$. By Proposition 4.1, $W(u, x) = P_\omega(u, u, x) = \ln(u)$ for $u \le x$. Note that the same static social welfare function would be obtained if Γ

were the largest possible set of profiles, that is, $\Gamma(u, x) = \{\delta \in \Sigma \mid \delta(\omega) \in S_\omega(x)$ for a.e. ω, $u_\delta = u\}$ for all u and x.

To calculate the optimal control, we use the Bellman equation written as follows: $\inf_{0<u\leq M} \big(\ln M - \ln(\min(x, u)) + w'(x) \cdot (-\max(0, u - rx))\big) = \xi \cdot w(x)$, where w is the value function.

We have a sequence of transformations:

$$\inf_{0<u\leq M} \big(-\ln(\min(x, u)) - w'(x) \cdot \max(0, u - rx)\big) = \xi \cdot w(x) - \ln M,$$

$$\sup_{0<u\leq M} \big(\ln(\min(x, u)) + w'(x) \cdot \max(0, u - rx)\big) = \ln M - \xi \cdot w(x),$$

$$\max\left(\sup_{0<u\leq rx} \ln u, \sup_{rx<u\leq x} \big(\ln u + w'(x) \cdot (u - rx)\big)\right) = \ln M - \xi \cdot w(x),$$

$$\max\left(\ln(rx), \sup_{rx<u\leq x} \big(\ln u + w'(x) \cdot (u - rx)\big)\right) = \ln M - \xi \cdot w(x),$$

$$\sup_{rx\leq u\leq x} \big(\ln u + w'(x) \cdot (u - rx)\big) = \ln M - \xi \cdot w(x),$$

$$\sup_{r\leq s\leq 1} \big(\ln(sx) + w'(x) \cdot (s - r)x\big) = \ln M - \xi \cdot w(x).$$

Let us fix some $x > 0$. For what s is the supremum attained?

First let's check $s = r$. If the supremum was attained at r, the Bellman equation would be equivalent to $\ln(rx) + w'(x) \cdot (r - r)x = \ln M - \xi \cdot w(x)$, which is equivalent to $\ln(rx) = \ln M - \xi \cdot w(x)$; therefore, $w(x) = (-\ln(rx) + \ln M)/\xi$.

We now check whether, for every $r < s \leq 1$, $\ln(sx) - r(1/(rx\xi))(s - r)x \leq \ln(rx)$. This is equivalent to $\ln(sx) - \ln(rx) \leq 1/\xi(s - r)$ and, consequently, $(\ln s - \ln r)/(s - r) \leq 1/\xi$. Since $s > r$, the left side of this inequality represents the right-side difference quotients of the logarithm at r. Since the logarithm function is strictly concave, those terms are less than the derivative. Therefore, $(\ln s - \ln r)/(s - r) < 1/r$ and, consequently, $(\ln s - \ln r)/(s - r) \leq 1/\xi$ for all $r < s \leq 1$ if and only if $\xi \leq r$.

(a) To complete the proof of optimality in case (a), it is enough to check the boundary condition $\liminf_{t\to\infty} w(X(t)) \cdot \Psi(t) = 0$. Indeed, we have

$$\liminf_{t\to\infty} w(x_0) \cdot e^{-\xi t} = \liminf_{t\to\infty} \frac{-\ln(rx_0) + \ln M}{\xi} \cdot e^{-\xi t} = 0.$$

To prove the uniqueness, let's assume that there exists an optimal control U such that the set of all t, such that $U(t) \neq \overline{U}(t)$, has positive measure. Let X denote the corresponding trajectory. It is obvious that $rX(t) \leq U(t)$. Therefore, we have $\dot{X}(t) = rX(t) - U(t)$ and, analogously, $\dot{\overline{X}}(t) = r\overline{X}(t) - \overline{U}(t)$.

We denote $(1/2)\overline{U} + (1/2)U$ by $\widetilde{U}$ and $(1/2)\overline{X} + (1/2)X$ by $\widetilde{X}$. Since $r\widetilde{X}(t) \leq \widetilde{U}(t) \leq \widetilde{X}(t)$ for almost every t, $\widetilde{X}$ is the trajectory corresponding to $\widetilde{U}$ and $\widetilde{U}(t) \in Y(\widetilde{X}(t))$. The function W

is strictly concave on the set $\{(u,x) \mid 0 \le u \le x, 0 \le x \le M\}$, and for all t, the pairs $(U(t), X(t))$, $\left(\overline{U}(t), \overline{X}(t)\right)$ and $\left(\widetilde{U}(t), \widetilde{X}(t)\right)$ belong to this set. Hence, for all t such that $U(t) \neq \overline{U}(t)$, we have $(1/2)W\left(\overline{U}(t), \overline{X}(t)\right) + (1/2)W\left(U(t), X(t)\right) < W\left(\widetilde{U}(t), \widetilde{X}(t)\right)$. This implies $\int_0^\infty W\left(\overline{U}(t), \overline{X}(t)\right)\Psi(t)dt < \int_0^\infty W\left(\widetilde{U}(t), \widetilde{X}(t)\right)\Psi(t)dt$, which contradicts the optimality of $\overline{U}$. This completes the proof of (a).

(b) In this case we first assume that the optimal control is given by the equation $u = sx$ for some $r < s \le 1$. Then the corresponding trajectory is $X(t) = x_0 \cdot e^{-(s-r)t}$. Therefore $w(x) = \int_0^\infty \left(\ln M - \ln\left(sx \cdot e^{-(s-r)t}\right)\right) \cdot e^{-\xi t}dt = (\ln M - \ln(sx))/\xi + (s-r)/\xi^2$.
We have

$$w'(x) = -1/\xi x \quad \text{and} \quad \operatorname{Argmax}_{r<s\le 1}\left(\ln(sx) + w'(x)\cdot(s-r)x\right)$$
$$= \operatorname{Argmax}_{r<s\le 1}\left(\ln(sx) - \frac{(s-r)}{\xi x}x\right) = \min(\xi, 1).$$

We now check a sufficient optimality condition, at first whether our function w fulfills the Bellman equation. Let $s = \min(\xi, 1)$. It is evident that

$$\ln M - \ln(sx) + \frac{1}{\xi x}(s-r)x = \xi\left(\frac{\ln M - \ln(sx)}{\xi} + \frac{s-r}{\xi^2}\right).$$

Finally, we check the boundary condition

$$\liminf_{t\to\infty} w(X(t))\cdot\Psi(t) = \liminf_{t\to\infty} w(x_0 e^{-(s-r)t})\cdot e^{-\xi t}$$
$$= \liminf_{t\to\infty}\left(\frac{\ln M - \ln\left(sx_0\cdot e^{-(s-r)t}\right)}{\xi} + \frac{s-r}{\xi^2}\right)\cdot e^{-\xi t}$$
$$= 0.$$

The proof of the uniqueness is exactly as in case (a).

(c) Now we have a static game: at some fixed time $\bar{t}$, the state of the system is $\bar{x}$. We want to enforce the aggregate extraction $\bar{u} = s\bar{x}$ with some measurable nonprofit, anonymous, tax-subsidy system such that $T_{\cdot}(\bar{u}, \bar{u}, \bar{x}) \equiv 0$. Since the players are of the same type, however, an anonymous, self-financing tax-subsidy system with this property is defined; one of the possible results is $\bar{\delta} \equiv \bar{u}$. Therefore, nobody pays taxes, and nobody gets subsidies.
The profile $\overline{\Delta}$ defined by $\overline{\Delta}(\cdot, t) = \bar{\delta}$ is optimal in the sense of property and in the sense of Pareto by Theorem 6.1.

If we want to enforce $\overline{\Delta}$ with a nonprofit, measurable tax-subsidy system such that $\left|T_\omega(\overline{U}(t), \overline{U}(t), \overline{X}(t))\right|$ is essentially bounded, then by Theorem 6.2, the aggregate aftertax/subsidy payoff equals the maximal aggregate payoff in the original game. □

Example 7.2 (Unstable rainforest problem). There is just one feature that makes this example different from the previous one. The system behaves the same way as in that of the stable rainforest unless its state is less than $x_{\min} = 2$. If, at some time t, the state of the system $X(t)$ is less than 2, the system cannot be saved from complete destruction in finite time whatever the subsequent actions of the players are.

Formally, we have the same assumptions on all the objects besides the regeneration function. The regeneration function φ is a function such that, for every admissible control function, there exists a unique corresponding trajectory. Let φ fulfill the following conditions:

$$\begin{aligned}\varphi(x,u) &= -\max(0, u - rx) \quad \text{for } x \geq 2;\\ \varphi(x,u) &\leq -2\sqrt{2-x} \quad \text{for } 1 \leq x < 2;\\ \varphi(x,u) &\leq -2\sqrt{x} \quad \text{for } x < 1;\end{aligned}$$

and $\varphi(x,\cdot)$ is strictly decreasing for $x \geq u \geq rx$. Because otherwise the example would become trivial, we take the initial state $x_0 > 2$. For the game of this example, the following proposition holds:

Proposition 7.2. (a) *No dynamic profile yielding finite payoffs for players in a set of positive measure is an equilibrium.*

(b) *Every dynamic profile leading to the destruction of the system in a finite time [i.e., $X(t) = 0$ for $t \geq \bar{t}$] is an equilibrium.*

(c) *Assume that players constituting a measurable set K decide to extract only a small part $0 < \varepsilon < r$ of admissible x in order to avoid the destruction of the unstable rainforest, while the others maximize their instantaneous utility. If $\mu(K) \leq 1 - r$, then, independently of decisions of the individuals in K, the system will be destroyed in a finite time.*

Proof. (a) Let Δ be a dynamic equilibrium such that players in a set of positive measure get finite payoffs, let U be the aggregate of the profile Δ, and let X be the trajectory corresponding to U. The static best response of every player to $U(t)$ at time t and state $X(t)$ is $d = X(t)$; therefore, the aggregate of the resulting profile is equal to $X(t)$. Thus, $U(t) = X(t) > rX(t)$.

We now prove that the system will reach 0 in a finite time.

For t such that $X(t) \geq 2$, $X(t) = x_0 \cdot e^{-(1-r)t} \overset{t\to\infty}{\longrightarrow} 0$. Thus, the subsequent trajectory will reach 2 at a finite time t'. After that moment, the system becomes self-destructing and for t such that $1 \leq X(t'+t) < 2$, we have $X(t'+t) \leq 2 - t^2$. After a period of $t'' \leq 1$, X reaches 1, that is, $X(t'+t'') = 1$. From that moment, $X(t+t'+t'') \leq (1-t)^2$; therefore, after $t''' \leq 1$, X reaches 0.

For $t \geq \tilde{t} = t' + t'' + t'''$, $X(t) = 0$. Therefore, $S_\omega(X(t)) = \{0\}$, so for $t \geq \tilde{t}$, every admissible profile fulfills the condition $\Delta(\omega,t) \equiv 0$. Hence, for $t \geq \tilde{t}$, $P_\omega(u, \Delta(\omega,t), x) = \ln 0 = -\infty$. Therefore, the player's payoff equal $-\infty$, so there exists no equilibrium with finite payoffs.

(b) Let Δ be a dynamic profile destructing the system in a finite time, let U be the aggregate of the profile Δ, and let X be the trajectory corresponding to U. The trajectory X fulfills the condition $X(t) = 0$ at $t > \bar{t} \in \mathbb{R}$. Since the only possible strategy of player ω for $t \in (\bar{t}, +\infty)$ is $d = 0$, his only possible payoff for the control function U is $-\infty$, regardless of the admissible dynamic strategy he chooses. Therefore, $\mathbf{B}_\omega(U) = \{D : \mathbb{T} \to \mathbb{S} \text{ measurable} \mid D(t) \in S_\omega(X(t))\} \ni \Delta(\omega, \cdot)$, which completes the proof of (b).

(c) Again we prove that the system will reach 0 in finite time.

The aggregate extraction is given by $u(x) = \mu(K)\varepsilon x + (1 - \mu(K))x = cx > rx$. For t such that $X(t) \geq 2$, we have $X(t) = x_0 \cdot e^{-(c-r)t} \overset{t\to\infty}{\longrightarrow} 0$, thus the trajectory will reach 2 at a finite time t'. The further reasoning is analogous to that in the proof of (a). □

8 Conclusions

The tragedy of the commons game presented in this paper is a model of the exploitation of a common ecosystem by many players. Every player is insignificant, but aggregate decisions affect the system. That is why games with a continuum of players are appropriate to model such situations.

Dynamic games with a continuum of players capture many phenomena that happen in the real world but are difficult or even impossible to model by games with finitely many players. One of them is equivalence between a dynamic equilibrium and a family of corresponding static equilibria. This fact explains why individually rational decisions of players may lead to destruction of the basis of their maintenance (which happened, e.g., in the whaling industry; a similar situation is illustrated by Example 7.2). It also reduces the problem of finding a dynamic equilibrium to a simpler problem of finding equilibria in a parametrized family of static games.

Besides theoretical results on dynamic games with a continuum of players, the paper also suggests a way to solve the problem of nonoptimality in the tragedy of the commons game. The solution is by the introduction of a social planner maximizing a dynamic social welfare function of a specific form and enforcing the aggregate extraction being the optimal control of his optimization, in static games modified by tax-subsidy systems. Under some assumptions on admissible tax-subsidy systems as well as on the parameters defining the social welfare function, the enforced profile suits some or all of three optimality concepts considered in this paper: it is Pareto optimal; it maximizes aggregated payoff; or it is at least as good as a profile arising after privatization (defined in this paper as *optimality in the sense of property*, specific to the problem of common property).

Acknowledgments

I would like to acknowledge Andrzej Wieczorek for helpful comments. This research has been supported by KBN Grant no. 1 H02B 015 15.

REFERENCES

[1] Aumann, R. J. Integrals of set-valued functions, *J. Math. Anal. Appl.*, **12**, pp. 1–12, 1965.

[2] Aumann, R. J. *Measurable Utility and Measurable Choice Theorem*, La Decision, Colloque Internationaux du C.N.R.S., Paris, pp. 15–26, 1969.

[3] Balder, E. A unifying approach to existence of Nash equilibria, *Intl. J. Game Th.*, **24**, pp. 79–94, 1995.

[4] Carlson, D. and Haurie, A. A turnpike theory for infinite-horizon open-loop competitive processes, *SIAM J. Cont. Opt.*, **34**, pp. 1405–1419, 1996.

[5] Carraro, C. and Filar, J. A. eds., Control and game-theoretic models of the environment, *Ann. Intl. Soc. Dyn. Games*, Birkhäuser, Boston, 1995.

[6] Coase, R. The problem of social cost, *J. Law and Econ.*, **3**, pp. 1–44, 1960.

[7] Dutta, P. and Sundaram, R. The tragedy of the commons?, *Econ. Th.*, **3**, pp. 413–426, 1993.

[8] Fisher, R. D. and Mirman, L. J. A strategic dynamic interaction. Fish wars, *J. Econ. Dyn. Cont.*, **16**, pp. 267–287, 1992.

[9] Folmer, H. and van Ierland, E. eds., *Valuation Methods and Policy Making in Environmental Economics*, Elsevier, New York, 1989.

[10] Hardin, G. The tragedy of the commons, *Science*, **162**, pp. 1243–1248, 1968.

[11] Hildenbrand, W. *Core and Equilibria of a Large Economy*, Princeton University Press, Princeton, 1974.

[12] Kaitala, V. Game theory models in fisheries management: A survey, pp. 252–266, in T. Başar, ed., *Dynamic Games and Applications in Economics*, Lecture Notes in Economics and Mathematical Systems, Springer-Verlag, New York, 1986.

[13] Kwiatkowski, M. Application of games with continuum of players in models of monetary economics (in Polish), Thesis, Faculty of Mathematics, Warsaw University, Warsaw, 1995.

[14] Levhari, D. and Mirman, L. J. The great fish war: An example using a dynamic Cournot–Nash solution, *Bell J. Econ.*, **11**, pp. 322–334, 1980.

[15] Mas-Colell, A. On the theorem of Schmeidler, *J. Math. Econ.*, **13**, pp. 201–206, 1984.

[16] Schmeidler, D. Equilibrium points of nonatomic games, *J. Stat. Phys.*, **17**, pp. 295–300, 1973.

[17] Van Ierland, E. *International Environmental Economics. Developments in Environmental Economics*, Elsevier, New York, 1994.

[18] Wieczorek, A. *Simple Large Games and Their Applications to Problems with Many Agents*, Report **842**, Inst. Computer Science, Polish Academy of Sciences, 1997.

[19] Wieczorek, A. and Wiszniewska (Wiszniewska-Matyszkiel), A. A game-theoretic model of social adaptation in an infinite population, *Applicationes Mathematicae*, **25**, pp. 417–430, 1999.

[20] Wiszniewska (Wiszniewska-Matyszkiel), A. Elements of mathematical theory of extraction of common resources (in Polish), Thesis, Faculty of Mathematics, Warsaw University, Warsaw, 1995.

[21] Wiszniewska-Matyszkiel, A. *Dynamic Game with Continuum of Players Modelling "the Tragedy of the Commons,"* *Game Th. Appl.*, **5**, 163–187, 2000.

[22] Wiszniewska-Matyszkiel, A. *Existence of Pure Equilibria in Games with Continuum of Players*, *Topological Methods in Nonlinear Analysis*, **16**, pp. 339–349, 2000.

[23] Wiszniewska-Matyszkiel, A. *Static and Dynamic Equilibria in Games with Continuum of Players*, Report **RW 00-3(70)**, Inst. Applied Mathematics and Mechanics, Warsaw University, Warsaw, 1999.

[24] Wiszniewska-Matyszkiel, A. *Static and Dynamic Equilibria in Games with Continuum of Players: Discrete Time Case*, Report **904**, Institute of Computer Science, Polish Academy of Sciences, 2000.

Annals of the International Society of Dynamic Games

Series Editor
Tamer Başar
Coordinated Science Laboratory
University of Illinois
1308 West Main Street
Urbana, IL 61801
U.S.A.

This series publishes volumes in the general area of dynamic games and their applications. It is an outgrowth of activities of "The International Society of Dynamic Games," *ISDG*, which was founded in 1990. The primary goals of *ISDG* are to promote interactions among researchers interested in the theory and applications of dynamic games; to facilitate dissemination of information on current activities and results in this area; and to enhance the visibility of dynamic games research and its vast potential applications.

The *Annals of Dynamic Games* series will have volumes based on the papers presented at the society's biannual symposia, including only those that have gone through a stringent review process, as well as volumes of invited papers dedicated to specific, fast-developing topics, put together by a guest editor or guest co-editors. More information on this series and on volumes planned for the future can be obtained by contacting the Series Editor, Tamer Başar, whose address appears above.

We encourage the preparation of manuscripts in LaTeX using Birkhäuser's macro.sty for this volume.

Proposals should be sent directly to the editor or to: Birkhäuser Boston, 675 Massachusetts Avenue, Cambridge, MA 02139, U.S.A.

Volumes in this series are:

Advances in Dynamic Games and Applications
Tamer Başar and Alan Haurie, Eds.

Control and Game-Theoretic Models of the Environment
Carlo Carraro and Jerzy A. Filar, Eds.

New Trends in Dynamic Games and Applications
Geert Jan Olsder, Ed.

Stochastic and Differential Games
Martino Bardi, T.E.S. Raghavan, and T. Parthasarathy, Eds.

Advances in Dynamic Games and Applications
Jerzy Filar, Vladimir Gaitsgory, and Koichi Mizukami, Eds.

Advances in Dynamic Games and Applications
Eitan Altman and Odile Pourtallier, Eds.